Laser Systems for Photobiology and Photomedicine

NATO ASI Series

Advanced Science Institutes Series

A series presenting the results of activities sponsored by the NATO Science Committee, which aims at the dissemination of advanced scientific and technological knowledge, with a view to strengthening links between scientific communities.

The series is published by an international board of publishers in conjunction with the NATO Scientific Affairs Division

A	**Life Sciences**	Plenum Publishing Corporation
B	**Physics**	New York and London
C	**Mathematical and Physical Sciences**	Kluwer Academic Publishers
D	**Behavioral and Social Sciences**	Dordrecht, Boston, and London
E	**Applied Sciences**	
F	**Computer and Systems Sciences**	Springer-Verlag
G	**Ecological Sciences**	Berlin, Heidelberg, New York, London,
H	**Cell Biology**	Paris, Tokyo, Hong Kong, and Barcelona
I	**Global Environmental Change**	

Recent Volumes in this Series

Series B: Physics

Laser Systems for Photobiology and Photomedicine

Edited by

A. N. Chester

Hughes Research Laboratories
Malibu, California

S. Martellucci

The Second University of Rome
Rome, Italy

and

A. M. Scheggi

IROE-CNR
Florence, Italy

Plenum Press
New York and London
Published in cooperation with NATO Scientific Affairs Division

Proceedings of a NATO Advanced Study Institute/15th Course of the
International School of Quantum Electronics on
Laser Systems for Photobiology and Photomedicine,
held May 11–20, 1990,
in Erice, Sicily, Italy

Library of Congress Cataloging-in-Publication Data

Course of the International School of Quantum Electronics on Laser
  Systems for Photobiology and Photomedicine (15th : 1990 : Erice,
  Italy)
    Laser systems for photobiology and photomedicine / edited by A.N.
  Chester, S. Martellucci, and A.M. Scheggi.
      p. cm. -- (NATO ASI series. Series B, Physics ; v. 252)
    "Proceedings of a NATO Advanced Study Institute/15th Course of the
  International School of Quantum Electronics on Laser Systems for
  Photobiology and Photomedicine, held May 11-20, 1990, in Erice,
  Sicily, Italy"--T.p. verso.
    "Published in cooperation with NATO Scientific Affairs Division."
    Includes bibliographical references and index.

    1. Lasers in medicine--Congresses. 2. Lasers in biology-
  -Congresses. 3. Lasers--Therapeutic use--Congresses. 4. Lasers in
  surgery--Congresses.   I. Chester, A. N. II. Martellucci, S.
  III. Verga Scheggi, A. M. (Anna Maria) IV. North Atlantic Treaty
  Organization. Scientific Affairs Division. V. Title. VI. Series.
    [DNLM: 1. Laser Surgery--congresses. 2. Lasers--diagnostic use-
  -congresses. 3. Lasers--therapeutic use--congresses.
  4. Photochemotherapy--congresses.   WB 117 C861L 1990]
  R857.L37C68 1990
  610'.28--dc20
  DNLM/DLC
  for Library of Congress   91-3006
                                                         CIP

© 1991 Plenum Press, New York
A Division of Plenum Publishing Corporation
233 Spring Street, New York, N.Y. 10013
Softcover reprint of the hardcover 1st edition 1991

ISBN-13: 978-1-4684-7289-9   e-ISBN-13: 978-1-4684-7287-5
DOI: 10.1007/978-1-4684-7287-5

PREFACE

    This volume contains the Proceedings of a two-week NATO Advanced Study
Institute on "Laser Systems for Photobiology and Photomedicine", conducted
from May 11 to 20, 1990 in Erice, Italy. This is the 15th annual course of
the International School of Quantum Electronics (ISQE), organized under the
auspices of the "Ettore Majorana" Center for Scientific Culture.

    The application of lasers to medicine and surgery has made amazing
progress since the last ISQE Course on this subject in 1983. The present
Proceedings give a tutorial introduction to today's most important areas,
as well as a review of current results by leading researchers. Among the
possible approaches to a NATO Advanced Study Institute on Laser Systems for
Photobiology and Photomedicine, we chose to emphasize the scientific and
technological aspects of advanced laser systems when applied to laboratory
and clinical tests. Since it is the policy of the School to stress the
advanced scientific and technological achievements in the field of Quantum
Electronics, the Course broadly covers performance already achieved and
potential applications.

    Because of the great variety of applications of laser systems in biol-
ogy, medicine, chemistry, engineering and related branches of science, this
Institute addressed a subject of interdisciplinary interest. The formal
sessions were balanced between tutorial presentations and lectures focusing
on unsolved problems and future directions. In addition, considerable time
was provided for the participants to meet together informally for additional discussions on the forefront of current work. Therefore the character
of the Institute was a blend of current research and tutorial reviews.

    We have brought together some of the world's acknowledged experts in
the field to summarize both the present state of their researches and the
background behind them. Most of the lecturers attended all the lectures
and devoted their spare hours to stimulating discussions. We would like to
thank them all for their admirable contributions. The Institute also has
taken advantage of a very active audience; most of the students were active
researchers in the field and contributed with discussions and seminars.
Some of these student's seminars are also included in these Proceedings.

    For this book, the lectures have been arranged in sections according
to topic, with tutorial and review material placed near the beginning of
each section. The papers in these Proceedings give a fairly complete accounting of the Course lectures with the exception of the informal discussions. In editing this material we did not modify the technical content of
the original manuscripts. In the particular case of the Chinese manus-
cript, the author does not always use accepted medical terminology; for
example, medical authors do not usually speak of "killing" cells, but use
other words. We did not change the author's terminology, because the
subject matter (laser acupuncture) is not a universally accepted medical
procedure anyway.

The topical sections are as follows:

A. "Physical and Biological Foundations", providing an overview of the field;
B. "Surgical and Ophthalmological Applications", covering the use of a variety of different laser types;
C. "Low Power Laser Therapy", presenting several important papers which give a firm scientific foundation to this area of photomedicine;
D. "Photodynamic Therapy: Mechanisms and Dosimetry";
E. "Photodynamic Therapy: Experimental Results";
F. "Diagnostic Techniques and Instrumentation"; and,
G. "Laser Safety and System Design".

These 1990 Proceedings give considerable coverage to "Low Power Laser Therapy" and "Photodynamic Therapy", recognizing the progress in these areas during recent years. The reader may also wish to consult the 1983 Proceedings ("Laser Photobiology and Photomedicine", edited by S. Martellucci and A. N. Chester, Plenum, E. Majorana Series N. 22). These earlier papers give more emphasis to "Fundamentals of Lasers and Tissue Interactions", "Laser Surgery", and "Photodermatology". Thus, the 1983 and 1990 Proceedings are complementary in their content.

This volume would, of course, not be possible without the considerable efforts put forth by the authors represented here; all of them are cordially acknowledged. We also wish to mention with sincere thanks Maria Teresa Petruzzi and Margaret Hayashi, secretaries to the ISQE Directors (S. M. and A. N. C.). We are also grateful to Mrs Vanna Cammelli for her skilful assistance in the organization of the course and to Miss Janie Curtis, editor at Plenum Press London, for her outstanding professional support. Before concluding, we acknowledge the organizations who sponsored the School, especially the generous financial support of the NATO A.S.I. programme.

The Directors of the NATO Advanced Study Institute:

A. N. Chester
Vice-President and Director
Hughes Research Laboratories
Malibu, California, U.S.A.

S. Martellucci
Professor of Physics
The Second University
Rome, Italy

A. M. Scheggi
Director of Research
I.R.O.E. - C.N.R.
Florence, Italy

October 30, 1990

CONTENTS

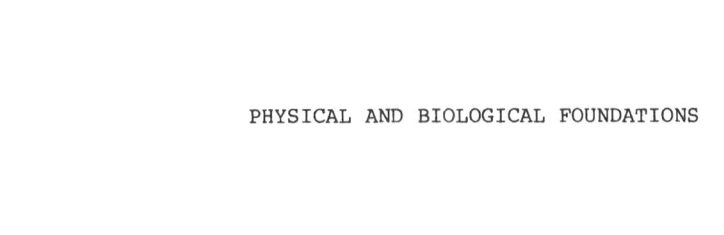

PHYSICAL AND BIOLOGICAL FOUNDATIONS

LASER PHOTOMEDICINE - AN OVERVIEW: PAST, PRESENT, AND FUTURE

D. H. Sliney (*) and M. L. Wolbarsht (**)

(*) US Army Environmental Hygiene Agency
Aberdeen Proving Ground, Maryland, USA
(**) Department of Psychology, Duke University
Durham, North Carolina, USA

I. INTRODUCTION

The Directors of this NATO-A.S.I. on "Laser system for photobiology and photomedicine" have asked us to provide some opening remarks, as we are the only two foreign faculty members of the previous Laser Photomedicine course held in Erice in 1983. It does not seem to us that seven years have passed since that last summer school, but there have been many new developments in laser applications in medicine and surgery. Nevertheless, most fundamental concepts remain unchanged and the basic lectures on lasers, tissue interactions and applications remain current. For this reason, many of the students (readers of the proceedings) may be interested in reviewing the proceedings (edited by S. Martellucci and A. N. Chester) of the previous school . In the previous school, emphasis was placed firstly upon tissue interactions, then studies at the cellular level, the photochemotherapy, photobiology, surgery, safety and diagnostics[1-26].

One of the fundamental difficulties encountered in any multidisciplinary field is the lack of effective communication due to a lack of common knowledge and common terminology. For example, the physicists concentrate only upon the initial events in laser-tissue interactions: absorption, transmission and scattering of energy, fluorescence, heat transfer and acoustic transients and ablation or vaporization. Design engineers were interested in providing safe, reliable lasers and practical delivery systems, and generally were enthusiastically hopeful that "their laser" could find more medical applications. The life scientists and clinicians were interested in primarily the morphological (structural) changes and the biological sequellae following laser exposure. In other words, physicists must be careful not to ignore the chain of biochemical and biological events that occur after energy absorption and dissipation, and the life scientist and clinicians must recognize the important optical and thermal properties of tissue and the variations of tissue vaporization and coagulation that can be achieved by varying wavelength (penetration depth, etc.) and exposure duration. J. A. Parrish and A. J. Welch emphasized these points in the previous school[2-4].

Seven years ago, there was an increasing number of physicians, surgeons and basic scientists who had achieved a sufficient level of multidisciplinary knowledge to communicate effectively on laser-tissue interactions. Although several investigators produced evidence for photobiological reactions induced by red light[5÷7], the consensus remained that low-power ("soft") laser effects referred to as "laser biostimulation", were most likely placebo effects.

The photophysical and photochemical interactions at the atomic and molecular level were a point of interest as photochemotherapy was a new subject of interest[8÷11]. The photobiological effects of short-wavelength light and ultraviolet radiation (UVR) were well accepted[12÷15]. Clinical experience in laser surgery was also a major point of interest[16÷20].

In the previous course, attention was paid to the use of proper radiometric terminology and interaction mechanisms[21]. Unfortunately, the misuse of radiometric terms such as fluence (for radiant exposure) and fluence rate (for irradiance) continue in the literature. These two terms should only be used for flux densities in tissue.

Laser safety was not ignored in the previous course[22÷24]. At that time, specific standards for laser safety were predicted, but had not been developed. Since then, medical laser safety standards are now an accepted fact.

Several papers concerned new technologies, such as light-emitting diodes[25], and fiber optics in laser medical applications[25÷27]. The previous proceedings provide an excellent insight into the state of progress of laser applications in medicine and surgery at that time. Since that time, a far better insight into the limits of laser applications in surgery has been realized[28]. It should now be evident that optimization of a laser surgical procedure requires a tradeoff between maximizing the competing, undesired interactions that produce unwanted side effects. For example, phototoxicity may often be present during coagulation, and destructing shock or acoustic warts may occur during ablative surgery. For example, phototoxicity may be present even during coagulation.

II. FUTURE CONSIDERATIONS

There will be a great host of new applications of lasers in medicine and surgery. However, it is our fervent hope that the new applications will be developed with an eye toward what tissue interactions are really needed, and do not continue as in the past depending only on a hit or miss empirical approach; in other words, design a laser with a specific wavelength and exposure characteristics rather than "Here is a laser - What can we do with it?" approach.

REFERENCES

1. S. Martellucci and A. N. Chester, Eds., "Laser Photobiology and Photo-medicine", E. Majorana International Science Series, vol. 22 Physical Sciences Plenum Press, New York (1985)
2. J. A. Parrish, "Photobiology and Photomedicine", in Reference 1, pp. 3-16
3. J. A. Parrish, "Effects of lasers on biological tissue: options for specificity", in Reference 1, pp. 17-28
4. A. J. Welch and M. Motamedi, "Interaction of laser light with biological tissue", in Reference 1, pp. 29-56
5. T. I. Karu and V. S. Letokhov, "Biological action of low-intensity monochromatic light in visible range", in Reference 1, pp. 57-66
6. S. Passarella, E. Casamassima, E. Quagliarello, I. M. Catalano and A. Cingolani, "Certain aspects of He-Ne laser irradiation on biological systems in vitro", in Reference 1, pp. 67-74

7. G. Delfino, S. Martellucci, J. Quartieri and E. Quarto, "A brief discussion on problems arising in the irradiation of whole biological specimens by laser beams, in Reference 1, pp. 75-84

8. F. Docchio, R. Ramponi, C. A. Sacchi, G. Bottiroli and I. Freitas, "Time resolved fluorescence Microscopy: Examples of applications to biology", in Reference 1, pp. 85-100

9. M. Aricò, M. Barcellona, M. S. Giammarinaro and S. Micciancio, "Time-resolved Spectrofluorometry of Melanin in Human Melanoma", in Reference 1, pp.101-106

10. A. Andreoni and R. Cubeddu, "Photophysical Properties of Hematoporphyrin Derivative and Related Components", in Reference 1, pp.109-116

11. L. Tomio, F. Calzavara and G. Jori, "Photoradiation Therapy with Hematoporphyrin as a Selective Technique for the Treatment of Malignant Tumors", in Reference 1, pp. 111-126

12. G. Palumbo, R. Massa, I. Vassalo, R. Bruzzese, S. Solimeno and S. Martellucci, "N_2 Laser Radiation Effects in Iodoaminoacids", in Reference 1, pp. 129-144

13. H. Malak, "The Influence of Oxygen-radicals on the Mechanisms of Conformational Changes and Denaturation of Hemoglobin Induced by UV Irradiation", in Reference 1, pp. 145-150

14. I. A. Magnus, "Aspects of Dermatological Photopathology", in Reference 1, pp. 151-162

15. R. Pratesi, "Phototherapy of Neonatal Hyperbilirubinemia", in Reference 1, pp. 163-208

16. C. Personne, A. Colchen and L. Toty, "Five years of Laser Applications in Broncology (1637 Endoscopy Resections, 921 Patients)", IN Reference 1, pp. 211-218

17. A. V. Fasano, G. F. Lombard, P. Martinetto, S. Tealdi and R. Tealdi, "Early Experiences with CO_2 Laser for the Treatment of Septic Pathology in Neurosurgery: Experimental Model with Other Sources", in Reference 1, pp. 219-226

18. R. Pariente, "Plastic and Reconstructive Surgery by Laser", in Reference 1, pp. 227-240

19. A. Musajo-Somma, "Laser Surgery in Hemophylia", in Reference 1, pp. 241-248

20. C. Balacco-Gabrieli, "Microsurgery by Means of a Nd-YAG Laser" in Reference 1, pp. 249-258

21. D. H. Sliney, "Laser Dosimetry and Damage Mechanisms", in Reference 1, pp. 261-270

22. D. H. Sliney, "Safety Aspects of Anterior Segment Laser Photodisrupters", in Reference 1, pp. 271-274

23. D. H. Sliney, "Laser Safety in Laser Surgery", in Reference 1, pp. 275-280

24. E. Righi, "Laser Safety in the Use of Laser Systems", in Reference 1, pp. 281-290

25. R. Pratesi, "Potential Use of (LED's) Incoherent and Coherent Light-émitting-diodes in Photomedicine", in Reference 1, pp. 293-308

26. A. M. Scheggi, "CO_2 Laser Radiation Delivery Systems", in Reference 1, pp. 309-316

27. A. M. Scheggi, M. Brenci and A. G. Mignani, "Optical Fiber Temperature Sensors for Medical Use", in Reference 1, pp. 317-323

28. D. H. Sliney, M. L. Wolbarsht, "Future applications of lasers in surgery and medicine", 82:293-296 (1989).

EXOGENOUS CHROMOPHORES FOR LASER NON-SURGICAL PHOTOMEDICINE

L. Goldman

Dermatology Department
Naval Hospital
San Diego, California, USA

I. INTRODUCTION

Chromophores in tissue are important for use with lasers in the visible light range. Chromophores may be endogenous, as with melanin and hemoglobin or exogenous, induced into tissues. Exogenous chromophores are used primarily for diagnostics in laser non-surgical medicine. However, the same exogenous chromophores may be used in laser photosurgery as aid in the initial diagnostics. This chapter is limited to exogenous chromophores for laser photomedicine.

The lists of exogenous chromophores truly increases almost daily. These chromophores change the imagery, as it were, for the incident or reflected or even absorbed laser beams. The beams induce fluorescence in the target. In this relatively new technology of exogenous chromophores, exogenous chromophores should have, briefly, the following characteristics: 1) Non-toxic in tissue, toxic dyes may be in laser systems; 2) Stability for the respective application; 3) Ease of administration [a. Parenteral: (i) Intravascular, and (ii) Peritoneal Cavity Absorption; b. Topical]; and, 4) Correlated to wavelength of a specific Laser System List of some chromophores.

Some of the current applications of this relatively new technology also which has a great future are: cancer diagnostic, organism applications, flow cytometry, enzymology, cardiovascular diagnostics, laser interferometry for cardiac and pulmonary pathology, individual cellular for Doppler, multicolor holography, transillumination diagnostics, diaphanographic diagnostics, histochemistry diagnostics, applications in dentistry, and spectroscopy in clinical medicine and clinical dentistry.

As one sees this list, it is evident that exogenous chromophores are an important technology for laser surgical medicine[1], and laser non-surgical medicine[2]. Laser medicine[3] differs from laser surgery. Laser non-surgical medicine includes laser photobiology, details of laser tissue interactions, laser spectroscopy in clinical medicine, laser Doppler flowmetry, laser cardiovascular diagnostics, laser immunology, laser treatments of arthritis, laser medicine in gastro-enterology, gynecology, head and neck surgery, perinatology, dermatology and urology. Also included in laser medicine are orthopedics, holography, the AIDS program, dentistry and veterinary medicine. For further details on the modern concepts of laser medicine and future developments see Reference n.2.

Although laser photobiology and laser diagnostic spectroscopy were the early beginnings of laser medicine, laser diagnostics in oncology is the real stimulus for initiating the progressive development of laser medicine of today. Because laser diagnostics of cancer are such an integral part of the laser treatment of cancer, laser cancer diagnostics will be considered in detail in laser exogenous chromophores in laser photosurgery.

Laser photobiology is the basis of all laser applications in biology, medicine and laser. Laser photobiology is also the basis of all laser safety programs. As new laser photons are introduced in biology, medicine and surgery, laser photobiology for laser tissue interactions is developed first.

Some exogenous chromophores and their wavelengths are:

1. Dihematoporphyrin ether or ester   643 nm
2. Rhodamine 1, 2, 3   500 nm
3. Rose Bengal   543 nm
4. Nile Blue A   623 nm
5. Crystal Violet   620 nm
6. Methylene Blue   587-643 nm
7. Tetracycline   730-770 nm
8. Metallo-phthalocyamines   760 nm

Others used recently include chlorophyll and hypericin.

The common lasers used often with exogenous chromophores include:

1. Ruby   614.3 nm
2. Argon   488-514.5 nm
3. Flash Pumped Dye   577-585   nm
4. Copper Vapor   577   nm
5. Gold Vapor   630   nm
6. Nd YAG   1064   nm
7. 2nd Harmonic Nd YAG   532   nm

An early application of exogenous chromophores, long before the advent of lasers, was tetracycline induced fluorescence for detection of the head of klarva migrans in the dermis, and on the teeth and in calcium of atheromas; for the head of larva migrans, deep laser impacts by Argon laser as compared to topical chemotherapy of Thiabendazole. Today, it is the fascinating test tube experiments of the induced fluorescence of the viruses on the envelope of leucocytes in the blood and their consequent destruction without apparent interference with the function of the blood. These experiments with PDT, described in the phototherapy of cancer and done at Baylor have included also, the exogenony chromophores for Candida and the malaria parasites. With this technique, many more pathogens will be included in this program.

Flow cytometry can make for "rapid measurements on individual biological cells as they pass in single file through a laser beam"[4]. The argon ion laser is used. The exogenous chromophore is a fluorochrome which binds, stoichiometrically, to DNA. In addition to individual cells, this technique uses flow cytometer instruments for such studies as: 1) studies of double stained cells; 2) single chromosome analysis; 3) measurements of scattered light with multiple detectors; 4) studies of polarized fluorescence emission from single cells; 5) scanning and imaging of cells. This has been important also for the diagnosis of cells of leukemia. This technique , "has created a need for multidisciplinary teams of biologists, chemists, physicists, engineers and statisticians."[4].

An experimental laser microfluorometer was developed[5], "based on (a) a subsecond-pulsed tunable dye laser, (b) a microscope system for fluorescence microscopy (Leitz MPV) with a PMT, and (c) a digital signal averager." Acridine dyes, important now in some phases of PDT and selective for induced papilloma studies were used, as well as fluorescent dyes, acridine DNA complexes, chromosomes and fluorescence decays. Many studies have developed since that initial reports.

Enzymology has recently become of interest in laser medicine. Topical enzymes were in the debridement of burns[6]. Coupling with fluorochromes could speed their action (stipulation? destruction?) in burn tissue. It has been reported that "carcinogen producing enzyme (AHH) absorbs blue light and fluorescences in the ultraviolet. This suggests therefore, that tumor discrimination may be achieved and could be used as a means of cancer detection"[7].

With the current interest in laser enzymology, there will be increasing applications for laser medicine.

Laser cardiovascular diagnostics relate to laser spectroscopy of normal vascular tissue and early and later developments of atheromas for laser angioscopy. Also, diagnostics of the various tissues of atheromas preparatory to atherolysis are important. Also, Sudan dyes for the fat; and persistent tetracycline induced fluorescence for the calcium of atheromas have been known for some time. Laser illumination of microscopic stained cardiac tissue is also under investigation.

Moire' patterns on the skin with the pulsed ruby laser, done originally at TRW of Redondo Beach, California, show the value of laser interferometry for the study of pulmonary pathology. Similar studies can be done over the precardium for cardiac pathology. Whether aerosol fluorochromes and intravascular fluorochromes would accentuate, as exogenous chromophores, and would amplify such patterns is not known. Also, whether such intravascular fluorochromes would delineate the reflectance patterns on ground glass screens, patterns of HeNe scanning of the precardium surface for arrhythmia is also not known.

Again, whether multicolor holography with exogenous chromophores would change the imagery of holography is not known, especially with the use of multiple laser systems. Synthetic wave-front interferometry (SWI) have been developed for ophthalmology[8]. Again, there is a possibility of developing this ophthalmologic technique for respiratory and cardiac functions.

With UV laser and selective staining of individual cells of the circulation, laser Doppler velocimetry of individual cells can be determined. With baseline studies of individual normal cells and normal circulation dynamics, changes in abnormal cells and abnormal circulatory changes can be studied.

Exogenous chromophores can be studied by laser transillumination with HeNe or Kr laser systems. Fixed tissue chromophores as with tattoos, including debris tattoos, can be located at relative depths in tissues. Areas of different tissue densities, as with cancers of the nose and breast, often may not require exogenous chromophore imagery. However, certain pathology reactions, as vascular changes, mucinoses, amyloidoses, myxedema and the like may require such. Laser colposcopy is another instrumentation which may be used for the vagina and cervix with topical exogenous chromophores. Similarly, laser, nasal, oral and laryngeal instrumentation may be used.

The procedure of recording transillumination field on film is known now, as diaphanography: "A diffuse, low intensity light transmitter with built-in flash is the source, and sensitive film is used for photographic registration." This is supposed to differentiate neoplastic, brown black, from benign tissue, red. Whether short-pulse laser system (Nd YAG-Q switched?) and exogenous chromophores will improve diaphanography is not known. Nd YAG low put with IR film and also with an image converter and photography; often such images have been studied in human tissue cancer phantoms. Nd YAG transillumination of the chest with IR film has been tried with unsuccessful results. Laser microscopy illumination with chromophore staining of cardiac tissue has been mentioned. Microscopic tissue illumination with HeNe and Kr laser systems and protective glasses was tried also with routine hematoxylin eosin staining and initial studies did not find this of value. Such preliminary studies should be continued with many more controls, more selective chromophores and remote control TV screen for laser safety.

For a detailed review in dentistry of laser luminescence reflection and diagnostics see Reference n.9. Luminescence of enamel with Ar and HeCd lasers in normal and pathologic enamel has been done. Laser reflection has been used for studies of abrasives in toothpaste. Doppler has been used for gingival blood flow. Laser transillumination has been used to detect cracks in teeth. "UV induced fluorescence spectroscopy has been used to detect residual tissue in canals during treatment"[9].

Holography has been used extensively in dentistry with holographic storage of x-ray images, holographic interferometry in molding materials, measurement of incisor extrusion, studies of orthodontic forces, tooth mobility, deformation of complete maxillary denture skull bone displacements by extra-oral forces and for forensic tooth models. From this long list, it is evident that holography has been used more in dentistry than in any other medical field. Combinations of red, green and blue laser lines have been used. Measurements, as shown, can be made of movemet, stress deformation and surface changes. "Computer assisted video holometry systems now permit observation of test subjects in real time with actual movements"[9]. Fiberoptics are now adapted for holography. Solid state small diode lasers "with sufficient energy output for holography are making their appearance"[9]. From early days microscopic holography has also been used.

From this review, it is evident that holography should be used more for laser diagnostic in medicine following the lead of dentistry. Initial holographic studies have started with the imagery by magnetic resonance imaging (MRI). Studies have been done in bone stress, but orthopedists should use holography more in both investigative and clinical studies.

This review of dental holography has stimulated a repetition of the section on Moire' patterns again in relationship to exogenous chromophores since lasers in the visible range are used in holography, and chromophores with respective wavelengths for these laser systems can be considered. These chromophores can bring out more surface changes and make for more diagnostic details, and stress formation changes in tissues may also show. More, even submicroscopic detail, arrhythmias may be detected earlier. With their greater experience, perhaps dentists could initiate the studies of exogenous chromophores in holography.

One of the early fields of laser non-surgical medicine was analytical laser spectroscopy. A plea is made for the practical value of spectroscopy in the clinical laboratory; these diagnostic analysis procedures are not only for toxicology but also for routine procedures. The LAMMA microscope is an important instrument for spectroscopy for trace metals and for spectroscopy of such microscopic structures. For chromophores like eosin B, laser analytical spectroscopy based on optical phase conjugation has been used[10].

The main use of exogenous chromophores in veterinary medicine at present has been PDT for the veterinary oncologists.

II. DISCUSSION

This chapter introduces the progress of the new technology in laser medicine and laser dentistry with greater potentials both in investigative and clinical studies. These applications include cancer diagnostics mainly and diagnostic potentials in many areas of biology and clinical medicine and dentistry. These vary from studies of micrograms, viruses, other organisms, flow cytometry enzymology, cardiovascular diagnostics including laser interferometry for pulmonary and cardiac function, the extensive developments in laser medicine. Also, included are Doppler, including individual velocimetry for individual cells in the circulation, transillumination and diagnostic diaphanography, histochemistry diagnostics, diagnostics in dentistry other than holography.

So, the future is great for the rapidly developing exogenous chromophores technology in biology, medicine and dentistry. When new diagnostic procedures are suggested for laser applications, as chronocoherent imaging (CCI)[11], the question of exogenous chromophores for better imagery should be considered.

Another example of exogenous diagnostics is MRI for both imagery and organ function[12]. For imagery also, there is research in holography for brain tumor diagnostics. With exogenous chromophores and with special non-metallic probes, there are great possibilities for deep interstitial tissue diagnostics for tumors diagnostics with MRI[13]. With the current laser scanning scope, this could be applied to MRI holographic brain tumor diagnoses. Chromophores with MRI could be used also with the fluorescence spectroscopy techniques. So, MRI is added to the great world of Photomedicine of the future.

Table 1. Exogenous Chromophores for deep interstitial tissue diagnostics.

1.	SPECIAL INTERSTITIAL FIBER OPTICS PROBES
2.	DEEP TISSUE TRANSILLUMINATION LASER TECHNIQUES
3.	CHROMO-COHERENT IMAGING (CCI) OF TISSUE
4.	OPTICAL PHASE CONJUGATION
5.	MAGNETIC RESONANCE IMAGING (MRI) WITH SPECIAL NON-METALLIC PROBES
6.	MAGNETIC RESONANCE IMAGING (MRI) WITH HOLOGRAPHY

III. ACKNOWLEDGEMENTS

The work reported herein was performed in part by the Navy Clinical Investigation Program reports 84-16-1968-224 and 84-16-1968-225. The views expressed in this article are those of the author and do not reflect the official policy or position of the Department of the Navy, Department of Defense, or the United States Government.

REFERENCES

1. L. Goldman, "Medical and Surgical uses for the lasers", <u>New Scientist</u>
    21:280-286 (1980)
2. L. Goldman, Ed. "The Biomedical Laser, Technology and Clinical
    Applications", Springer Verlag, New York (1981)
3. L. Goldman, Ed., "Laser Non-Surgical Medicine", <u>Technomic Publishing</u>
    <u>Co. Inc.</u>, (1990)
4. G. C. Salzman, "Flow cytometry: the use of lasers for rapid analysis
    and separation of single biological cells", 33-53, in Ref. 2
5. A. Andreoni, A. Congoni, C. A. Sacchi, O. Svelto, "Laser fluorescent
    microirradiation: a new technique", 69-83, in Ref. 2
6. J. P. Fidler, "Techniques of laser burn surgery": 209-218, in Ref. 2
7. D. B. Greenberg and M. D. Tribbe, "Tumor diagnosis by laser transil-
    lumination and diaphanographic methods" 284-291, in Ref. 2
8. L. S. Horwitz, Wm. H. Haas, and J. A. Householder, personal communi-
    cation
9. G. C. Willenborg, "Laser and electro-optics in dentistry", in Ref. 3
10. Z. Q. Wu and W. G. Tng, "Laser analytical spectrometry based on opti-
    cal phase conjugation by degenerate four wave mixing in a flowing
    liquid analyte cell", <u>Anal. Chem.</u> 1:61(9):998-1001 (1989)
11. K. G. Spears, J. Serapin, N. H. Abramson, X. Zhu, and H. Bjelkhagen,
    "Chrono-coherent imaging", IEEE Transactions on Biomedical Engi-
    neering 36 No. 12:1210-1221 (1989)
12. D. J. Castro, R. E. Saxton, L. J. Layfield, H. R. Fetterman, P. B.
    Tartell, J. D. Robinson, S. Y. O. To, E. Nishimura, R. B. Lufkin,
    P. H. Ward, "Interstitial laser phototherapy assisted by magnetic
    resonance imaging: a new technique for monitoring laser tissue
    interaction", American Society for Laser Medicine and Surger,
    Tenth Annual Meeting, Nashville, Tenn., April 6 (1990)
13. D. J. Castro, R. E. Saxton, R. B. Lufkin, L. J. Layfield, P. B.
    Tartell, J. D. Robinson, P. H. Ward, "Interstitial laser photo-
    therapy for treatment of deep tumors assisted by MRI", American
    Society for Laser Medicine and Surgery, Tenth Annual Meeting,
    Nashville, Tenn., April 8 (1990)

MODELLING AND MEASUREMENTS OF LIGHT PROPAGATION IN TISSUE

FOR DIAGNOSTIC AND THERAPEUTIC APPLICATIONS

B.C. Wilson

Hamilton Regional Cancer Centre, McMaster University
Ontario Laser and Lightwave Research Center
Hamilton, Ontario, Canada

I. INTRODUCTION

The propagation of light (ultraviolet, visible and infrared) in tissue is fundamental to many diagnostic and therapeutic applications in photomedicine. The former are based on the effect of tissues on light, as in, for example, diffuse transmittance or reflectance spectroscopy, and fluorescence spectroscopy. Therapeutic applications, which depend on the effect of light on tissue involve the absorption of light energy by tissue chromophores, leading to specific or non-specific photophysical/photochemical changes. In both cases, the spatial (and temporal) distribution of light fluence is important, and is determined essentially by the optical properties of the tissue.

Light distributions in tissue may be described either by the propagation of electromagnetic waves through a dielectric, governed by Maxwell's equations, or by the transport of discrete photos, described by the Boltzmann equation. In the former it is possible, in principle, to account for wave-dependent phenomena such as interference, refraction and reflection, and to determine both amplitude and phase dependencies. However, solutions have been developed for only very restricted cases[1]. More progress has been made using transport theory to describe the local energy fluence (rate) distribution, which is usually the primary quantity of interest.

The relevant optical properties are then the absorption (extinction) and elastic scattering coefficients and the differential scattering coefficient (phase function). The absorption can be related directly to the tissue dielectric properties. The scattering has been associated with microscopic fluctuations in refractive index. The average tissue refractive index, n, accounts for specular reflection at tissue boundaries and for the speed of light in tissue.

II. LIGHT ABSORPTION AND SCATTERING IN TISSUE

The absorption of light tissue is due to specific chromophores. At low fluence rates, the local rate of energy absorption, $P(\underline{x},\lambda)$ is

$$P(\underline{x},\lambda)= \sum_i \epsilon_i(\lambda) \times C_i(\underline{x}) . \quad \phi(\underline{x},\lambda) = \mu_a(\lambda) \times \phi(\underline{x},\lambda) \tag{1}$$

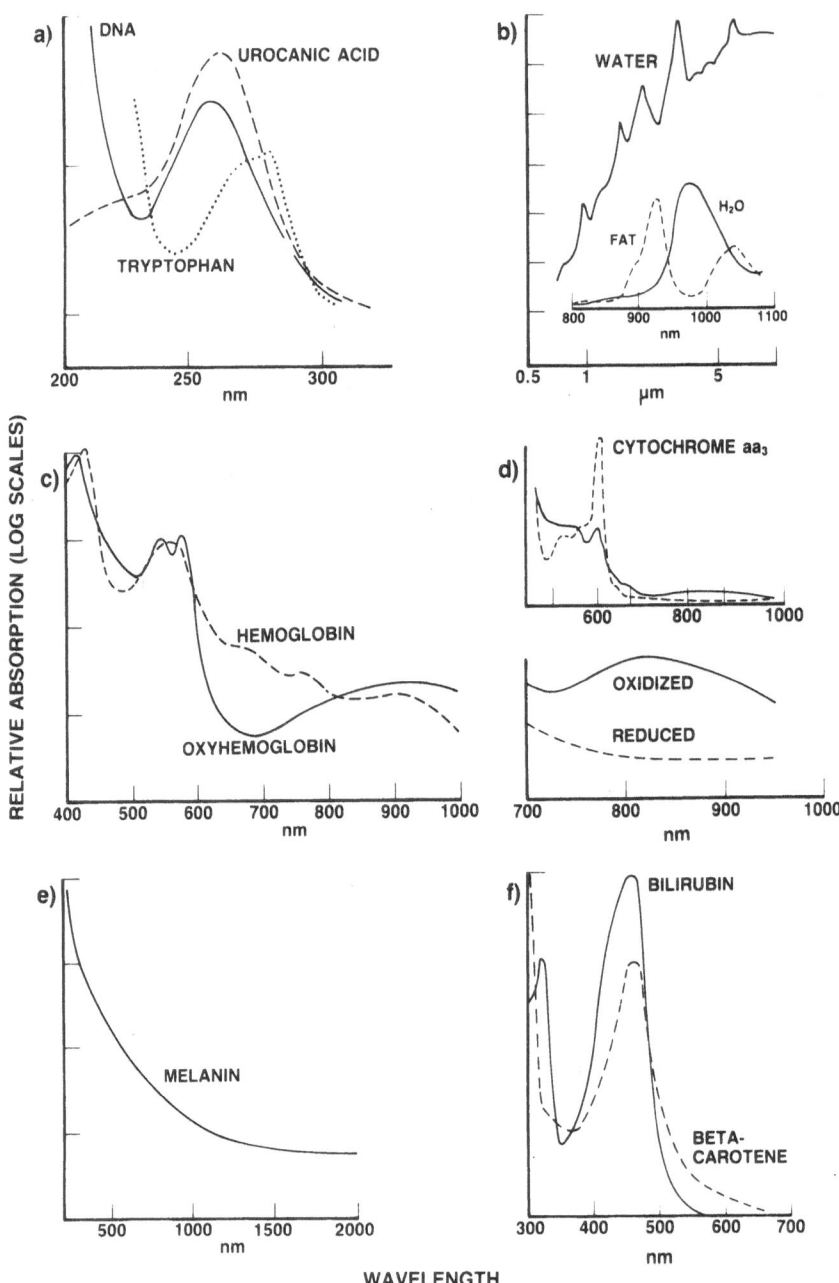

Fig. 1. Absorption (extinction) spectra of various tissue chromophores.

where C_i is the local concentration of the i-th chromophore of extinction coefficient ε_i and ϕ is the local energy fluence rate. The total tissue absorption coefficient is μ_a. With short-pulse lasers generating very high energy fluence rates, non-linear absorption processes may occur, so that Eq. (1) no longer holds and the dependence of R on ϕ may be of higher order.

Many of the important chromophores in tissue have distinctive absorption spectra, as illustrated in Figure 1. Different chromophores are dominant in different spectral regions. At mid- and far-infrared wavelengths (> ~1500 nm), water is the main chromophore in soft tissues: it has its maximum of about 1300 mm^{-1} at 2,9 µm, so that the fluence rate for an incident beam falls by 99% in less than 10 µm depth. In the UV-B and UV-C regions, biomolecules such as nucleic acids dominate.

In the visible and near-infrared, hemoglobin and other pigments such as melanin and bilirubin are strong absorbers, while other chromophores are also important in specific applications, as for example cytochrome-aa$_3$. Note that the absorption spectrum of hemoglobin, and hence of blood, is markedly different between the oxygenated and reduced states, but there are some wavelengths ("isobestic points") at which the extinction coefficients are equal. In whole blood the absorption in the visible/near-IR is due essentially to the hemoglobin in the erythrocytes, and is proportional to the hematocrit: at the 815 nm isobestic point the absorption coefficient for a 50% hematocrit is about 1 mm^{-1}. In the visible/near-IR, μ_a is typically ~ 0,05 - 0,5 mm^{-1} for soft tissues[2-5]. The spectral "signature" of individual chromophores may be apparent, such as the 500-600 nm Hb peaks or the H_2O near-IR peaks in Figure 2b.

Scattering in tissue may be regarded as due to reflection and refraction from micro-inhomogeneities in the refractive index. In the radiation transport model the scattering is assumed to occur at discrete, independent scattering centres. In scattering from such "particles", the scattering coefficient and phase function depend on the difference in refractive index between the particle and its surrounding medium and on the ratio of particle size to wavelength. For small particles (Rayleigh scattering), the coefficient varies as $1/\lambda^4$ and the phase function is roughly isotropic, i.e. independent of scattering angle: for particles large compared to the wavelength (Mie scattering), the cross-section varies approximately as $1/\lambda$ and is forward peaked.

In the visible/near-IR, where most measurements have been made, scattering in tissue appears to be Mie-like: see, for example, Figure 2. The scattering coefficients of soft tissue, μ_s, are typically ~10-100 mm^{-1} and decrease slowly with increasing wavelength. The scattering is generally forward-peaked, with g ($<\cos \theta>$) in the range 0.7-0.95. If the phase functions are considered as due to independent spherical particles, then the equivalent "scattering particle" size in tissue is of the order of microns. However, it is not clear that this interpretation is physically correct, since tissue is not merely a suspension of cells. The scattering coefficient is independent of oxygen saturation, and shows a more complex dependence on hematocrit than does the absorption, due to loss of independent scattering between cells at higher concentrations (~H{1-H}{1.4-H})[6]. The coefficient falls slowly with increasing wavelength, being approximately 500 mm^{-1} for 50% hematocrit at 815 nm. As would be predicted by Mie theory, the scattering is highly forward-peaked (g ~ 0.98 in the visible).

III. LIGHT FLUENCE DISTRIBUTIONS WITHIN AND FROM TISSUE

The spatial distribution of light within tissue, and the distribution of diffuse (i.e. multiply scattered) reflected and transmitted light depend

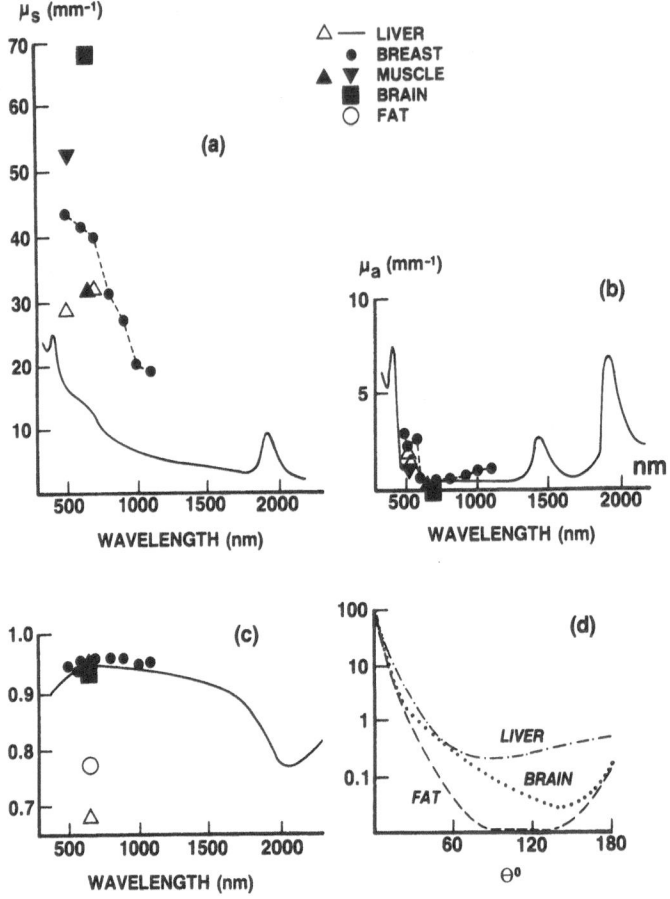

Fig. 2. Optical properties of typical soft tissue in the visible/
near IR. The solid curves in (a)-(c) are in rat liver, after
Parsa et al.[2]; those in (d) are from Flock et al.[3] (brain,
fat; 633 nm) and Marchesini et al.[4] (liver; 635 nm). Other
data are human glandular breast tissue[5], bovine muscle (▲)[3],
human muscle (▼)[4], pig brain and fat[3]. All data are from
measurements made in vitro. Some of the apparent spectral
structure in μ_s and g may be measurement artefact.

critically on the tissue optical properties. The ratio of absorption to
scatter is particularly important. Three main spectral zones may be identified: absorption-dominated, scattering-dominated and similar scatter and
absorption.

Consider a collimated beam incident on a semi-infinite tissue with
homogeneous absorption and scatter. For the absorption-dominated case
$(\mu_a \gg \mu_s)$, Beer's Law applies and the decrease of fluence with depth is
simply exponential (Figure 3a). There is no significant beam spreading
beyond the edges of the incident light field and little diffuse reflectance. The specular reflectance is determined by the refractive index,
according to the Fresnel's equations. For soft tissue, propagation is
essentially absorption-dominated at wavelengths ≲ 600 nm and ≳ 2000 nm.

For scattering-dominated propagation (Figure 3b), diffusion theory may be applied and, at large enough depth, the decrease of fluence with depth is also exponential: for a broad-beam the effective attenuation coefficient, $\mu_{eff} = [(3\,\mu_a\{\mu_a + \mu_s\,(1-g)\})]^{1/2}$.

Effective attenuation coefficients of soft tissues in the red/near-IR are typically[7,8] in the range 0.1-2 mm^{-1}. Thus, in this "optical window" the diffuse transmittance can be substantial. There may be a sub-surface peak in the fluence due to backscattered light, and the fluence can be several times larger than the incident fluence[9,10]. This depends on the boundary conditions at the surface: for example, at an air-tissue boundary, significant internal (specular) reflection occurs due to refractive index

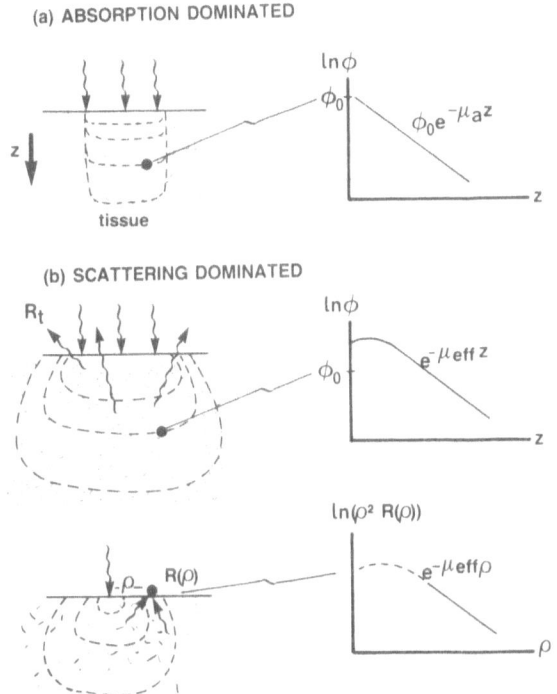

Fig. 3. Illustration of light fluence and diffuse reflectance
        distributions in tissue.

mismatch. There is substantial sideways spreading of the beam in the tissue. The total diffuse reflectance, R_t, depends primarily on the tissue transport albedo, $a' = \mu_s'/(\mu_a + \mu_s')$, where $\mu_s' = \mu_s(1-g)$. The local diffuse reflectance, $R(\rho)$, at radial distance ρ from an incident pencil beam varies approximately[11] as $\rho^{-2}\exp(-\mu_{eff}\rho)$ at large ρ. Thus, apart from the ρ^{-2}

factor, the radial dependence of local diffuse reflectance with distance from a point source at the surface is the same as the depth dependence of fluence within tissue for broad beam irradiation. At smaller radial distances the behaviour is more complex[12], depending both on μ_{eff} and μ'_s. The diffuse transmittance of light through a finite thickness of tissue can also be described by diffusion theory for the scatter-dominated case[13]. However, there is presently no corresponding simple analytic relationship between the total or local diffuse transmittance and the tissue optical properties.

In the intermediate spectral zones of comparable absorption and scatter, there are no straightforward descriptions for the fluence distributions or reflectance/transmittance. Higher-order solutions of the transport equation are required[14]. As with the scattering-dominated case, Monte Carlo computer modelling has proved to be a powerful tool for studies of light propagation in such media[10].

IV. MEASUREMENTS OF TISSUE OPTICAL PROPERTIES

In Vitro - To date, much of the information on tissue optical properties has been derived from measurements made in vitro. This may be done: i) by direct determination using optically thin ($<<1/[\mu_s + \mu_a]$) tissue sections[3,4], ii) indirectly as in measurement of the total diffuse reflectance, R, and transmittance, T, for optically-thick sections[2,5], or iii) by measurement of the change in μ_{eff} as known amounts of absorber are added to the tissue[15]. The indirect techniques require derivation of the "inverse solution" to obtain the optical coefficients from the measured data.

In Vivo - Until recently, in vivo measurements comprised mainly the mapping of fluence distributions in tissues using interstitial opticalfiber detector probes[7,16]. This has been used to determine parameters such as μ_{eff} and the radiance pattern of light at different depths within irradiated tissue, from which the basic coefficients can be derived. There are now, however, a number of promising non-invasive methods for determination of tissue optical properties. These are also directly relevant to applications of in vivo spectroscopy.

The two main classes of non-invasive techniques are radiometric and photothermal. The discussion here will be concerned mainly with the former. Photothermal methods include pulsed photothermal radiometry (PPTR) and photoacoustic spectroscopy (PAS). In PPTR the tissue is irradiated with a light pulse, at the wavelength of interest, of sufficient energy to cause an initial temperature rise of a few degrees distributed according to the local energy absorption. The subsequent temperature-time curve at the tissue surface resulting from the thermal diffusion is determined by measuring the IR emission. This technique has been mainly used for the absorption-dominated situation. Recently, studies have used PPTR in tissues at wavelengths where there is significant scattering[17].It is not clear how feasible it will be to separate the absorption and scattering properties by this technique. The same is true in the second photothermal technique, PAS, in which the signal measured is the acoustic wave generated by the induced thermal wave in the tissue heating an enclosed air volume at the tissue surface[18].

Purely radiometric techniques based on measurements of the diffuse reflectance or transmittance of light are intrinsically most sensitive and accurate in the scatter-dominated regime, i.e. in the red/near-IR spectral range, and may, therefore, be complementary to photothermal methods which are simplest in the absorption-dominated regions. As illustrated in Figure 4, there are essentially three different categories of measurement which can be made: reflectance or transmittance; total or local fluence; steady-state or time-resolved. The discussion here will be illustrated mainly by the diffuse reflectance, for which most work has been done. The measurable quantities are then R_t, $R(\rho)$, $R_t(t)$, $R(\rho,t)$, corresponding to total and local reflectance under steady-state (time-independent) and time-dependent conditions.

IV.1. Steady State Measurements

As discussed above, measurement of R_t yields the transport albedo, a', while the asymptotic slope of $\ln[\rho^2 R(\rho)]$ versus ρ yields μ_{eff}. The absorption coefficient, μ_a, and the transport scattering coefficient, μ'_s, can then be calculated. In practice, it is difficult to measure R_t accurately, but recent work[12] has shown that it may be possible to obtain both μ_a and μ_{eff} by measuring the shape of the complete R( ) curve, including the dependence at small ρ values. This has the advantage that only relative fluence measurements are required, not absolute values.

IV.2. Time-Resolved Measurements

Light photons take a finite time to propagate in tissue: speed, $c' = c/n \simeq 0.21$ mm/psec. Thus, the remitted light is spread in time following an incident short light pulse. Consider the example in Figure 5 in a typical soft tissue in the red/near-IR ($\mu'_s = 2$ mm^{-1}, $\mu_a = 0,02$ mm^{-1}) following an incident picosecond light pulse. At radial distance $\rho = 40$ mm, no photons will be detected for $t < 40/0.21 \sim 190$ psec, representing the direct line-of-flight. There is then a sharp rise, with the signal reaching a peak at about 1350 psec. The signal finally reaches an exponentially-decaying region. Note that photons delayed by 1350 psec have travelled 28 cm in the tissue, and (for g = 0.9) have undergone approximately 5,500 scattering interactions. The probability of a photon travelling this distance without being absorbed is equal to exp (0.02 × 280) = 0.4%. At 4 ns time delay, the corresponding values are 84 cm path length, 17,000 scatters and 5×10^{-6}%.

Under such conditions, where the light is fully diffused ($\rho \gg 1/\mu'_s$)

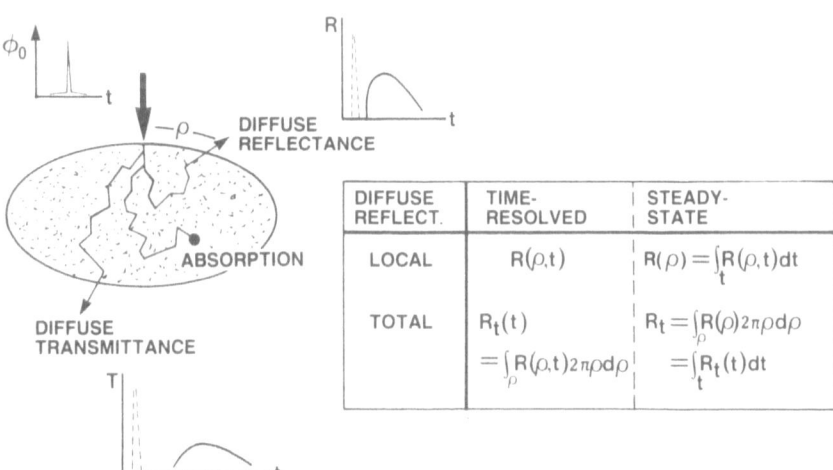

DIFFUSE REFLECT.	TIME-RESOLVED	STEADY-STATE
LOCAL	$R(\rho,t)$	$R(\rho) = \int_t R(\rho,t)dt$
TOTAL	$R_t(t)$ $= \int_\rho R(\rho,t)2\pi\rho d\rho$	$R_t = \int_\rho R(\rho)2\pi\rho d\rho$ $= \int_t R_t(t)dt$

Fig. 4. Steady-state and time-resolved optical remittance, as illustrated by diffuse reflectance.

the following observations have been made, both theoretically by time-dependent diffusion theory[13] and experimentally in tissues and tissue-like media[19÷21]:

i) $R_t(t)$ and $R(\rho,t)$ depend on μ_a and $\mu_s^{\,!}$ (and n);
ii) the final slope of $\ln[R(\rho,t)]$ equals $-\mu_a \cdot c'$, independent of the scattering coefficient and radial distance; and,
iii) the time-to-maximum value of $R(\rho,t)$ increases roughly linearly with $\mu_s^{\,!}$, and ρ, but is only weakly dependent on μ_a : $\mu_s^{\,!}$ can then be obtained from t_{max} at fixed ρ, given μ_a.

An alternative approach to such time-dependent light propagation measurements is to work in the frequency domain[22÷24]. Thus, rather than measuring the reflectance-time curve following a short input light pulse, the amplitude modulation and phase shift of the remitted light fluence are determined with reference to a continuously modulated input light source of known modulation frequency. In principle, the same information can be obtained as in the time-domain method[22].

There are many factors which enter into the choice between the steady-state and time-dependent techniques. These include equipment cost, complexity, wavelength tunability and clinical feasibility, as well as fundamental questions of accuracy of the optical property determination. With present technology, steady-state measurements are significantly cheaper and simpler, and clinical instrumentation is fairly straightforward[12]. Time-dependent techniques are presently confined to a major laboratory setting. However, the technology is changing rapidly and this approach should become more generally feasible in the next few years. In addition, hybrid methods, using a combination of steady-state spectroscopy and limited, but simpler, time-dependent measurements (e.g. gated photon time-of-flight) are being investigated[22]. As regards the fundamental accuracy, major factors

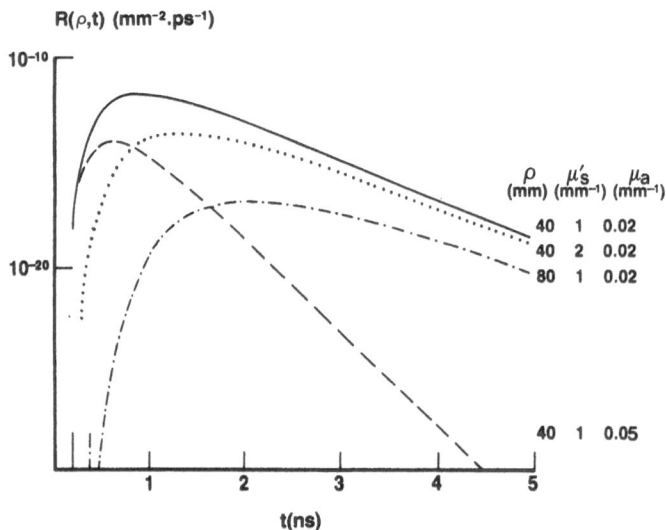

Fig. 5. Time-resolved local diffuse reflectance spectra for a semi-infinite, optically homogeneous medium. (Medium-air boundary, n = 1.4). The vertical lines mark the direct times of flight ρ/c'. These curves are predicted from time-dependent diffusion theory[13÷19].

include the effects of optical heterogeneity of tissue and the effects of finite tissue volume and tissue boundary conditions. The optimum technique may be different for different conditions: for example, it may be more accurate to use time resolved methods to measure μ_a (or changes in μ_a) for small tissue volumes using time-resolved measurements.

In the above, the diffusion regime has been assumed. It is also possible to examine the spatial and temporal distribution of light fluence and reflectance at small distances and short time-scales, say of the order of 1-100 psec, before a high degree of multiple scattering has occurred[25÷27]. As illustrated in Figure 6 a,b it is possible that the reflectance measurements, representing early back-scattered light, are sensitive to the scattering anisotropy, g, and not simply to the transport scattering coefficient. The fluence distributions within tissue (Figure 6 c,d) also show interesting behaviour at these short time ranges, with a propagating light "front" which has a complex dependence on the tissue optical properties.

V. APPLICATIONS

There are two main classes of application, therapeutic and diagnostic, in which knowledge of the optical properties of tissues and/or of the spatial/temporal distribution of light propagating in tissue is important.

V.1. Therapeutic Applications

Light is used therapeutically through a number of different light-tissue interaction mechanisms: photothermal, photochemical and photomechanical. In laser-based treatments, each of these may utilize either linear or non-linear (e.g. multi-photon) absorption processes, the latter being associated with high peak power, short pulsed irradiation.

In the linear regime the most extensive studies of light propagation in tissue have been for photodynamic therapy (PDT) of solid tumors in the red/near-IR. Ultimately, the goal is to determine prospectively the optimal treatment parameters. Recent investigations of quantitative diffuse reflectance in vivo have two main objectives as applied to PDT: 1) the determination of light fluence distributions at the treatment wavelength, and 2) the determination of photosensitizer concentration in target tissue. For the former, μ_a, μ_s' (and possibly also g and n) are required at the photoactivating wavelength. For the latter, the true absorption spectrum, $\mu_a(\lambda)$, is needed in order to discriminate the photosensitizer absorption spectrum from the background tissue spectrum, and hence to measure the photosensitizer concentration, correcting the reflectance signal for the effects of the intrinsic tissue absorption and scatter. Instruments have been developed for this based on steady-state spectroscopy[12,28]. The use of $R(\rho,\lambda)$ measurements to determine $\mu_a(\lambda)$ (tissue + photosensitizer) and $\mu_s'(\lambda)$ (tissue) have allowed absolute quantification of photosensitizer concentration in vivo. Some preliminary work has also been done to evaluate the feasibility of using time-resolved (time or frequency domain) measurements for this application[22]: this may have advantages for limited tissue volumes.

Spin-off from these studies are techniques to quantify the fluorescence yield spectra of photosensitizers in tissue in vivo, by determining the optical properties of the tissue at the fluorescence excitation and emission wavelengths, thereby accounting for the spatial distribution of the excitation light and the attenuation of the emitted light.

The spatial distribution of light in tissue is also important in other

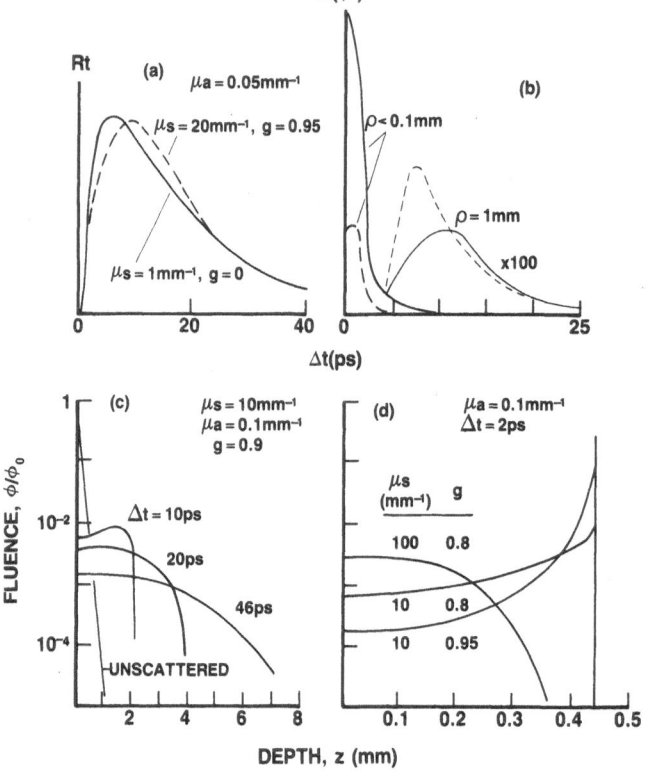

Fig. 6. Time-resolved reflectance (a,b) and fluence-depth (c,d) distributions in the short-time regime, illustrating the dependence on the scattering anisotropy g as well as on the scattering and absorption coefficients. These are smoothed curves based on Monte Carlo studies: a) Ref. 25; b) this laboratory, unpublished; c, d) Ref. 27.

therapeutic medical laser applications. Studies have included, for example: i) light propagation in optically-layered media, such as in laser irradiation of skin[29], ii) the changes in tissue optical properties arising from laser exposure[30], and, iii) optical-thermal modelling in photothermal treatments, such as laser hyperthermia with near-IR irradiadiation[31,32].

V.2. Diagnostic Applications

There is an increasing range of clinical and experimental diagnostic applications based on non-invasive reflectance and/or transmittance optical spectroscopy. Each involves effectively the determination of absolute or relative tissue absorption or, more rarely, scattering coefficients or spectra. The applications include:

i) measurements of endogeneous tissue chromophores, such as melanin and other skin pigments[33÷35], body fat content[36] cytochrome aa₃ and hemoglobin concentration and oxygenation in muscle and brain tissue[20,37÷39]
ii) measurement of exogeneous chromophores[12,28];
iii) transillumination spectroscopy for detection of breast cancer[40];

iv) monitoring of metabolic or electrical activity of tissues such as retina and cerebral cortex[41];

v)   the calibration of macroscopic or microscopic fluorescence spectra[12,42,43]; and,

vi) the analysis of laser Doppler blood flow measurements to account for multiple scattering by moving red blood cells[44].

Here, examples of the first class of techniques will be presented to illustrate the principles.

Non-invasive characterization of the endogeneous chromophores hemoglobin and cytochrome-aa$_3$ have been used widely in different organ systems, both for clinical applications and in physiological research. Brain and muscle have been studied extensively using surface illumination and detection[20,37-39], while endoscopic reflectance spectroscopy has been developed recently for internal tissues such as gastrointestinal mucosa[44]. Hemoglobin spectra give information on the oxygenation of the blood, while cytochrome-aa$_3$ oxygenation is related to tissue metabolic rate. Both chromophores have distinctly different spectra in the oxygenated and reduced states (see Figure 1). Thus, measurements of the diffuse reflectance or transmittance spectra yield information on the oxygenation status. Relative changes in tissue oxygenation, as in continuous monitoring, can be determined fairly easily, and contact oximetry using comparative measurements at isobestic and non-isobestic wavelengths is a well-established clinical tool.

The determination of absolute tissue oxygenation, or indeed of the absolute absorption of any chromophore, is more difficult, since the measured spectrum depends not only on the true absorption of the chromosphere but also on the tissue scattering, and on the tissue and source-detection geometries. In the general case, techniques such as those discussed above using spatially and/or temporally-resolved remittance measurements, are required. For situations where there is a well-defined and fixed geometry, absolute calibration might be achieved without a complete determination of the separate absorption and scattering coefficients. For example, in transillumination spectroscopy of the neonatal head to determine brain tissue oxygenation, a measurement of the average optical path length can be used to calibrate the Hb/HbO$_2$ signal. This could be obtained by a time-resolved measurement of the transmitted signal following a short input pulse. Alternatively, it may be possible to use the size of the steady-state absorbance signal from a chromophore whose tissue concentration is known, e.g. that of water at 970 nm[45]. These are, of course, only approximate calibration techniques, since they assume, for example, that the optical path length is wavelength independent (i.e. that the scattering coefficient does not vary with wavelength).

In skin, the chromophores of interest are hemoglobin, melanin, and bilirubin. In this organ, a major complicating factor is that the chromophores are not uniformly distributed in the tissue, but are essentially layered. The scattering properties of the stratum corneum, epidermis, dermis and subcutaneous tissues are also different. Again, in general, relative changes in chromophores are easier to determine than absolute concentrations. For example, changes in arterial and venous blood can be monitored, as in erythema following UV irradiation or vasoconstriction caused by topical steroids[48]. Little work has been done on determination of the spatial distribution of chromophores or of their absolute concentrations in skin. The feasibility of such measurements may be illustrated by the preliminary results shown in Figure 7, where good agreement was obtained between independent measurements of total diffuse reflectance spectra and the $R_t(\lambda)$ values calculated from measured $R(\rho,\lambda)$ values.

Finally, there has been recent interest in the possibility of regional in vivo spectroscopy and in vivo optical imaging, for example for measuring regional brain tissue oxygen levels or in detecting localized lesions such as tumors in the breast. It is likely that time-dependent methods will be essential for this, and, as illustrated in Figure 8, there is preliminary evidence[24,49,50] that diffuse reflectance/transmittance time spectra are significantly altered by the presence of localized optical inhomogeneities within tissues. Ultimately, this may lead to the optical analogue of computed tomography, with the mapping of the spatial distribution of optical properties. The major question is whether sufficient spatially-localized information can be obtained, either by measuring the complete tissue spectra of remitted light or by time gating by suppressing the scattered component in the remittance signal.

VI. CONCLUSIONS

The propagation of light in tissue, under conditions of linear absorption, has been described in terms of radiation transport models. Those models, and experimental studies of the corresponding optical absorption and scattering properties of tissues, have lead to an understanding of how to measure such properties non-invasively. This capability may be exploited in various phototherapeutic and photodiagnostic techniques.

These techniques involve fundamentally the determination of the true absorption (and/or scattering) spectra of the tissue. New technologies, particularly ultrafast (psec) lasers and corresponding time-resolved or time-gated detectors, may contribute significantly to in vivo spectroscopy. A major challenge will be to develop methods and instrumentation appropriate to specific clinical problems.

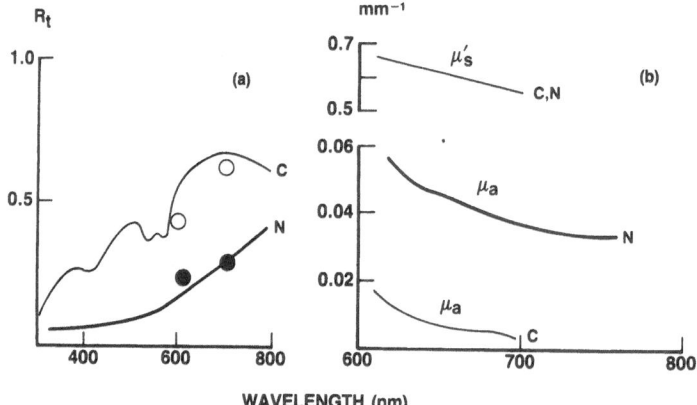

Fig. 7. Reflectance spectroscopy in human skin in vivo. a) The curves show the total diffuse reflectance spectra for Caucasian (C) and Negroid (N) skin, after Ref. 35; b) The absorption and transport scattering coefficient spectra, determined from $R(\rho,\lambda)$ measurements using the technique and instrument described in Ref. 12. The data points in a) were then calculated from these μ_a and μ_s values using diffusion theory[10] with n = 1.4: open symbols – Caucasian; closed symbols – Negroid.

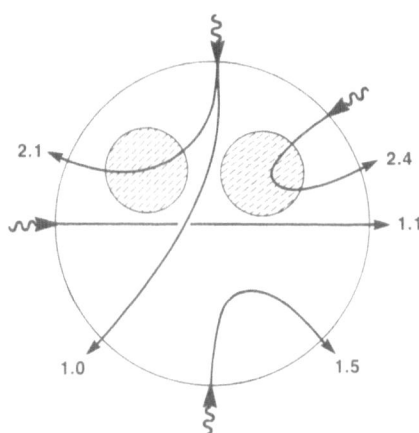

Fig. 8. The effect of optical inhomogeneity on time-resolved
        remittance spectra: 12 cm diameter cylinder of scattering
        medium with absorber added to two 3 cm diameter cylin-
        drical inserts. The numbers shown on each "optical path"
        between different source-detector distances represent the
        relative magnitude of the final slope of the remitted time
        spectra (= $-\mu_a c'$ for homogeneous medium). Data kindly
        supplied by Dr. B. Chance, University of Pennsylvania.

VII. ACKNOWLEDGEMENTS

Thanks are due to T. Farrell and K. Moon for providing data for Figures 6 b and 7 and to Dr. B. Chance, University of Philadelphia for data
in Figure 8. Support is provided by the National Cancer Institute of
Canada and the Ontario Laser and Lightwave Research Centre.

REFERENCES

1. A. Ishimaru, "Wave Propagation and Scattering in Random Media",
   Academic Press, New York (1978)
2. P. Parsa, S.L. Jacques and N.S. Nishioka, "Optical properties of rat
   liver between 350 and 2200 nm", Appl. Opt. 28:2325 (1989)
3. S. T. Flock, B.C. Wilson and M.S. Patterson, "Total attenuation coefficients and scattering phase functions of tissues and phantom
   materials at 633 nm", Med. Phys. 14:835 (1987)
4. R. Marchesini, A. Bertoni, S. Andreola, E. Melloni and A.E. Sichirollo,
   "Extinction and absorption coefficients and scattering phase functions of human tissues in vitro", Appl. Opt. 28:2318 (1989)
5. V. G. Peters, D.R. Wyman, M.S. Patterson and G.L. Frank, "Optical
   properties of normal and deceased human breast tissues in the
   visible and near infrared", Phys. Med. Biol (in press)
6. J. M. Steinke and A.P. Shepherd, "Diffusion model of the optical absorbance of whole blood", J. Opt. Soc. Am. 5:813 (1988)
7. B. C. Wilson, W.P. Jeeves and D.M. Lowe, "In vivo and post mortem
   measurements of the attenuation spectra of light in mammalian
   tissues", Photochem. Photobiol. 42, 153 (1985)
8. B. C. Wilson and M.S. Patterson, "The physics of photodynamic therapy"
   Phys. Med. Biol. 31:327 (1986)
9. W. M. Star, J.P.A. Marijnissen and M.J.C. van Gemert, "Light dosimetry
   status and prospects", J. Photochem. Photobiol. B1:149 (1987)

10. S. T. Flock, M.S. Patterson, B.C. Wilson and D.R. Wyman, "Monte Carlo modelling of light propagation in highly scattering tissues Parts I and II", IEEE Trans. Biomed. Eng. 36:1162 and 1169 (1989)

11. M. S. Patterson, E. Schwartz and B.C. Wilson, "Quantitative reflectance spectrophotometry for the noninvasive measurement of photosensitizer concentration in tissue during photodynamic therapy", Proc. SPIE 1065:115 (1989)

12. B. C. Wilson, T.J. Farrell and M.S. Patterson, "An optical fiber-based diffuse reflectance spectrometer for non-invasive investigation of photodynamic sensitizer in vivo", Proc. SPIE (in press)

13. M. S. Patterson, B. Chance and B.C. Wilson, "Time resolved reflectance and transmittance for the non-invasive measurement of tissue optical properties", Appl. Opt. 28:2331 (1989)

14. W. M. Star, "Comparing the P3-approximation with diffusion theory and with Monte Carlo calculations of light propagation in a slab geometry", SPIE Inst. Series IS5:146 (1989)

15. B. C. Wilson, M.S. Patterson and D.M. Burns, "Effect of photosensitize concentration in tissue on the penetration depth of photoactivating light", Lasers Med. Sci. 1:235 (1986)

16. J. P. A. Marijnissen and W.M. Star, "Quantitative light dosimetry in vitro and in vivo", Lasers Surg. Med. 7:235 (1987)

17. R. R. Anderson, H. Beck, U. Bruggemann, W. Farinelli, S. L. Jacques and J. A. Parrish, "Pulsed photothermal radiometry in turbid media: internal reflection of backscattered radiation strongly influences optical dosimetry", Appl. Opt. 28:2256 (1989)

18. U. Bernini, R. Reccia, P. Russo and A. Scala, "Quantitative photoacoustic spectroscopy of cateractous human lenses", J. Photochem. Photobiol. B4:407 (1990)

19. B. Wilson, Y. Park, Y. Hefetz, M. Patterson, S. Madsen and S. Jacques "The potential of time-resolved reflectance measurements for the noninvasive determination of tissue optical properties", Proc. SPIE 1064:97 (1989)

20. B. Chance, J. S. Leigh, H. E. Miyake, D. S. Smith, S. Nioka, R. Greenfield, M. Finander, K. Kaufmann, W. Levy, M. Young, P. Cohen, H. Yoshioka and R. Boretsky, "Comparison of time-resolved and unresolved measurements of deoxyhemoglobin in brain", Proc. Natl. Acad. Sci. 85:4971 (1988)

21. R. Nossal, R.F. Bonner and G.H. Weiss, "Influence of path length on remote optical sensing of properties of biological tissue", Appl. Opt. 28:2238 (1989)

22. M. S. Paterson, J. D. Moulton, B. C. Wilson and B. Chance, "Applications of time-resolved light scattering measurements to photodynamic therapy dosimetry", Proc. SPIE 1203 (in press)

23. B. Chance, M. Maris, J. Sorge and M. Z. Zhang, "A phase modulation system for dual wavelength difference spectroscopy of hemoglobin deoxygenation in tissues", Proc. SPIE (in press)

24. J. R. Lacowitz, K. W. Berndt and M. L. Johnson, "Frequency and time-domain measurements of photon migration in scattering media and tissue", Biophys. Soc., Baltimore (abstract) (1990)

25. B. C. Wilson, M. S. Patterson, S. T. Flock and D. R. Wyman, "Tissue optical properties in relation to light propagation models and in vivo dosimetry", in: Photon Migration in Tissue, B. Chance, ed., Plenum Publ. Corp., New York:25 (1990)

26. S. L. Jacques, "Time resolved reflectance spectroscopy in turbid tissues", IEEE Trans. Biomed. Eng. 36:1155 (1989)

27. S. L. Jacques, "Time resolved propagation of ultrashort laser pulses within turbid tissues", Appl. Opt. 28:2223 (1989)

28. W. R. Potter, "PDT dosimetry and response", Proc. SPIE 1065:88 (1989)

29. M. Keijzer, W. M. Star and P. R. M. Storchi, "Optical diffusion in layered media", Appl. Opt. 27:1820 (1988)

30. G. J. Derbyshire, D. K. Bogen and M. Unger, "Thermally induced optical

property changes in myocardium at 1.06 m", <u>Lasers Surg. Med.</u> 10:28 (1990)

31. L. O. Svaasand, C. J. Gomer and A. E. Profio, "Laser-induced hyperthermia of ocular tumors", <u>Appl. Opt.</u> 28:2280 (1989)

32. D. Wyman, C.-L. Swift, R. Siwek and B. C. Wilson, "Optimal temperature control in laser hyperthermia", <u>Proc. SPIE</u> 1201 (in press)

33. N. Kollias and A. Bager, "On the assessment of melanin in human skin in vivo", <u>Photochem. Photobiol.</u> 43:49 (1986)

34. J. W. Feather, M. Hajizadeh-Saffar, G. Leslie and J. B. Dawson, "A portable scanning reflectance spectrophotometer using visible wavelengths for the rapid assessment of skin pigments", <u>Phys. Med. Biol.</u> 34:807 (1987)

35. R. R. Anderson and J. A. Parrish, "Optical properties of human skin", <u>in</u> "The Science of Photomedicine", J.D. Regan and J.A. Parrish, eds., Plenum Press, New York: 147 (1982)

36. J. M. Conway, K. H. Norris and C. E. Bodwell, "A new approach for the estimation of body composition: infrared interactance", <u>Am. J. Clin. Nutr.</u> 40:1123 (1984)

37. S. Wray, M. Cope, D. T. Delpy, J. S. Wyatt and E. O. Reynolds, "Characterization of the near infrared absorption spectra of cytochrome aa_3 and hemoglobin for the non-invasive monitoring of cerebral oxygenation", <u>Biochim. Biophys. Acta</u> 933:184 (1988)

38. M. Cope and D. T. Delpy, "System for long-term measurement of cerebral blood and tissue oxygenation on newborn infants by near infra-red transillumination", <u>Med. Biol. Eng. Comp.</u> 26:289 (1988)

39. B. Chance, M. Maris, J. Sorge and M. Z. Zhang, "A phase modulation system for dual wavelength difference spectroscopy of hemoglobin deoxygenation in tissue", <u>Proc. SPIE</u> (in press)

40. R. L. Egan and P. D. Dolan, "Optical spectroscopy: pre-mammography marker", <u>Acta Radiol.</u> 29:497 (1988)

41. A. Grinwald, E. Lieke, R. D. Frostig, C. D. Gilbert and T. N. Wiesel, "Functional architecture of cortex revealed by optical imaging of intrinsec signals", <u>Nature</u> 324:361 (1986)

42. R. Anderson Engles, "Laser-induced fluorescence for medical diagnosis <u>Lund Rep. Atomic Physics</u>, 108 (1989)

43. R. Richards-Kortum, R. P. Rava, M. Fitzmaurice, L. L. Tong, N. B. Ratcliff, J. R. Kramer and M. S. Feld, "A one-layer model of laser-induced fluorescence for diagnosis of disease in human tissue: applications to atherosclerosis", <u>IEEE Trans. Biomed. Eng.</u> 36:1222 (1989)

44. R. J. Gush, T. A. King and M. I. V. Jayson, "Aspects of laser light scattering from skin tissue with application to laser Doppler blood flow measurement", <u>Phys. Med. Biol.</u> 29:1463 (1984)

45. F. W. Leung, T. Morishita, E. H. Livingston, T. Reedy and P. H. Guth, "Reflectance spectrophotometry for the assessment of gastroduodenal mucosal perfusion", <u>Am. J. Physiol.</u> 252:G797 (1987)

46. D. T. Delpy, M. Cope, P. van der Zee, S. Arridge, S. Wray, and J. Wyatt, "Estimation of optical path length through tissue from direct time of flight measurement", <u>Phys. Med. Biol.</u> 33:1433 (1988)

47. J. C. Hebden and R. A. Kruger, "A time-of-flight imaging system: simulations and preliminary experimental results", <u>Proc. SPIE</u> (in press)

48. J. W. Feather, K. S. Ryatt, J. B. Dawson, J. A. Cotterill, D. J. Barker and D. J. Ellis, "Reflectance spectrophotometric quantification of skin colour changes induced by topical corticosteroid preparations", <u>Br. J. Dermatol.</u> 106:437 (1982)

49. G. Jarry, S. Ghesquiere, J. M. Maarek, F. Fraysee, S. Debray, B. M. Hung and D. Laurent, "Imaging mammalian tissues and organs using laser collimated transillumination", <u>J. Biomed. Eng.</u> 6:70 (1984)

50. J. Hebden and R. A. Kruger, "Simulating the performance of a time-of-flight transillumination imaging system", <u>Proc. SPIE</u> 1305 (in press)

CHEMICAL LASER INTERACTIONS WITH HUMAN TISSUES

G. L. Valderrama, R. F. Menefee, B. D. Krenek and M. J. Berry

Laser Applications Research Center, HARC; and
Rice University, Department of Chemistry
Houston, Texas, USA

I. INTRODUCTION

Controlled chemical laser ablation of human tissues may offer a means
to perform precise microsurgical procedures such as laser keratomileusis[1]
and laser angioplasty [2,3]. Some of the potential advantages of chemical
lasers for these applications have been identified previously[4-6]. Repetitively pulsed (rp) hydrogen fluoride (HF) chemical laser interactions with
human corneal and cardiovascular tissues have been studied to understand
tissue ablation phenomenology, effects and mechanisms under well characterized laser irradiation conditions. RP HF chemical laser experiments have
been performed at two wavelengths (λ = 2.78 µm and 2.91 µm) over a radiant
exposure/fluence range of 0.05 to 10 J/cm^2 to determine ablation efficiencies and effective enthalpies of ablation (Q*) as a function of wavelength
and radiant exposure/fluence. The experimental results have been analyzed
to consider the physical and chemical processes associated with thermochemical ablation of human tissues by pulsed mid-infrared lasers. The
present chapter summarizes the nature of pulsed HF chemical lasers and
their match to tissue ablation requirements, presents quantitative tissue
ablation results obtained using an rp HF chemical laser operating in the
λ = 2.7 - 3.0 µm wavelength region, and discuss tissue ablation mechanisms.

II. PULSED HYDROGEN FLUORIDE CHEMICAL LASERS

Chemical lasers are lasers that operate on population inversions produced by chemical reactions. Several hundred elementary chemical reactions
yield chemical lasers based on, for example, vibration-rotation transitions
of hydrogen and deuterium halides. A very useful chemical laser is generated by the bimolecular reaction between atomic fluorine and molecular
hydrogen:

$$F + H_2 \rightarrow HF\dagger(v = 0\text{-}3) + H \tag{1}$$

wherein hydrogen fluoride (HF) is formed preferentially in vibrationally
excited states with inverted vibrational populations [7] . This bimolecular
chemical reaction can be initiated rapidly in a stable reactant mixture by
using pulsed input energy (for example, pulsed light or a pulsed electrical
discharge) to generate fluorine atoms from a suitable precursor such as
CF_3I[7] or SF_6[8].

Figure 1 shows the relative gains of HF chemical laser transition expected from the measured nascent vibrational population inversions of HF produced by the $F + H_2$ reaction[7]. In a particular chemical laser device, various vibration-rotation transitions (including a subset of those shown in Figure 1) are observed. The lasing transitions and the pulse durations and pulse energies of these lasing transitions depend sensitively on complex formation and decay rate processes associated with generation of the nascent population distribution of vibration-rotation states of HF, alteration of nascent populations by collisional and radiative relaxation, cavity resonator conditions, etc.[7‡10]. The relative importances of formation and decay rate processes are functions of (time-dependent) number densities of reactants, products, and other species (diluent gases, electrons, etc.)[7‡10]. Lasing transitions also depend on gas mixture temperature and pressure and on the presence or absence of wavelength-selective optics; for example, the R-branch transitions shown in Figure 1 can only be observed if a diffraction grating or other wavelength-selective optical resonator is used to prevent higher gain Pbranch transitions from lasing[7,9]. Hence, each HF chemical laser device tends to have its own output laser characteristics (which can be modified by changing reagent pressures, discharge conditions, optics, etc.).

In principle, pulsed HF chemical lasers can be engineered to operate on any of the transitions shown in Figure 1 (and on other transitions over an extended wavelength range using other pumping reactions such as $H + F_2 \rightarrow HF\dagger + F$ that generate even more highly vibrationally excited HF) at submicrosecond pulse durations and useful pulse energies (greater than

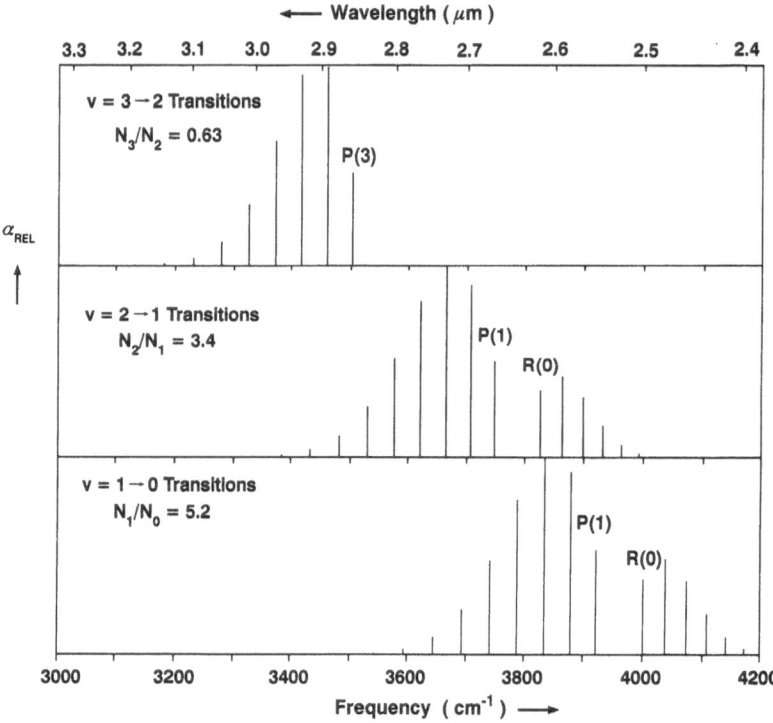

Fig. 1. Relative gain of HF chemical laser transitions expected from the nascent vibrational population inversions N_v/N_{v-1} of HF produced by the $F + H_2$ chemical reaction. Each panels shows $v \rightarrow v-1$ band transitions normalized to the highest gain transition within that band. Selected P- and R-branch transitions are identified within each panel.

100 mJ/pulse) at high pulse repetition frequency (greater than 100 Hz). With proper optical resonator design and suitable spatial filtering, the output beam quality can be engineered to near-diffraction limite performance. However, an "off-the shelf" pulsed HF chemical laser device from a commercial vendor usually has several undesirable characteristics (poor beam quality, multiple lasing transitions, etc.) that must be improved in order to conduct well characterized and meaningful experiments.

The "ideal" HF chemical laser has several potential advantages for tissue ablation: 1) optimal (tunable) irradiation wavelengths in the λ = 2.5 - 3.1 μm spectral region for altering the optical penetration depth in tissue from a few micrometers (for highly localized energy deposition) to several hundred micrometers; 2) optimal (short duration) pulse waveforms for minimizing thermal damage effects on remaining unablated tissue; 3) optimal pulse peak irradiances and radiant exposures/fluences for producing efficient tissue ablation; 4) the availability of fiber optic delivery systems (based on fluoride glass fibers); and, 5) the availability of compact "user friendly" devices with long lifetime, stable performance, ease of operation and modest cost. The quantitative tissue ablation experiments described below were undertaken to define "ideal" HF chemical laser characteristics for microsurgical applications, as well as to understand mechanistic aspects of tissue ablation.

III. QUANTITATIVE TISSUE ABLATION EXPERIMENTS

Advantages #1 through 3 above have been explored by means of quantitative corneal and cardiovascular tissue ablation experiments [1,3,11]. In order to achieve near-"ideal" rp HF chemical laser operation, a Lambda Physik Model EMG-204MSC excimer laser was converted into a chemical laser device using SF_6 as the fluorine atom source with H_2 as the "fuel", and He as a diluent gas. Under typical operating conditions, the spectral distribution of HF chemical laser vibration-rotation transitions included strong v = 2→1 and 1→0 band emissions and a few weak v = 3→2 band emissions. Two laser wavelengths (λ = 2.78 and 2.91 μm) were selected for initial tissue ablation studies. Infrared absorption spectra of corneal[1,12] and cardiovascular[2,3] tissues indicated that λ = 2.91 μm wavelength irradiation should produce much better energy localization than the corresponding λ = 2.78 μm wavelength irradiation. Controlled and well characterized quantitative tissue ablation experiments were therefore performed at these two wavelengths to investigate wavelengthdependent effects and tissue ablation mechanisms.

Figure 2 shows the rp HF chemical laser/tissue interaction apparatus that was used to irradiate human cornea and normal aorta specimens with single wavelength (λ = 2.78 and 2.91 μm), 200 ns pulse duration, 0.4 Hz pulse repetition frequency laser pulses. The apparatus components are: 1) the rp HF chemical laser device described above; 2) an optical delivery system including beam steering mirrors, an external grating and aperture mask for selection of single wavelengths, attenuators for variation of radiant exposure levels (at a constant spot size), and focussing lenses; 3) laser beam diagnostics that characterize wavelengths, pulse powers, energies, and temporal/spatial beam profiles; and, 4) a tissue specimen XYZ positioner plus alignment telescopes. Further discussion of the apparatus is presented elsewhere [1,3,11].

Human cornea and normal aorta tissue specimens were irradiated in air at normal incidence. Care was taken to measure HF chemical laser beam irradiance distributions on the tissue specimens and to use only the central, uniform, irradiance region (at ≥ 90% of the peak value) of ablation craters for quantitative measurements of tissue recession depths.

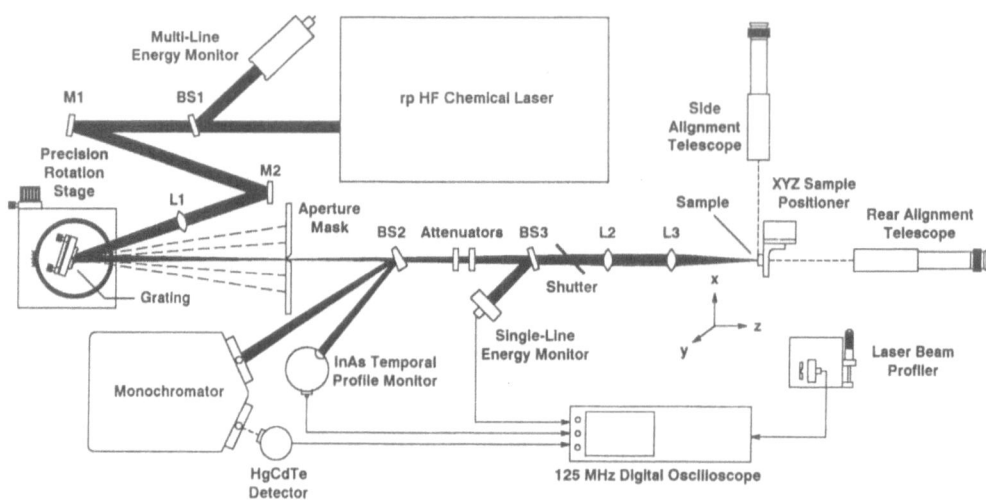

Fig. 2. RP HF chemical laser/tissue interaction apparatus. Components M1 and M2 are beam steering mirrors, BS1 through BS3 are beam-splitters, and L1 through L3 are beam focussing lenses.

A fixed number of laser pulses (typically 25) was used at each radiant exposure/fluence level to produce ablation. The resulting ablation crater depths were measured with a microscope to an accuracy of $\pm$ 5 µm, leading to a typical recession depth per pulse Δ_p measurement accuracy of $\pm$ 0.2 µm. Further discussion of materials and methods is presented elsewhere[1,3,11].

Figure 3 shows measured values of the recession depth per pulse (also termed the ablation rate) Δ_p [units: µm/pulse] as a function of peak laser pulse radiant exposure (also termed the peak fluence or peak pulse areal energy density) F_p [units: J/cm^2] for λ = 2.91 µm wavelength irradiation by the rp HF chemical laser. Linear least-squares fits to low radiant expo-sure ($F_p \leq$ 2.4 J/cm^2) data are also shown in Figure 3. The intercept pa-rameters of these fits correspond to values of the "threshold" radiant ex-posure F_0 of (0.67 $\pm$ 0.06) J/cm^2 and (0.37 $\pm$ 0.06) J/cm^2 for cornea and normal aorta data, respectively. These differences in "threshold" radiant exposures were confirmed qualitatively by direct experiments at very low (sub-"threshold") radiant exposure levels. The slope parameters of the linear fits shown in Figure 3 correspond to values of the effective en-thalpy of ablation Q* [units: kJ/g] of (3.66 $\pm$ 0.22) kJ/g and (3.83 $\pm$ 0.26) kJ/g for cornea and normal aorta data, respectively, calculated using a density ρ of 1.0 g/cm^2 for both tissues and the relation[1,3]:

$$Q^* = F_p/(\rho \Delta_p) \qquad (2)$$

Similar fits were obtained for λ = 2.78 µm wavelength Δ_p vs. F_p data[1,3,11]. Table 1 summarizes the resulting values of "threshold" radiant exposure F_0 and effective enthalpy of ablation Q* for all data set. The ablation rates for all cornea and normal aorta data "roll-off" slightly from the linear fits at higher values of radiant exposure extending to 10 J/cm^2.

Figure 4 shows corresponding effective enthalpy of ablation Q* data derived from recession depth per pulse Δ_p data using equation (2) above, together with fits to the "one-dimensional (1D) loss" form of Q* derived from an energy balance model[1,3]:

$$Q^*_{\lambda,1D} = H_{MR}/(\alpha_\lambda - F_0/F_p) \qquad (3)$$

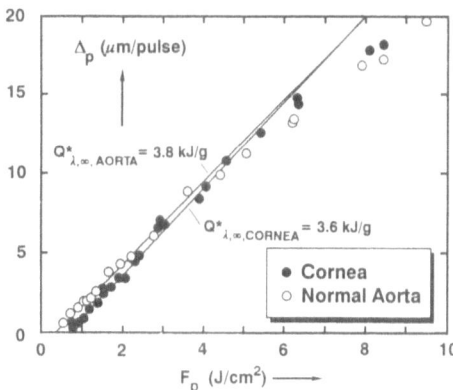

Fig. 3. Recession depth per pulse Δ_p as a function of peak
laser pulse radiant exposure F_p for λ = 2.91 μm wavelength irradiation of human cornea and normal aorta
tissues. Linear least-squares fits to low radiant
exposure ($F_p \leq$ 2.4 J/cm^2) data are shown.
(RP HF Ablation: λ = 2.91μm)

where H_{MR} is the enthalpy of mass removal [units: kJ/g] and α_λ is the wave-
length-dependent thermal coupling coefficient [dimensionless]. For opaque
tissues and transparent plumes, α_λ is equal to the spectral absorptance A_λ
and the spectral emittance ε_λ; all three quantities are related to the
spectral reflectance R_λ:

$$\alpha_\lambda = A_\lambda = \varepsilon_\lambda = 1 - R_\lambda \qquad (4)$$

Direct measurements of spectral reflectances[2,11] indicate that $\alpha_\lambda \gtrsim$ 0.09 at
HF chemical laser wavelengths. There are no direct measurements of H_{MR},
but the enthalpy of mass removal can be calculated from a thermochemical
ablation model[1,3] from thermochemical properties of tissue constituent materials (water and organic polymers) assuming that there is no significant
discrete mass removal (also termed spallation or "chunking"), leading to
estimates of $H_{MR} \cong$ ca. 3.1 kJ/g and 3.3 kJ/g for human cornea and normal
aorta tissues, respectively.

In the high coupled radiant exposure ($\alpha_\lambda F_p \gg F_0$) limit, equation (3)
reduces to:

$$Q^*\infty = H_{MR}/\alpha_\lambda \qquad (5)$$

where $Q^*\infty$ is the high coupled radiant exposure limit value of the effective
enthalpy of ablation; this quantity should be equal to the value of Q^* derived from linear least-squares fits to Δ_p vs. F_p data. From equation (5),

Table 1. Quantitative Tissue Ablation Parameters
($F_p \leq$ 2.4 J/cm^2, λ = 2.78 and 2.91 μm)

Tissue	$F_{0,2.78}$ (J/cm^2)	$Q^*_{2.78}$ (kJ/g)	$F_{0,2.91}$ (J/cm^2)	$Q^*_{2.91}$ (kJ/g)
Cornea	0.59 ($\pm$ 0.07)	3.60 ($\pm$ 0.23)	0.67 ($\pm$ 0.06)	3.66 ($\pm$ 0.22)
Normal Aorta	0.30 ($\pm$ 0.10)	3.65 ($\pm$ 0.33)	0.37 ($\pm$ 0.06)	3.83 ($\pm$ 0.26)

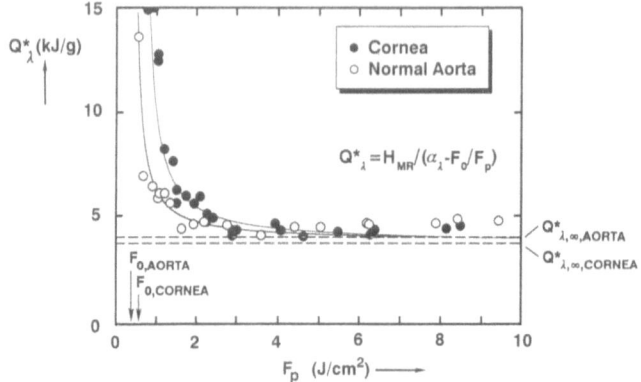

Fig. 4. Effective enthalpy of ablation Q* as a function of peak
laser pulse radiant exposure Fρ for λ = 2.91 μm wavelength
irradiation of human cornea and normal aorta tissues.
Equation (3) fits to the data are shown (RP HF laser).

combining the experimental lower limits for α_λ with the estimated values
for $H_{MR} \approx$ leads to thermochemical ablation estimates of $Q^*_\infty \cong 3.4$ and
3.7 kJ/g for cornea and normal aorta tissues, respectively, in close
agreement with measured experimental values in the range 3.60 - 3.83 kJ/g
(cf. Table 1).

The origin of the "threshold" radiant exposure is primarily the loss
of residual energy (also termed retained energy) between laser pulses at
low pulse repetition frequency (PRF). At higher PRF, the "threshold"
radiant exposure is expected to drop as residual energy between pulses
contributes to mass removal during subsequent pulses. Following energy
deposition during a laser pulse, the tissue pyrolyzes and ablates to a
recession depth governed by the energy balance equation for the coupled
pulse radiant exposure[13]. The residual areal energy density $\alpha_\lambda F_{p,res}$
(units: J/cm^2) is approximately equal to the threshold radiant exposure and
is given by[13]:

$$\alpha_\lambda F_{p,res} = \rho \delta_\lambda \int_{T_0}^{T_s} C_p(T) dT = \rho \delta_\lambda H_S \simeq F_0 \qquad (6)$$

where ρ is the tissue density, δ_λ is the optical penetration depth
(units: cm) at irradiation wavelength λ, T_S and T_0 are the surface temperature (units: °K) during ablation and between pulses, respectively, $C_p(T)$
is the temperature-dependent specific heat capacity [units: J/(g•K)], and
H_S is the total enthalpy required to heat the tissue from T_0 to T_s. As the
tissue cools between pulses, the residual energy is lost; this is the limit
that pertains to low PRF experiments.

Two cases for calculation of the optical penetration depth δ_λ from
equation (6) are treated[3]: 1) tissue properties and ablation energetics are
those of pure water vaporizing at T_S = 100 °C (as often assumed in ablation
models[14]) with ρ = 1.0 g/cm^3 and H_S = 313 J/g[15]; and, 2) tissue properties
and ablation energetics are those of a dehydrated surface layer of residual
organic solids with a fractional density ρ_f that matches the organic solid
content of the unablated tissue and with $H_S \simeq 2$ kJ/g (probably an upper
limit to the true enthalpy requirement).

Case 1 leads to the conclusion that the optical penetration depth is ~10-12 μm for normal aorta at $\lambda = 2.78$ and 2.91 μm, respectively, and ~19-21 μm for cornea at the corresponding wavelengths. Case 2 leads to the conclusion that δ_λ is ~ 5-7 μm for normal aorta and ~ 13-15 μm for cornea. Both of these cases (or any combination thereof) lead to inferred optical penetration depths that are markedly different from those obtained from room temperature infrared spectra of nonablated tissues[1,3,12].

IV. DISCUSSION

Quantitative tissue ablation experiments indicate that: a) the ablation of both human cornea and normal aorta tissues match those predicted by a thermochemical ablation model, with similar values of the effective enthalpy of ablation $Q*\infty$ for both tissues at both irradiation wavelengths studied; and, b) optical penetration depths δ_λ inferred from "threshold" radiant exposures F_0 are larger for cornea ablation compared to normal aorta ablation; δ_λ values for $\lambda = 2.78$ and 2.91 μm and for both sets of tissues are much larger than expected on the basis of room temperature infrared absorption spectra of "virgin" nonablated tissues.

Since the "threshold" radiant exposures F_0 at $\lambda = 2.78$ and 2.91 μm are very similar for each tissue type (Table 1), the inferred optical penetration depths δ_λ are also similar. This result is not expected from measured values of room temperature infrared absorption spectra[13] which indicate that $\delta_{2.78} \simeq 2$ μm and $\delta_{2.91} \leq 1$ μm. It is very likely that "bleaching" of both cornea and normal aorta tissues occurs during laser irradiation and/or from pulse to pulse as the water content and chemical composition of residual tissue change from virgin to irradiated values.

The relatively large optical penetration depths δ_λ inferred from F_0 measurements suggest that pulsed HF chemical lasers can only provide energy localization within ~ 5-10 μm thickness of tissue. This degree of localization may still be suitable for many laser microsurgical procedures.

V. ACKNOWLEDGMENTS

This work was sponsored by the Robert A. Welch Foundation under Grant Numbers C-0002 and C-0812 and by the National Institutes of Health under Grant Number 2 R44 EY07037-02 through Helios, Inc. and Grant Number 1 R01 HL36894-01 through Baylor College of Medicine (BCM). The authors thank the sponsoring organizations and Dr. Michele P. Sartori (BCM) and Mr. Robert A. Fort (Lions Eye Bank of Texas) for tissue specimens.

REFERENCES

1. G. L. Valderrama, R. F. Menefee, B. D. Krenek and J. M. Berry, "Chemical laser interactions with human corneal tissue", SPIE Proc. Thermal and Optical Interactions with Biological and Related Composite Materials 1064:135 (1989)
2. M. P. Sartori, P. D. Henry, G. L. Valderrama, R. F. Menefee, B. D. Krenek, L. G. Fredin and M. J. Berry, "Chemical laser interactions with human cardiovascular tissues", SPIE Proc. - Laser Interaction with Tissue 908:34 (1988)
3. G. L. Valderrama, R. F. Menefee, B. D. Krenek, M. J. Berry, M. P. Sartori and P. D. Henry, "Chemical laser interactions with human cardiovascular tissue", SPIE Proc. - Laser/Tissue Interaction 1202:149 (1990)
4. M. L. Wolbarsht, "Laser Surgery: CO_2 or HF", IEEE J. Quantum Electron. QE-20:1427 (1984)

5. H. Loertscher, S. Mandelbaum, R. K. Parrish II and J.-M. Parel, "Preliminary report on corneal incisions created by a hydrogen fluoride laser", <u>Am. J. Ophtalmol</u>. 102:217 (1986)

6. T. Marshall, S. Trokel, S. Rothery and R. Krueger, "The potential of an infrared hydrogen fluoride (HF) laser (3.0 µm) for corneal surgery", <u>Lasers in Ophtalmol</u>. 1:49 (1986)

7. M. J. Berry, "The $F + H_2$, D_2, and HD reactions: chemical laser determination of the product vibrational state populations and the $F + HD$ intramolecular kinetic isotope effect", <u>J. Chem. Phys</u>. 59:6229 (1973).

8. A. Ben-Shaul, G. L. Hofacker and K. L. Kompa, "Characterization of inverted populations in chemical lasers by temperaturelike distributions: gain characteristics in the $F + H_2 \rightarrow HF + H$ system", <u>J. Chem. Phys</u>. 59:4664 (1973).

9. M. J. Berry, "Chemical laser studies of energy partitioning into chemical reaction products", <u>in</u> R.D. Levine and J. Jortner (Editors), "Molecular Energy Transfer", Wiley, New York (1976)

10. A. Ben-Shaul, Y. Haas, K. L. Kompa and R. D. Levine, "Lasers and chemical change", Springer-Verlag, New York (1981)

11. G. L. Valderrama, "Laser/Tissue Interactions", Ph. D. Dissertation, William Marsh Rice University, Houston, TX (1990)

12. L. M. A. Levine, L. G. Fredin and M. J. Berry, "Infrared absorption spectra of human corneal tissue and cured epoxy resin at temperatures up to 450 °C", <u>SPIE Proc. - Thermal and Optical Interactions with Biological and Related Composite Materials</u> (1064:131 (1989)

13. A. Bailey, J. Baker, C. Rollins, E. Pugh, L. Popper and P. Nebolsine, "Simple models for laser-material interaction", Report PSI-9445/SR-420, Physical Sciences Inc., Andover, MA (1989)

14. F. Partovi, J. A. Izatt, R. M. Cothren, C. Kittrel, J. E. Thomas, S. Strikwerda, J. R. Kramer and M. S. Feld, "A model for thermal ablation of biological tissue using laser radiation", <u>Lasers Surg. Med</u>. 7:141 (1987)

15. M. W. Chase, Jr., C. A. Davies, J. R. Downey, Jr., D. J. Frurip, R. A. McDonald and A. N. Syverud, "JANAF thermochemical tables, 3rd edition", American Institute of Physics, New York (1986)

SURGICAL AND OPHTHALMOLOGICAL APPLICATIONS

MID-INFRARED LASER SURGERY

M. L. Wolbarsht and D. Shi

Department of Biomedical Engineering
Duke University
Durham, North Carolina, U.S.A.

I. INTRODUCTION

Present CO_2 laser instruments for general surgery offer many advantages over conventional methods with scalpels and cauterizers yet there are still many limitations. A particularly critical one is the damage to tissue adjacent to that ablated by the laser energy. Laser energy applied to the tissue immediately produces thermal or mechanical damage to a volume surrounding the impact zone[1÷3].   Part of this is due to scattering[4,5], but other mechanisms also act on the tissue. Steam formation, thermal expansion, and acoustic shockwaves, with much of the energy in the ultrasonic region, have an effect termed physical amplification. In a living system, there is also a secondary and delayed amplification of the size of the exposure site. This delayed change, the reaction of the living system to the physical trauma, has been termed biological amplification[1,2]. Some of the important factors of biological amplification are: cell death and histamine or toxin release, inflammation and edema, immune responses, hemorrhage, muto/carcinogenesis, and interference with respiration and cardiac function. For surgery to be successful, biological amplification must be avoided or, at least, minimized.

In CO_2 laser surgery, the chief absorber is water, and the laser radiation and its resultant heat leave a layer of dead or altered tissue adjacent to the ablation, mostly by thermal denaturation and steam formation but also by convection of heated and liquified tissue. This is the physical part of the amplification process, while the biological portion is often accompanied by some tissue poison (perhaps histamine) released by the injured cells and/or hemorrhage and edema. In tissue ablation procedures, it is desirable to eliminate or, at least, minimize biological amplification by removal of all tissue within the physical amplification zone.

Often, the CO_2 laser is not suitable for surgery on tissues abutting highly sensitive and important structures to which damage must be avoided at all costs. The use of excimer lasers in the ultraviolet (193 nm) to remove plastics to close tolerances without grossly visible damage in adjacent regions[6,7] has suggested that this method may also be helpful in surgery[8÷10]. However, certain features of this type of laser exposure, especially tissue effects of short wavelength UV and the shockwave associated with ablation, remain to be investigated. The present paper will suggest that a laser in another spectral region, in the infrared around 3 μm, is similar to the ArF excimer laser at 193 nm in its precise tissue removal without, however, its possible deleterious physical implication.

In CO_2 laser surgery, the tissue evaporates or decomposes when some critical or threshold temperature is reached, at which point the tissue components, mostly water, have become gaseous or, at least, are broken up into small particles which acquire sufficient kinetic energy to be blown away. However, some thermal diffusion from this impact zone takes place into the remaining material. To minimize the opportunity for thermal diffusion, it is desirable to limit the laser absorption depth to the thinest layer nearest the surface in order to vaporize or remove only the superficial material without disturbing anything underneath. The depth of cut for each exposure can be controlled by variations in the laser power and the duration of application to any given spot as compared with conventional CO_2 laser surgery where the depth of cut is controlled mostly by the speed of the laser beam traversing across the tissue. Deeper cuts are made by successive applications of the laser beam until the required depth is reached. An analysis of this type of cutting shows that at least three factors are important in limiting the conduction of heat from the laser exposure site to the adjoining material: the pattern of energy deposition which is dependent on the absorption and scattering proportion of the tissue; thermal diffusion (and possibly convection); and change of state or vaporization of tissue elements - usually water. Surface cooling by convection or mixing may also make a contribution if the thermal gradient lasts sufficiently long.

An analysis of ablative laser surgery indicates that heat diffusion from the site of laser exposure may be minimized by proper selection of wavelength and exposure duration. A model for tissue removal with a laser is developed taking account of the threshold phenomena of ablation as a function of wavelength. In the model, radiation at 2.94 µm by an Er:YAG (or 2.95 µm by an HF) laser with short duration pulses is compared with that from a CO_2 (10.6 µm) laser as well as that from a 193 µm ArF excimer laser. Thermal diffusion is minimized by taking advantage of the large amount of heat removed by the phase change of water into steam. This model suggests that for deep cuts at a wavelength with a shallow absorption depth, many short pulses are preferable to a single long duration exposure. Optical energy scattering within biological tissue may cause internal temperatures which exceed those on the surface at threshold ablation energy values. In this situation, much of the material is removed by a sub-surface explosion from water converted into steam. For coagulation (hemostasis control), as well as ablation, two or more simultaneous wavelengths may be required, a very heavily absorbed one (near 3 µm) for cutting and another with deeper penetration (such as 1.06 µm from Nd:YAG or 488-514 nm from argon lasers) for more penetration and heating of deeper layers and blood vessels. The selection of midIR (around 3 µm) lasers allows the use of commercially available low absorption flexible glass optical fibers adaptable for many specialty delivery systems such as tracheal and GI endoscopy, cardiovascular catheterization, and other percutaneous procedures.

II. PATTERN OF ENERGY DEPOSITION

The pattern of energy deposition in tissue is a function of absorption and scattering. Both of these vary in each substance with wavelength, and together they define the limit for minimizing thermal diffusion in most types of cases. In CO_2 laser surgery, the only important absorber is water. Its absorption and scattering properties at 10.6 µm determine the pattern of energy deposition. The absorption depth of water for radiation at the 10.6 µm wavelength emitted by the CO_2 laser is often given as 16 mm in tissue, although other considerations indicate[3,11] that it may be closer to 20 mm. In the following analysis, a value of 20 mm ($\alpha \sim 500$ cm^{-1}) has been selected as shown in Table 1.

The literature[11÷15] is ambiguous as to the exact wavelength for the highest specific absorption of water in the near and mid-IR range, g<1 mm. However, recent measurements[16], made especially to clarify this issue, indicate an absorption peak for pure water at 2.94 ± 0.01 μm and a specific absorption depth of 1.0 mm with $\alpha \sim 10,000$ cm^{-1}. The results of the measurements of the absorption of liquid water also indicated that increases in the temperature resulted in a shift of the maximum absorption toward shorter wavelengths. Also, an increased temperature increased the value of the maximum absorption coefficient. Based on these findings, the optimum laser for clinical practice may be one whose wavelength shifts from 2.94 μm to 2.70 μm during the pulse so as to always match the peak absorption for the tissue as it becomes heated.

In the visible and near IR, water has a very low absorption. The values are given in Table 1 for selected laser wavelengths (based on data from References 11 and 16). The values given in Table 1 assume that at all wavelengths, water is the significant tissue absorber. Thus, any other absorber will modify these values. However, discrete tissue components such as hemoglobin and melanin may absorb heavily, often in a strongly wavelength dependent way. Scattering generally increases in the blue and in the near UV region. The rise in scattering and the increased effective absorption of many specific molecular components of various tissue types which act as localized targets will be discussed in detail in Section VI.

III. THERMAL DIFFUSION

The removal of material while minimizing thermal damage to adjacent zones is limited by thermal conductivity and the time required for heat diffusion of the material involved. An analysis[1,2] similar to that used for photocoagulation of the retina with melanin granules as hot spots shows that the high temperatures needed for the phase change from water to steam

Table 1. Tissue Absorption Parameters

Laser	Wavelength (nm except as indicated)	α for H_2O OH (cm^{-1})	H_2O Specific Absorption Depth at 1/e (mm)	α for Tissue (Assuming 70% Water) H_2O OH (cm^{-1})	Tissue Specific Absor.Depth at 1/e (mm)
CO$_2$	10.60 μm	500	20	350	29
Er:YAG	2.94 μm	10,000	1	7,000	1
HF	2.71 μm	1,667	6	1,167	9
Ho	2.09 μm	33	300	23	433
Tm:YLF	1.95 μm	100	100	70	143
Er:YLF	1.73 μm	40	250	28	360
Nd:YAG	1.32 μm	7	1430	5	2,044
Nd:YAG	1.06 μm	1.430	7000	1	10,000
Ruby	694.3	8	1250	5.6	1,790
HeNe	632.8	7	1430	5	2,040
2×/Nd:YAG	532	20	500	14	714
Argon	514.5	23	435	16	621
Argon	488	33	303	20	433
XeF	351	67	150	47	213
XeCl	308	333	30	233	43
KrF	248	1,000	10	700	14
ArF	193	667	15	467	21

without appreciable heating of adjacent tissues are reached only when the exposure radiation is shorter than the thermal relaxation time. This effect is ultimately limited by the depth absorption properties of the tissue at the selected wavelength. Equation (1) gives the fraction of the absorbed energy (F) available for heating a given volume of material in terms of its thermal characteristics, T_h, and the duration of time exposure, T_e:

$$F = [T_h/T_e] \times [1-\exp(-T_e/T_h)] \tag{1}$$

The longer the exposure, the greater the loss of heat by diffusion. This loss is limited by the ratios between T_e, the duration of the exposure, and T_h, the thermal relaxation time of the material. The thermal relaxation time, T_h, is controlled by the absorption depth of the radiation at the selected wavelength, d, and the thermal diffusivity, k, of the material in question as given in the following equation (2):

$$T_h = d^2/4k \tag{2}$$

In the simplest case, we can use Lambert's law or Beer's law to describe the distribution of the optical radiation along the path.

$$-dI_x/dx = \alpha\, I_x \tag{3}$$

where I_x is the intensity of radiation at a point in the medium, x is the path length of the light up to that point in the medium, α is the wavelength dependent absorption constant of the medium and is the reciprocal of d. Upon integration of this equation with respect to x along the values from 0 to d (the absorption path length), the following expression is obtained:

$$\ln (I_d/I_0) = -\alpha d \tag{4}$$

or

$$I_d = I_0 \exp(-\alpha d) \tag{5}$$

The rate of change of the depth of energy deposition, which varies in each substance with wavelength, defines the limit for minimizing thermal diffusion. To take full advantage of the high α, it is important to make the laser pulse width less than the characteristic thermal relaxation time. This value for a particular wavelength can be calculated from

$$T_h = (\delta S/K)\, L \tag{6}$$

As water is by far the major tissue component, it determines the thermal diffusion rate in general. For water, δ, the mass density is 1 g cm^{-3}; S, the specific heat, is 4.186 J/g °C. L is the specific absorption depth in cm (or the reciprocal of L), and k, the thermal conductivity, is 6.27 cal/sec °C cm^2. The expression in seconds can be calculated approximately as 7 times the absolute value of L. As L is a function of wavelength, each wavelength will have a different thermal relaxation time. For a typical tissue with 70% H_2O content, equation (2) shows that the CO_2 laser at 10.6 μm with L at a nominal value of 20 mm, gives a thermal relaxation time of 2.8 ms. The Er:YAG laser at a wavelength of 2.94 μm has an absorption depth near 1 mm corresponding to nominal 1 ms thermal relaxation time. At 2.94 μm, the free running pulsed Er:YAG laser for each flash lamp excitation has a pulse train of 75 ms with individual pulses or spikes, about 2 ms in length. For each individual spike which exceeds the ablation threshold, the above pulse width limit condition of 1 ms is essentially satisfied. Accordingly, more efficient ablation with the least surrounding thermal damage could be expected to be approached by a short pulse or Q-switched Er:YAG laser, if dielectric break down could be avoided.

The percentages of absorbed energy available for heating with a variety of exposure durations for a CO_2 laser at 10.6 μm and an Er:YAG laser at 2.94 μm can easily be calculated. For example, when the exposure duration is 100 ns, which is about 1/10 the T_h of 1 ms for water at a wavelength of 2.94 μm, essentially all the energy (95%) will remain right where it is absorbed (in the top several mm) long enough to heat and vaporize just the surface portion of the material. An even shorter duration, 10 ns, would allow less than 1% of the energy to diffuse into the surrounding tissues before the temperature rises to a peak and vaporization takes place. Since only the duration of energy deposition can be selected by the operator, the need for reliable exposure duration selection and control can be appreciated. However, as shown below, even the elilination of heat flow for all practical purposes by selecting exposure durations in the nanosecond range does not eliminate the problem of thermal damage.

The above analysis is overly simplicistic as it omits several sizable contributions from other factors, although, the absorption depth and thermal diffusion are very important. A useful approximation based on this analysis yields, at least, a minimum volume of tissue thermally damaged by heating over the 50 °C level which is left behind following thermal ablation. This is a volume more or less equal to the portion of the top layer of tissue which is vaporized. Another factor to be considered is the additional heat required for the change of state for water (liquid to gas), as this will enlarge the damage zone even more. Thus, if a unit volume of tissue is ablated at threshold, a volume of tissue at least five times as large will be damaged at any given wavelength. The energy and temperature parameters of this effect are given in Section IV.

It is not enough to minimize heat flow, or rather maximize temperature peaks, by using short pulses. Even for very short pulses, where heat flow is negligible, the greater the absorption depth, the less steep the temperature profile, with correspondingly greater relative amounts of unablated tissue attaining lethal temperature levels. For this reason, to minimize adjacent absorption by deeper tissues, a wavelength should be selected with a specific absorption depth as small as possible. To confine the thermal effects to the shallowest surface layer possible, the laser wavelength must be matched closely to the absorption peak of the material in question. A laser at the absorption peak of water, 2.94 μm, will have the maximum absorption in the shallowest tissue depth. This offers a great advantage for minimizing thermal absorption outside of the vaporized or ablated zone. From this consideration alone, the best choice for surgery may be to shift the laser wavelength from 10.6 μm to 2.94 μm to decrease both the absorption (ablation) depth and the size of the adjacent damage zone.

IV. CHANGE OF STATE (VAPORIZATION)

The high proportion of water in the body tissues allows the use of a simple model for the short duration exposures as described in Section III. Indeed, the best approach is to consider tissue as a solid with all the thermal properties of water except that there are no thermal convection currents. Water has a heat of vaporization of 2,260 J/g. This is the amount of energy that the liquid phase absorbs as part of its change of state into steam. This value is much higher than for most other substances and is also much greater than the amount required to heat the tissue from 37 °C to 100 °C. Although the steam takes with it most of the large amounts of heat which went into its creation, heating of the adjacent tissue is still a problem. In the short pulse case, the heat flow is less, but energy required is greater than that given in the simple model, as this extra heat required for vaporization heats the deeper layers even more as discussed in Section IV. This analysis is valid only when scattering of the radiation is ignored. The possible corrections due to scattering are discussed in Section VI.

The deposition of additional energy into tissue required to vaporize the water and is over and above that required to heat it to 100 °C. This additional energy widens the zone of killed and thermally stressed tissue surrounding the ablation site. Its size can be best described by, first, giving the pattern of energy deposition and, secondly, translating this into an instantaneous temperature rise by depositing the energy within that period of time sufficiently short to effectively make thermal diffusion a negligible quantity. We will continue with the model introduced in Section III. in which the surface layer (about 1 mm thick) is heated to approximately 100 °C. Then, we add to it sufficient energy to vaporize it. The distribution of energy is given by equation (4). This surface, of course, will remain at 100 °C while it is absorbing the sufficient energy to vaporize the water contained within it. However, during this process, more energy will be put into the lower layers, heating them still further. We will then have a temperature distribution which can best be plotted by merely acting as if the surface has absorbed sufficient energy to heat it instantaneously to approximately 640 °C. The actual depth values of absorbed energy, however, are different according to the absorption characteristics of the material at the wavelength selected. The curves found in this way correspond to the energy which would have been absorbed had the surface been heated to 640 °C with no change of state taking place, perhaps as would be approximated under very high pressure.

The damage zone is now much wider than calculated in Section III, and corresponds more to the experimentally observed values. This then represents the approximate volume of damaged tissue to be expected with a short pulse. It should be noted that the zone of thermally stressed tissue still depends upon the specific absorption depth at the particular wavelength of the laser radiation. Even very short durations cannot get around the differences in zones of damage between, for example, an Er:YAG and a CO_2 laser, or a CO_2 laser and a Nd:YAG.

The above information is suitable if there is no scattering and also if there is no plasma formation. These points will be considered separately in later sections, but in the simple model available, it can be seen that this is the minimum amount of tissue damage for thermal ablation. Of course, the damage zone can be smaller if not all the water is vaporized, for example, if there is an explosion or a tissue is weakened and breaks apart along a natural fracture line. Thus, with particles of any being ejected, all the water does not break the tissue in the vapor phase.

At first, the continuous wave (CW) case can be examined to determine if it is possible to continue this type of evaporation or ablation without short pulse delivery.

From considerations of the thermal input, it appears possible that in certain cases, a microplasma is formed which acts as a black body absorber and radiator. Within the plasma, the electron cloud would absorb strongly and intercept most of the energy from the laser beam and act as a secondary source for continued ablation. Obviously, such a plasma must be continuously sustained to be effective. An extended discussion of this mechanism of ablation as applied to CW situations is given in Reference 17. From that discussion, a general approach to the steady state condition can be formulated. It should be emphasized that the steady state case applies only to plasma generation, and thus differs considerably from the purely thermal situation.

44

V. STEADY STATE CASE

The exact formulation of the steady state or continuous energy input case depends upon the material and the characteristics of the laser exposure and experimental measurements may be needed for. Perhaps, the simplest way to express the CW case is as the limiting case of a series of individual pulses of constant duration in which the inter-pulse interval becomes smaller and smaller - that is, the pulse repetition frequency (PRF) increases. When the pulses are far apart, there is no interaction between the pulses. As the effect of each single pulse is independent of the proceeding pulses(s), the discussions given in Section III and IV above are valid. However, when PRF increases sufficiently, a complex interaction occurs at the surface of the material where the heat evaporated material absorbs a significant fraction of the energy. In the present situation, it can be assumed that the material is water and that the ablation is governed by the change of phase of tissue water from liquid to gas. As the PRF increases, the threshold power level for microplasma formation at the surface will decrease. This suggests that in the CW case, the surface has a continually renewed microplasma which is itself the effective source for tissue ablation.

The main objection to the supposition that CW ablation depends on the generation of the microplasma is that, in many cases, no plasma phenomena such as a spark or pronounced shielding by the ablated material are observed. In these cases, a pure thermal boiling or explosive removal of tissue seems to be the best model. On the other hand, the microplasma model may fit very well the data from some UV lasers, such as the 193 nm ArF excimer laser, particularly when direct ionization by the laser photon energy itself is considered as an important part of the input, and most other lasers when the pulse duration is under 50 ms. The plasma process appears to be less efficient than the direct heating seen in the purely thermal model because of poor coupling between the plasma and the material. The plasma would stay close to the surface as the majority of the atoms contributing new electrons would only enter from that direction and will leave going in the other direction due to their kinetic energy. Thus, a stable, thin plasma is formed with the majority of its radiation losses in the direction of the laser beam. Efficiency would increase as the plasma began to cover the surface significantly. However, a new assessment of the damage in the unablated material adjacent to the ablative zone would be required as radiation from the plasma would be in the ultraviolet rather than the infrared. By the Wien displacement law, a plasma in the 3,000 to 5,000 °K range would have its dominant wavelengths between 575 nm and 825 nm, which are poorly absorbed by tissue.

Although, the majority of the absorbed energy goes into evaporation and acceleration of the vapor from the ablated surface layer, some undetermined fraction, dependant on the material, is responsible[6] for thermally or photonically mediated chemical changes such as bond breaking, etc. As the energy is applied, an equilibrium is established with a constant rate of vaporization and ablation and a constant thickness heated zone remaining with the CW case described above as a limiting case.

VI. SCATTERING AND EXPLOSIVE MODEL

From the preceding discussion, it seems that minimizing thermal denaturation of the tissue adjacent to the site of exposure will require short durations and high power pulses at the water absorption peak near 3 μm, but that the pulse duration should remain long, and the peak power low enough to minimize dielectric breakdown and plasma formation. Possibly, the optimal solution might be any of a group of crystal lasers (erbium in a variety of host materials; YAG at 2.94 µm; YSGG at 2.87 µm; YALO at 2.79 µm) or a gas laser such as HF. A series of lines available from 2.70 to 2.95 µm will be required for experimental purposes. At just above threshold levels, the amount of tissue removed at each pulse would be very small, and many pulses will be needed to make cuts of any appreciable depth. In a clinical situation, the peak powers and repetition rate would have to be optimized to maximize the phase change of water from liquid to steam. Although the factors that govern the thermal loading of the tissues adjacent to those absorbing laser energy are not completely known, a start, at least, has been furnished for maximizing tissue removal while minimizing the thermal damage in the adjacent area. Further investigation is certainly required to reveal the optimum set of conditions with minimal damage in the unablated neighboring zone suitable for the use of a pulsed or CW laser in 3 µm region.

The literature on the distribution of light or energy flux in the skin and other tissues of the body resulting from surface application is voluminous. Attempts to give an analytical description of the distribution have been particularly stimulated by the requirement to assess the dose in photodynamic therapy for cancer with such light activated molecules as HPD. It is interesting to note the strong dependance of the calculated flux distribution upon the model selected. The two most widely accepted models are those by Kubelka-Munk[18] and Welch[5]. As expected, both emphasize the scattering contributions but are sketchy on any problems raised by inhomogeneities in absorption properties or time dependent changes in each factor. The Welch model itself is derived from the Hallderson and Langerholc[19] formulation which emphasizes the dependence of the flux distribution upon the ratio between scattering and absorption. When the scattering coefficient begins to approach the absorption coefficient in magnitude, the energy flux at the surface becomes higher than the incident flux from the source. Furthermore, below the surface, a peak flux is found. If these energy fluxes are translated into instantaneous relative temperatures corresponding to the energy absorption sites prior to any heat flow, then the highest flux will have the highest temperature, and the relative temperature distribution will show a peak in temperature below the surface. Although the peak temperature will always be below the surface, the exact dimensions and importance of the displacement will depend upon the values of the various factors governing any particular situation. The quantitative aspects of the instantaneous temperature distribution remain to be determined for the individual cases. For any short pulse (<1 ms) situation, the exact temperature distribution will be a function of the laser wavelength, as this defines both the effective absorption distance and the scattering. This analysis shows that in any real situation, ablation an threshold energy values will take place as an explosion. That is, the initial change of phase of water into steam, with its corresponding expansion, will occur inside that tissue, and the surface layer will be blown off as solid particles. Such an effect must must always be present to some extent, but whether this explosion makes a significant contribution than the total ablation is unknown and requires an experimental determination of the relation between scattering and absorption in each individual case.

The scattering characteristics of tissue are strongly related to temperature. An extreme example is egg white which scatters almost not at all at 37 °C and at 100 °C quickly becomes a very strong scatterer, as the protein constituents are thermally denatured. Less extreme changes in the molecules may, however, cause almost as great changes in scattering. Tissue transparency depends upon a quite different mechanism that the extreme regularity of spacing the atomic molecules in a crystal must have to produce constructive (non-destructive) interference which allows an unscattered image to be transmitted through a medium. Biological tissue optics depends upon a delicate balance between molecular sizes, the spacing between adjacent molecules and tissue elements such as microtubules, microfilaments, organelles, etc. Benedek[20] has analyzed this problem with respect to the cornea and the sclera, but his conclusions apply equally as well to all other tissues. Changes either in the average spacing of the optical elements or in their relative indices of refraction will change the relation between constructive and destructive interference, and the scattering properties will alter. The molecular spacing may not change during short pulse (1 ms) laser exposure, but the index of refraction of the various components can change. This may increase the scattering and, as in a Christiansen filter, the absorption will increase. This increased absorption will make the temperature rise non-linear in a run-a-way fashion. This change of indices of refraction will also change a low scattering case into a high scattering case with the resulting sub-surface temperature peak leading to an explosion.

The subsurface explosion described above also becomes an important consideration when the energy needed for ablation is expressed in quantitative terms. For example, if an explosion occurs, the amount of tissue ablated per pulse will be much less than that calculated on the basis of the energy required to vaporize all of the water contained in an equivalent volume of tissue. Stated in another way, less laser energy would be required than would be calculated for the total ablation of tissue based only on its water content. This phenomenon usually has been neglected in previous analyses - obviously, more research is needed. Also, there may be unwanted tissue change which accompany any subsurface explosion. For example, exposure conditions which maximize the explosive effect could also result in destruction from the mechanical or acoustic shockwave. Thus, caution must be taken in order that any positive advantages of a sub-surface explosion are not accomplished with undesirable side-effects.

Tests with long pulses (~ 1-2 ms) HF and Er:YAG lasers around 3 μm have not shown any shockwave effects in the tissues; possibly the peak power levels of the individual pulses in the train were below the level for plasma generation. Almost certainly, a Q-switched Er:YAG laser with 5-10 ms pulses generates a plasma at the threshold for tissue ablation. For the long pulses, the high absorption coefficient in this case certainly means that any explosion cannot be very far below the surface. However, particles are observed in the ejected material which suggests that there is some type of sub-surface explosion. The situation is different for the 193 nm ArF excimer laser. Here the scattering is extremely high, and although the absorption may be somewhat lower than at 3.0 μm it is predictable that the 193 nm exposure would cause a subsurface explosion. Also, perhaps of greater importance is the almost certain plasma generation that takes place at the tissue surface because of the much shorter pulse duration (< 20 ns). This sub-surface explosion will also facilitate the formation of the disruptive microplasma, and both together would certainly cause extensive and distinct mechanical damage. Even if there is not any microplasma formation, the explosion itself should lead to some form of shockwave damage, possibly seen most strongly where there are reflective and interference effects at membranes or tissue discontinuities.

VII. DELIVERY SYSTEMS

For most surgery in enclosed body cavities by endoscopic (or similar) techniques, it is necessary to have a delivery system compatible with the other surgical tools. In general, this puts an outside diameter of the delivery system or catheter of 2 mm as the upper limit, with 1.0 mm or less more desirable.

Clinically usable delivery systems for the traditional surgical laser,

CO_2 at 10.6 μm, have been difficult to achieve, especially for any kind of endoscopic or catheter situation where the necessary flexibility almost always requires optical fibers. The available optical fibers are mainly single or poly-crystalline material such as CsBr, KCl [21,23], AgBr[24,25] or TlCl[26]. The latest fibers, a polycrystalline solution of AgCl and AgBr[27] seem the most promising. Although many of these crystalline fibers can carry quite sizeable amounts of power without degradation, and the poisonous ones can be protected from tissue contact by diamond or sapphire windows, nevertheless, none achieve the types of cflexibility required to achieve a bending radius below 5 cm , and all have rather high attenuation losses in clinically usable fiber lengths. Although this latest type[27] seems to avoid many problems, it has not been tested to any large extent in clinical use for durability or efficiency, nor is it known whether it can be incorporated in a system suitable for microsurgery. Thus, although the CO_2 laser cuts most types of tissue with a rather narrow zone of thermal damage around the ablated region, the lack of a suitable delivery system has severely limited applications in many surgical situations.

The same type of problem is found at the other end of the spectrum for the ArF excimer laser at 193 nm. Again, the tissue removal is satisfactory - perhaps, the best of any presently known system, and thermal damage to adjacent tissue is virtually non-existant. However, a flexible optical fiber has not even been suggested, much less achieved.

When the ablative wavelength is in the 3 μm region, construction of the flexible glass optical fiber delivery system is somewhat more difficult than in the visible and near infrared, but it is, at least, possible. For example, in the 2.5 - 3.0 μm spectral band, comparatively flexible very low loss zirconium fluoride (ZrF_4) glass fibers of conventional cladded construction (i.e., 50-500 μm core diameter) are available[28,29]. Many other types have been suggested[30]: graded index profile fibers, germanium oxide glass fibers, and 10% B_2O_3 doped silica glass fibers. Indeed, sapphire fibers are also available with low losses in this spectral region. Although transmission losses of all fiber types in this spectral region are often sizeable due to bound water and metallic impurities, nevertheless, ZrF_4 glass fibers with an attenuation of 1 dB/m are already available in short lengths. Even presently available fibers have sufficiently low loss for the 10 or more meters transmission length necessary to allow placement of the laser outside of the operating room for safety or greater convenience. Thus, if a short pulse laser emitting around 2.94 μm, the water absorption peak, is too large or hazardous for operating room use, it can, nevertheless, be placed at a remote location with the power delivered to the patient through a very flexible optical fiber system. This could allow not only conventional surgery but also interchangeable fiber ends for special endoscopic and percutaneous techniques.

As both the ZrF_4 and sapphire optical fibers are sufficiently transparent in the visible region of the spectrum, it is easy to pass an auxiliary visible laser beam through the same fiber to serve as a marker which will remain in precise optical alignment with the invisible intrared beam which is used for surgery. This wide band transparency will also give the possibility of adding other wavelengths for rapid tissue ablation, coagulation, and tissue bonding as discussed in the following section.

VIII. AUXILIARY WAVELENGTHS FOR COAGULATION, TISSUE BONDING, ETC.

In many types of surgery, the coagulation effect of the conventional CO_2 surgical laser is sufficient for homostasis. However, at laser wavelengths near the peak water absorption band at 2.94 μm, the absorption depth is very shallow, and all the laser tissue interaction is very close

to the surface. Where ablation of surface tissues must be combined with a
more penetrating laser action for coagulation or tissue bonding, auxiliary
wavelengths are required. Significant penetration into the remaining unablated tissue is required for coagulation of large blood vessels as well as
for generalized control of bleeding or seeping, and for more gentle, but
even, heating of the cut surfaces of blood vessels, nerves, and other
tissues as part of tissue anastomosis or bonding procedures. Also, some
surgical procedures may require removal of a large volume of tissue as rapidly as possible; this will require deeper penetration of the energy over
a wide area. All of the above procedures require a laser operating at a
wavelength away from the water absorption peak.

A combination of laser effects can be accomplished in several different ways. For example, the Er:YAG (2.94 μm) or the HF (2.95 μm) laser
might be used with other lasers which are optimal for these secondary functions. Table 1 shows tissue absorption parameters at various laser wavelengths. This suggests that an argon laser at 488514 nm can be used where
the blood itself must be the absorber, while a Nd-YAG laser at 1,064 nm can
be used for deep penetration, if desired. Indeed, many other less familiar
wavelengths might also be picked. The concept of the multi-wavelength
laser delivery system will certainly be more popular in the future whenever
such instrumentation becomes sufficiently available for its utility to be
widely recognized.

IX.   SURFACE COOLING AND FUME CONTROL

Cooling by the loss of the surface layer may be understood better by
considering a water jar (or canteen) covered with a damp cloth. As warm
air blows over the surface, the surface is not warmed to the air temperature much less above by atmospheric friction on the outer surface. Rather,
the surface layer of water vapor molecules are blown away. This surface
layer is composed of the more energetic water molecules, thus, the net effect of this process is cooling of the surface of solid as the more energetic, or "hotter", molecules are lost preferentially from it. In this
way, the surface evaporation that goes along with the purging action of the
gas used for fume removal, or moisturizing at least, can assist in preventing dangerous temperature rises for repetitive exposures in the tissue
remaining after ablation. The energy transfer by the vapor, in the case of
water, can be both rapid and sufficient enough to prevent conduction from
the heated zone into the unablated adjacent tissue. When the energy flux
is sufficiently rapid to initiate evaporation, then heat diffusion is
terminated where the evaporation takes place, and the surface temperature
is controlled by the factors influencing evaporation.

Delivery system utilizing optical fibers or wave guides are all suitable for endoscopic or similar usage within the body. In such a location,
the removal of the ablation by-products becomes very important. An additional consideration in same cases is to maintain a clear pathway from the
exit window of the delivery system to the ablation site. It is, however,
possible to have the delivery system in direct contact with the tissue and
eliminates this problem. Where applicable for sub-surface tissue cooling,
a gas jet, as moist as possible, can be blown at the site of the laser application at all times except during the laser pulse. The gas jet could be
mixed with a cooling spray of water (or saline) applied as soon as possible
after the initial vaporization to minimize any further heat diffusion or
convection. If the ablation is accomplished by pushing an optical fiber
against a piece of tissue in a blood or Ringer's solution environment, as
in a blood vessel, when cooling is less of a problem, of course. However,
a high repetition rate for the laser exposures may lead to heat build up in
the ablation site. This will have significant thermal diffusion and, in

some cases, convection currents, both of those factors may produce significant thermal damage to adjacent tissues. When ablation takes place in an enclosed cavity, pressure surges due to the rapid thermal expansion must be minimized by some type of escape valve. The thermal expansion could also help with fume or debris removal, supplementing the normal inflow/outflow type of flush system.

X. CONCLUSION

An analysis of the use of a laser in the 3 µm region for surgery indicates that absorption and heat flow, as well as an adaptable delivery system are the limiting factors. Nevertheless, it seems obvious that in the near future, the technology will be developed to the point that laser system suitable for handling delicate types of surgery in which heat effects in adjacent tissues must be minimized. The selection of laser wavelengths to minimize the energy absorption depth in water, such as the Er:YAG laser at 2.94 µm with short pulses (<1 ms), may make possible new types of precision surgery in all parts of the body and, even scarfree plastic surgery.

Although the material presented has been detailed only as regards minimizing the depth of injured tissue adjacent to the ablation site, the other factors involved in surgery such as heating for simultaneous and precise hemostasis control are equally as important. The analysis of laser energy deposits in tissues suggests this end point can be accomplished with a similar approach based on a second laser wavelength that penetrates more deeply than the ablating one, and heats the adjacent blood vessels or surrounding tissues sufficiently to prevent bleeding or effect bonding. Thus, it is proposed that the optimal surgical system will incorporate several laser types which are available for simultaneous use. The group that appears to meet the boundary conditions best is the Er:YAG or HF in the 3 µm region for cutting and ablation; an (Ar^+, or Nd:YAG), for deeper penetration, coagulation, and tissue bonding. A wavelength that can be combined for faster tissue removal, however, such as the CO_2 laser remains a problem. What is needed is a dependable laser in 2.6 µm or 3.4 µm region.

REFERENCES

1. J. R. Hayes and M. L. Wolbarsht, "A Thermal Model for Retinal Damage Induced by Pulsed Lasers", Aerospace Med. 39:474-480 (1968)
2. J. R. Hayes and M. L. Wolbarsht, "Models in Pathology-Mechanisms of Action of Laser Energy With Biological Tissues", pp. 255-274, In: M. L. Wolbarsht (ed.), "Laser Applications in Medicine and Biology", vol.1. Plenum Press, New York (1971)
3. M. L. Wolbarsht, "Laser Surgery: CO or HF", IEEE J. Quant.Electron. QE-20, 1427-1432 (1984)
4. F. Partovi, J. A. Izatt, R. M. Cothren, C. Kittrell, J. E. Thomas; S. Strikwerda, J. R. Kramer and M. S. Feld, "A Model for Thermal Ablation of Biological Tissue Using Laser Radiation", Laser Surg. Med. (1987)
5. A. J. Welch, "The Thermal Response of Laser Irradiated Tissue", IEEE J. Quant. Electron., QE-20, 1471-1481 (1984)
6. R. Srinivasan and V. Mayno-Banton, "Self Developing Photoetching of Poly (Ethene Terephthalate) Films by Far Ultraviolet Laser Radiation", Appl. Phys. Lett., 41:576-578 (1982)
7. R. Srinivasan, "Ablation of Polymers and Biological Tissue by Ultraviolet Lasers", Science, 234:559-564 (1986)
8. R. J. Lane, R. Linsker, J. J. Wynne, A. Torres and R. G. Geronemus, "Ultraviolet-Laser Ablation of Skin and Other Tissue", Conference on Lasers and Electro-Optics, IEEE/OSA CLEO, Baltimore, MD. (1984)

9. S. L. Trokel, R. Srinivasan and B. A. Bodil Brare, "Excimer Laser Surgery of the Cornea", Amer. J. Ophthalmol., 96:710-715 (1983)

10. W. S. Grundfest, I. F. Litvack, L. Morgenstern, J. S. Forrester, I. S. McDermid, J. Pacala, D. M. Rider and J. B. Laundenslager, "The Effect of Excimer Laser Irradiation on Human Atherosclerotic Aorta: Amelioration of Laser Induced Thermal Damage", Conference on Lasers and Electro-Optics, IEEE/OSA CLEO, Baltimore, MD (1984)

11. L. Esterowitz and C. Hoffman, "Laser-Tissue/Water Interaction of the Erbium 2.9 µm Laser, pp. 196-197", In "Lasers in Medicine", Soc. Photo. Instru. Engineer. (SPIE), 112 (1986)

12. J. G. Bayly, V. B. Kartha and W. H. Stevens, "The Absorption Spectra of Liquid Phase H_2O, HDO, and D_2O from 0.7 µm to 10 µm.", Infrared Physics, 3:211-223 (1963)

13. M. A. Bramson, "Infrared Radiation - A Handbook for Applications" (Trans. R. B. Rodman), Plenum Press, New York (1968)

14. M. Centeno, 5th, "The Refractive Index of Liquid Water in the Near Infrared Spectrum", J. Opt. Soc. Amer. 31:244-247 (1941)

15. C. W. Robertson and D. Williams, "Lambert Absorption Coefficients of Water in the Infrared", J. Opt. Soc. Amer., 61:1316-1320 (1971)

16. A. V. Lukashev, Personal Communication (1989)

17. O. N. Krokhin, "Generation of High Temperature Vapors and Plasmas by Laser Radiation", pp. 1371-1407, In: F. T. Arecchi, and E. D. Schultz-Dubois (eds), "Laser Handbook", North Holland Publishing Co., Amsterdam (1972)

18. P. Kubelka, "New Contributions to the Optics of Intensity Light Scattering Materials", J. Opt. Soc. Amer., 38:448-457 (1948)

19. T. Hallderson, and J. Langerholc, "Thermodynamic Analysis of Laser Irradiation of Biological Tissue", Appl. Opt. 17:3948-3958 (1978)

20. G. B. Benedek, "Theory of Transparency of the Eye", Appl. Opt., 10:459-473 (1971)

21. Y. Mimura and C. Ota, "Transmission of CO_2 Laser Power by Single Crystal CsBr Fibers", Appl. Phys. Lett., 40:774-775 (1982)

22. Y. Mimura, Y. Okamura and C. Ota, "Single Crystal CsBr Infrared Fibers", J. Appl. Phys., 53:5491-5497 (1982)

23. D. A. Pinnow, A. L. Gentile, A. G. Standlee, A. J. Timper and L. M. Holbrook, "Polycrystalline Fiber Optical Waveguides for Infrared Transmission", Appl. Phys. Lett., 33:28-29 (1978)

24. T. J. Bridges, C. K. N. Patel, A. R. Strnad, O. R. Wood, E. S. Brewer and D. B. Karlin, "Syneresis of Vitreous by Carbon Dioxide Laser Radiation", Science, 219:1217-1219 (1983)

25. T. J. Bridges, J. S. Hasiak and A. R. Strnad, "Single Crystal AgBr Infrared Optical Fibers", Opt. Lett. 5:85-86 (1980)

26. E. Garmire, T. McMahon and M. Bass, "Low-Loss Optical Transmission Through Bent Hollow Metal Waveguides", Appl. Phys. Lett. 31:92-94 (1977)

27. V. Artjeshenko, Personal Communication (1989)

28. D. C. Tran, "Advances in Mid-Infrared Fibers", Proc. Tech. Digest, 5th Int. Conf. Integrated Optics and Optical Fiber Communications, 2:1320 (1985)

29. P. W. France, S. F. Carter, M. W. Moore and C. R. Day, "Progress in Fluoride Fibres for Optical Communications", Br. Telecom. Technol. J., 5(2) (1987)

30. S. Mitachi, T. Miyashita, and T. Kanamori, "Fluoride-Glass-Cladded Optical Fibers for Mid-Infrared Ray Transmission", Electron. Lett., 17 (1981)

THE Nd:YAG LASER - APPLICATIONS IN SURGERY

R. A. Kirschner

The Institute for Applied Laser Surgery, Inc.
Suburban General Hospital
Bala Cynwyd, Pennsylvania, U.S.A.

I. INTRODUCTION

The Nd:YAG laser that is used in surgery today is a CW instrument that has an output of 1,064 nm in near-infrared light spectrum. This modality has become the second most important tool in the laser surgeon's armamentarium, the first being the CO_2 laser. The light emitted from the Nd:YAG laser may be transmitted through a flexible quartz fibreoptic. This ability frees the surgeon from cumbersome articulated arms, when the laser is used in the hand-held or macro mode. It also permits laser energy to be transmitted through flexible endoscopes. This latter ability was responsible for the rapid establishment of this tool in surgery. The Nd:YAG laser is a poor cutter and an excellent coagulator. It has a propensity for increased absorption by darkly pigmented tissue.

"When Nd:YAG laser light encounters tissue, the result is a combination of back scatter (reflection), forward scatter, and absorption. The scattering effect around the incident laser beam within the tissue heats up a large volume and causes tissue coagulation and necrosis over a large volume of tissue without its removal. Tissue vaporization can be achieved by further heat generation of the coagulated and dessicated tissue with high energy density over time."[1] In most instances this laser is not employed to vaporize tissue. This will only occur after tissue has been carbonized. The depth of thermal change due to the increased absorption coefficient of carbonized tissue produces a situation that would be quite hazardous. We must remain constantly cognizant of the fact that energy of focused laser radiation may be stronger than the sun over a very small area.

II. ENDOSCOPIC Nd:YAG LASER SURGERY

The qualities of the Nd:YAG laser have made it an excellent tool for many areas of endoscopic application. One of the earliest proponents of the Nd:YAG laser in endoscopic surgery was J.-F. Dumon. He brought this technology to the U.S. when we held the first American workshop in endobronchial lader surgery in 1981. Many obstructive lesions of the tracheobronchial tree may be removed with the aid of the Nd:YAG laser. This treatment is often not of a curative nature but, in cases of cancer, it allows the patient to enjoy a very high quality of life, often for a much longer period of time than that which would have been possible without this technology.

Laser Systems for Photobiology and Photomedicine
Edited by A. N. Chester *et al.*, Plenum Press, New York, 1991

Nd:YAG laser radiation is transmitted through a flexible quartz fibre. This fibre may be introduced through a rigid endoscope or through a flexible scope. The fibre tip can be used bare, with its end polished or in concert with a contact tip. We will discuss the contact tip in greater detail later. However, at this time let us say that the contact tip allows actual contact with the lesion that is being ablated.

The operator will focus the aiming beam of the laser on the lesion and apply power through the fibre. He must not allow the polished tip of the fibre to contact the tissue, while the power is being transmitted through the fibre. If this occurs, superficial tissue will adhere to the tip and it will become carbonized. If this occurs a new fibre must be utilized while the old fibre must be cleaved and repolished.

The Nd:YAG laser is essentially a coagulator. It is not used to vaporize tissue safely. Once tissue has been coagulated through an endoscope it must be removed. This is usually accomplished with a biting cup forceps. This portion of the procedure usually takes the longest to perform. If a flexible endoscope is utilized the working channel is quite small. This limits the size of the forceps that can be used. Use of this small forceps is a very tedious proposition. If a rigid scope is used, larger forceps may be employed and this portion of the procedure progresses quickly.

There is a continuous controversy amongst surgeons regarding whether rigid endoscopy or flexible endoscopy is superior for the ablation of endobronchial lesions. A comparative list is offerred that contrasts the relative advantages and disadvantages of both procedures.

III.1. Rigid bronchoscopy

III.1.i. Advantages

a) Airway is maintained through the endoscope
b) Local or General Anesthesia may be utilized
c) Rigid viewing fibreoptics are significantly better than flexible viewing fibreoptics
d) Large surgical instruments may be employed through the scope
e) Easier to control bleeding than with flexible technique
f) Airway fire less likely because the scope cannot ignite

III.1.ii. Disadvantages

a) Difficult technique to properly master - requires special training
b) More uncomfortable than flexible techniques with local anesthesia
c) Difficult and sometimes impossible to perform with certain anatomy - the "No-Neck monster".

III.2. Flexible bronchoscopy

III.2.i. Advantages

a) Easy technique to master
b) Comfortable for patient
c) Facilitates endoscopy of the "No-Neck Monster"
d) Can traverse greater angulation

III.2.ii. <u>Disadvantages</u>

a) The procedure usually takes longer because only smaller instruments fit through the scope
b) It is easier to start an airway fire
c) Viewing Optics are inferior to rigid ones
d) Airway more difficult to maintain
e) Scopes cannot be autoclaved - must be cleaned with chemicals or gas sterilized.

IV. UROLOGY

The Nd:YAG laser is finding other endoscopic applications. In Urology, surgeons are using this modality to resect strictures and condylomata within the urethra. In our Institute a large series of patients had the prostate resected with the aid of the Nd:YAG laser[2]. The Nd:YAG laser is ideal to utilize through a resectascope. The distal manipulating bridge of the instrument allows accurate positioning of the fibre tip.

V. LASER LAPAROSCOPY

The Nd:YAG laser has been employed through a laparoscope for lysis of adhesions and ablation of endometriosis. Most recent applications include the Laser Laparoscopic Cholecystectomy. This procedure allows for many of these patients to be done on an out-patient basis.

VI. CUTANEOUS SURGERY

VI.1. <u>Focusing lens</u>

The facultatively selective absorption of Nd:YAG laser radiation by dark colors (chromophores) such as hemoglobin may be used to the surgeon's advantage. The laser may be used with a focusing handpiece for the treatment of cavernous hemangiomas and cavernous types of port-wine nevi. With the contact tip in place the laser may be used in the ablation of spider veins and capillary type port-wine nevi.

We have found the Nd:YAG laser with a focusing handpiece to be an excellent tool for the removal of tattoo pigment. The tattoo pigment is exposed at the time of surgery. The patient removes pigment and coagulum with hydrogen peroxide several days postoperatively.

VI.2. <u>Contact tips</u>

Contact tips are used in the free-hand mode as well as on the end of a flexible fibre. When these tips are used free-hand they are mounted on the end of a handle which is hollowed out to allow the passage of the fibre to the crystal.

The scalpel type tips are currently enjoying increasing applications in free hand surgery. A number of individuals are performing mastectomies, radical head and neck dissections and cosmetic surgery with these tools. These scalpels function in the place of conventional scalpels with the added advantage of some degree of innate hemostasis. These tips also cause a much lower degree of thermal necrosis than that experienced with electro-dissection instruments.

AlO_2 (artificial sapphire) is the material that is currently utilized for contact tips. Many other substances are being evaluated for use in manufacturing contact tips. I feel that the tight lattice of a diamond would produce an enduring tip. Other possibilities would include a diamond coating on another crystal substrate.

REFERENCES

1. R. P. Apfelberg, T. Smith, H. Lash, D. N. White and M. R. Maser, "Preliminary Report on Use of the Neodymium YAG Laser in Plastic Surgery, Laser Surg. Med. 7:189-198, (1987)
2. L. Finkelstein et al., "Lectures at Urology Symposia", The Institute for Applied Laser Surgery, (1986)

LASER IN PLASTIC SURGERY AND DERMATOLOGY

J. S. Nelson

Beckman Laser Institute and Medical Clinic
University of California
Irvine, California, USA

Since the laser was first developed in 1960, it has found many uses in the treatment of cutaneous and superficial lesions in plastic surgery and dermatology. Moreover, lasers are now the treatment of choice for several clinical entities for which no reliable or effective modality was previously available.

Port-wine stain (PWS) is a congenital vasculopathy consisting of an abnormal network of capillaries in the upper dermis with an overlying normal epidermis[1]. It occurs most commonly on the face and neck, but may be found anywhere on the body as an isolated finding or in association with systemic syndromes. PWS has been treated in the past with an array of therapeutic modalities including skin grafting, ionizing irradiation, dermabrasion, cryosurgery, tattoing, and electrotherapy[2], which have often left cosmetically unacceptable secondary scarring.

The introduction of the argon laser in the early 1970's represented the first major advance in therapy for PWS. The bluegreen light (488, 514 nm) produced by the argon laser is preferentially absorbed by hemoglobin in the dilated ectatic capillaries in the upper dermis. There, the photon energy is converted to heat causing thermal damage and thrombosis in the targeted vessels[3÷6].

Unfortunately, after initial enthusiasm, the mechanism of action described above was found to be overly simplistic in that the epidermis is not totally spared (due to undesired absorption of energy therein by melanin and other dermal components, including collagen, and to dissipation of heat from the injured vessels) and suffers some irreversible damage. Histopathology shows nonspecific coagulation necrosis of the upper reticular dermis to a depth of 0.45 mm. In the subjacent zone, extending to a depth of 0.75 mm, specific damage to chromophore-containing organelles is noted[7,8]. Results of clinical studies are encouraging but scarring remains a worrisome complication even in the hands of the most skilled practitioner.

Continued improvement in treatment results, with reduction in scarring, will depend upon the ability to use lasers to induce selective injury of only the abnormal blood vessels in the dermis while sparing the normal overlying epidermis. Recent increases in our understanding of the optical characteristics of skin have made it possible to concentrate not only on the effects of any particular laser system but on the basic biological and physical principles of laser-tissue interactions.

If the clinical objective is to cause selective destruction of dermal blood vessels, the wavelength chosen should match the high absorption of the targeted HbO_2 molecule relative to other optically absorbing molecules. Choice of wavelength also determines the depth to which the light will penetrate with sufficient energy density to effect tissue change. The wavelengths suitable for consideration are the HbO_2 Soret absorption band at 418 nm and the absorption bands at 542 and 577 nm. Despite the higher extinction coefficient of the Soret band, this wavelength can be rejected for clinical use on the basis that penetration of these photons into the dermis is insufficient to produce blanching of vessels deeper than 0.1 mm from the surface. Furthermore, absorption by melanin is higher at this wavelength leading to nonspecific thermal injury of the epidermis. However, if one can take advantage of the longer wavelength HbO_2 absorption band at 577 nm where tissue penetration is increased and melanin absorption is reduced, less heating of the epidermis should occur and more incident light energy will be transmitted to the blood vessels. Also, the extinction coefficient for HbO_2 is higher at this longer wavelength than at the blue-green (488, 514 nm) wavelengths produced by the argon laser[9,10]. Newer laser systems now employ a wavelength of 585 nm based on a recent study which demonstrated that a greater depth of penetration can be achieved (1.20 mm at 585 as opposed to 0.50 mm at 577) at this longer wavelength while maintaining almost the same degree of specificity for vascular and perivascular injury as that previously described after exposure to 577 nm irradiation[11].

In addition, optical factors such as wavelength and tissue factors such as absorption and penetration are not the only criteria for successful laser therapy. Given that one goal of treatment is the precise control of thermal energy, equally as important as optical and tissue factors is the pulse duration of laser irradiation. One way to maximize the spatial confinement of heat is to use a short pulsed laser with a pulse duration on the order of the thermal relaxation time of the tissue. The latter constant is defined as the time required for the heat generated by the absorbed light energy within the target chromophore to decrease to 50% of its initial value immediately after exposure to the laser. Longer pulse durations offer a more generalized heating, and therefore, less spatial selectivity resulting in non-specific thermal damage to adjacent structures regardless of how carefully one has chosen a wavelength since the absorbed energy is invested almost uniformly in heating of the tissue during exposure, despite its origin in the target structure. However, if the laser pulse is suitably brief, its energy is invested in the target chromophore before much heat is lost by thermal diffusion out of the exposure field. Shorter pulse durations confine the laser energy to progressively smaller targets with more spatial selectivity. A maximum, transient temperature differential between the target and adjacent structures will then be achieved. The calculated thermal relaxation time for dermal blood vessels, typical of PWS in children, is 190 micro- to 3 milliseconds[12,13]. In commonly available argon lasers, the shortest available pulse duration is 0.05 seconds which accounts for the non-specific damage produced by these devices.

The flashlamp-pulsed dye laser (FLPDL) at 585 nm, a wavelength well absorbed by the targeted oxyhemoglobin (HbO_2) molecule relative to other optically absorbing structures, causes selective thermal damage to dermal blood vessels while minimizing the epidermal melanin absorption. Furthermore, the 450 microseconds pulse duration produced by this laser closely matches the thermal relaxation time for dermal blood vessels thereby confining the laser energy to the targeted HbO_2 molecule before much heat is lost by thermal diffusion out of the exposure field.

The safety of this form of therapy has now been demonstrated by the successful management of many patients with PWS [14÷17]. The immediate post-treatment purpura, edema, and erythematous flare resolve within 24 hours to several days. The scaling and/or crusting that occurs occasionally is also transient and resolves without any significant sequelae (Figure 1).

In contrast to other laser systems, minimal (less than 2%) or no scar formation occurs following the use of the FLPDL. The majority of scars have occurred in areas accidentally traumatized soon after laser therapy. Hyperpigmentation, which has been reported in up to 57% of patients[17], is usually temporary and will resolve spontaneously over 6-12 months or with the subsequent use of hydroquinone bleaching preparation.

A wide range of other vascular abnormalities such as facial telangiectasias, telangiectasias of Osler-Weber-Rendu disease, spider angiomas, "post traumatic red nose syndrome", cherry (senile) angiomas, angiokeratomas, and venous lakes (Figure 2) have all benefited from treatment with FLPDL or argon laser[18]. Therapy for large nasolabial fold telangiectasia is more difficult due to the large diameter of the vessels. More recently, a technique using a continuous wave, mechanically pulsed, yellow (577 nm) tunable dye laser with a 100 μm spot diameter hand piece and pulse duration of 0.05 to 0.10 sec, has been shown effective (Figure 3). Pulses were delivered sequentially along the blood vessel to produce the clinical response of blanching, leaving normal epidermis and skin appendages between consecutive pulses. Power and energy densities were 89-127 W/nm^2 and 4.5-13 J/mm^2; respectively[19]. Trunk and upper extremity teleangiectasias respond less favorably. Unfortunately, treatment of superficial lower extremity teleangiectasias is disappointing, often resulting in hyperpigmentation, atrophic scarring, or patches of unimproved or even more noticeable teleangiectasias. Less than 25% of such patients obtain good results.

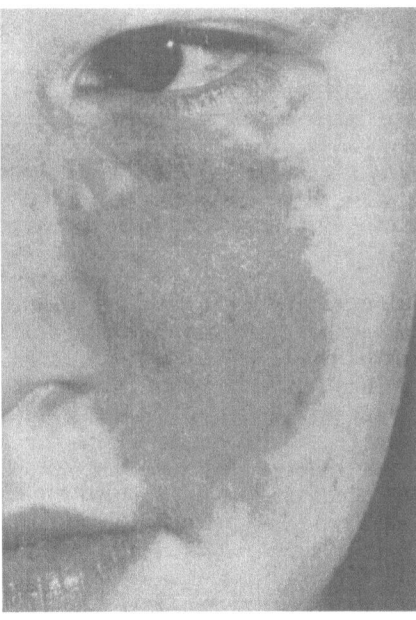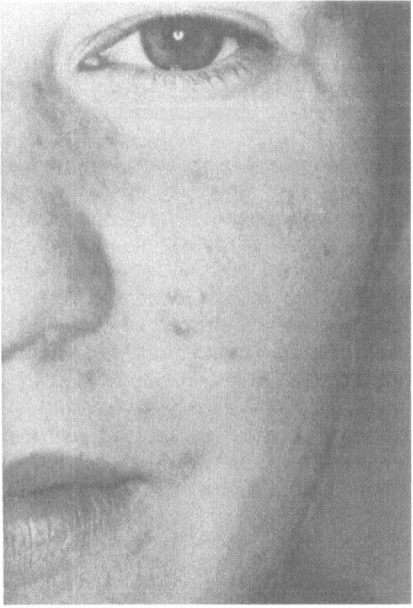

Fig. 1. Twelve year old female child with PWS, a) prior to laser therapy; and, b) after four treatments with the flashlamp-pulsed dye laser.

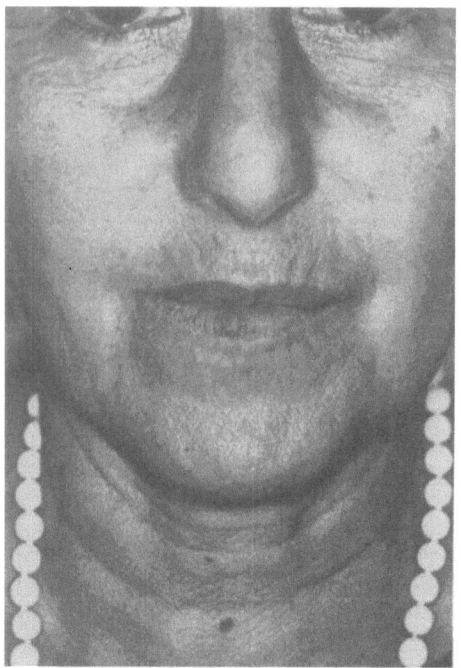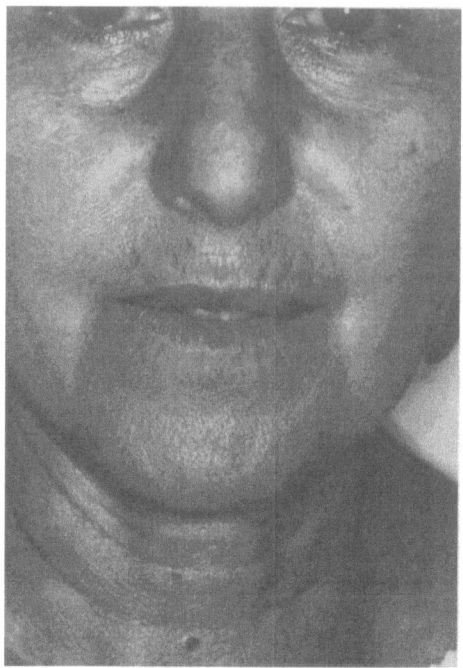

Fig. 2. Sixty-seven year old female adult with venous lake of the
lower lip, a) prior to laser therapy; and, b) after one
treatment with the flashlamp-pulsed dye laser.

Capillary hemangiomas of infancy (strawberry hemangiomas), which make
up 65% of hemangiomas, are composed of small superficial vascular channels.
Cavernous hemangiomas, which comprise 15% of hemangiomas, have deeper and
larger vascular spaces and may present as flesh-colored tumors. The remaining hemangiomas have components of both and are termed mixed hemangiomas. Spontaneous involution occurs in most capillary hemangiomas, resulting in normal-appearing skin or, slight atrophy. Cavernous hemangiomas and
the deep component of mixed hemangiomas usually involute incompletely,
leaving behind folds of atrophic teleangiectatic skin[2].

In most cases no treatment of any kind is required as the hemangiomas
resolve with good cosmetic results. Active treatment may be dictated by
situations such as extensive involvement; repeated ulceration, hemorrhage,
or infection; obstruction of vital organs such as the eye, nose, mouth,
genitals, or anus. Treatment with the laser may stop growth and induce
resolution or aid in control of these complications. Laser treatment of
these lesions should be undertaken only for specific indications and not
used for what is generally a self-limited process.

Argon laser therapy (Figure 4) offers clinical advantages over surgical forms of therapy for certain hemangiomas,including its ease of use,
absence of intraoperative hemorrhage, and satisfactory results[20][22]. The
Nd:YAG laser is most commonly used to treat larger, boggy cutaneous hemangiomas and vascular tumors of mucosal surfaces and the tongue[23,24]. Penetration into the tissue with this laser is deep, resulting in a larger
volume of coagulated tissue, substantially larger than with other lasers.
Profound thermal damage can be produced unknowingly by an inexperienced
operator making the Nd:YAG laser a potentially dangerous tool.

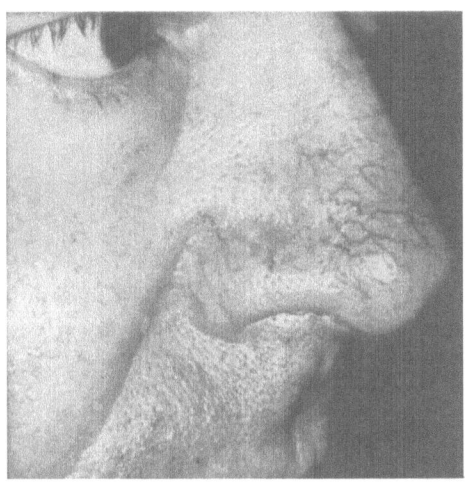

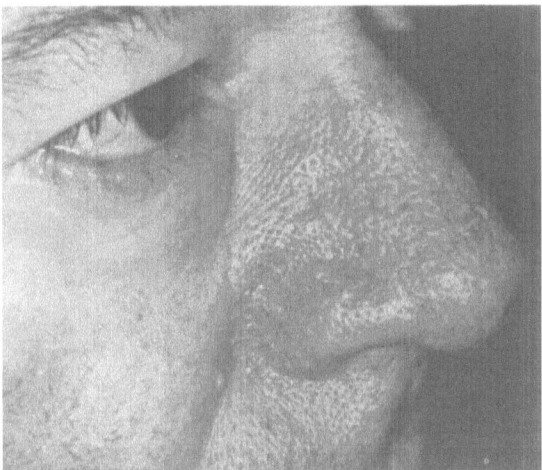

Fig. 3. Fifty-three year old male adult with teleangiectasias of
the nose, a) prior to laser therapy; and, b) after two
treatments using the continuous wave, mechanically pulsed,
yellow (577 nm) tunable dye laser with a 100 μm spot diameter
handpiece

Both the argon and CO_2 lasers have been used to treat the decorative
tattoo. Successive layers of skin are removed to expose the intradermal
pigment which is subsequently vaporized and the wound allowed to heal by
re-epithelialization from adjacent skin and undamaged dermal appendages.
Because there is no color selective absorption and therefore no epidermal
sparing, the treated skin never completely returns to normal and 30-40% of
patients have hypertrophic scarring. Patients must be told what to expect
but most are relieved that the original tattoo has been removed[25÷27].

The Q-switched ruby laser has also been proposed for the removal of
tattoos based on target selectivity, lack of vascular injury, and optical
penetration into the dermis. This laser emits nanosecond pulses with extremely high peak powers (megawatts). The red 694 nm wavelength penetrates
several millimeters into the dermis and is well absorbed by carbon, india
ink and organometallic dyes typically found in dark blueblack amateur and
professional tattoos, relative to other optically absorbing structures.
Furthermore, the nanosecond pulse duration produced closely matches the

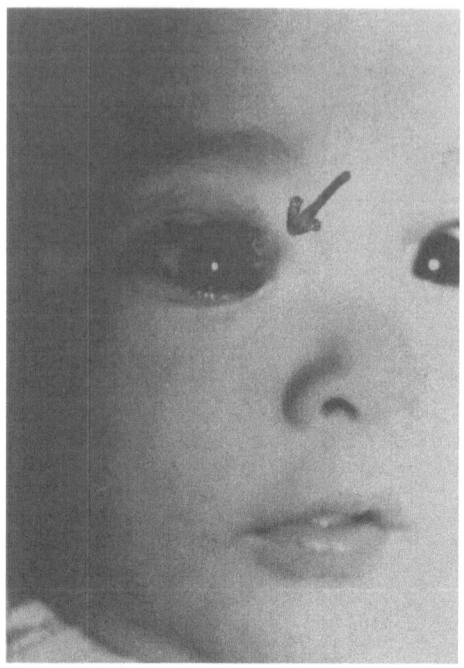

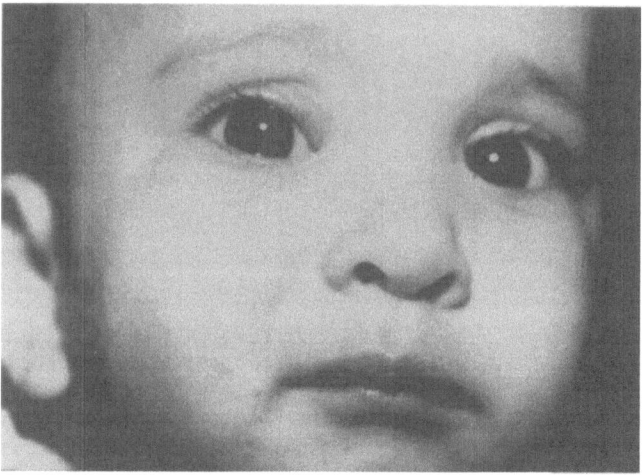

Fig. 4. Eighteen month old male child with strawberry hemangioma
         of left upper eyelid, a) prior to laser therapy; and,
         b) after one treatment using the argon laser

thermal relaxation time for tattoo pigment thereby confining the laser
energy to the targeted pigment before much heat is lost by thermal diffusion. The treatment, which is described as feeling like an elastic band
snapped against the skin is well tolerated. If patients complain of discomfort, a local anesthetic can be used.

Energy densities typically used for treatment of dark blue-black
tattoos range from 2-8 J/cm^2. Immediately after treatment, a white-ash
discoloration is seen. This opaque white appearance in the skin is caused
by the formation of vacuoles in the dermis probably due to the production
of steam [28].

62

In the ensuing 48-72 hrs, the whitening of the skin subsides at which time the normal-slightly faded preoperative tattoo will reappear. Additional fading of the tattoo will be noted over the next 6-8 weeks. The fading and removal of ink is brought about by two mechanisms. First, the dye absorbs the light energy and converts it rapidly to heat causing the production of steam with probable chemical alteration of the ink into colorless compounds. Second, there is an increase in macrophage activity in the subsequent weeks and months postoperatively. The large deposits of ink are broken up into smaller particles which are then removed by phagocytosis[29].

Complete removal of the tattoo after one treatment is rare. The great majority of patients require multiple treatments to the same area to obtain optimal fading. Amateur tattoos generally require fewer treatments as compared to those professionally done due to the higher concentration of pigment in the latter. Darker colors, especially blue and black, respond best (Figure 5). Yellow and orange respond less well and red not at all.

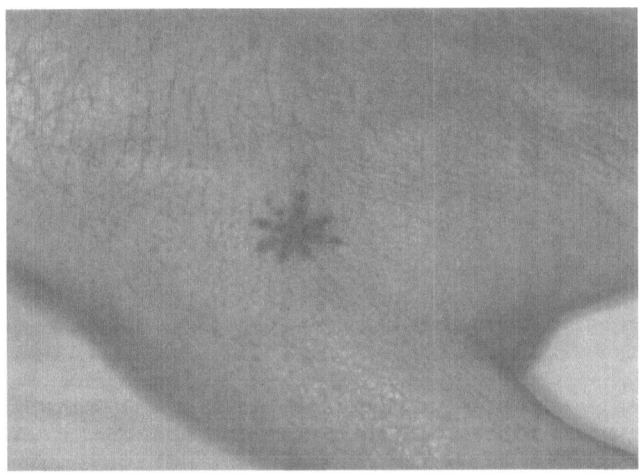

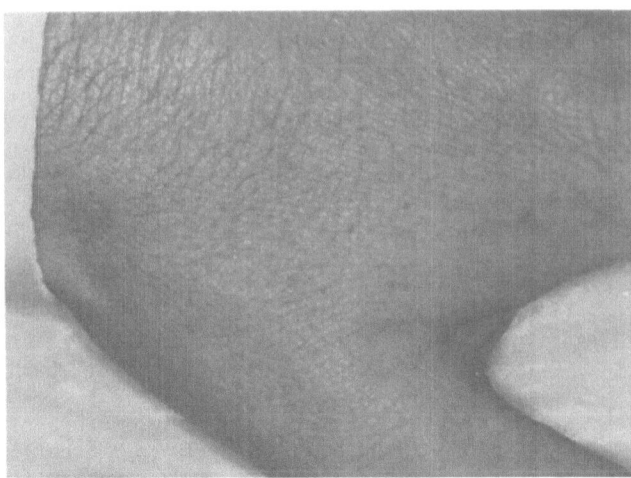

Fig. 5. Thirty one year old male adult with decorative hand tattoo, a) prior to laser therapy; and, b) after four treatments with the Q-switched ruby laser.

In contrast to other methods of tattoo removal, no scar formation or skin textural changes were observed following the use of the Q-switched ruby laser. Hypopigmentation, which can occur in up to 20% of patients, is usually temporary and will resolve spontaneously over 6-12 months[30]. These findings, which confirmed the efficacy of the Q-switched ruby laser in the treatment of tattoos, facilitated FDA approval for this clinical application in November, 1989.

Because all living tissue contains 70-90% water, the CO_2 laser has multiple applications in plastic surgery and dermatology. The CO_2 laser may serve as a "light scalpel" to accomplish incisional or excisional procedures involving infected or highly vascular lesions or in lesions located in a highly vascularized anatomic region. In these situations, the beam diameter is reduced to its minimum possible size, usually in the range of 0.1 to 0.2 mm. The power output is relatively high, resulting in irradiance values in the range of 25,000 watts/cm^2. The actual depth of the incision will depend upon the power and the rate of movement of the beam across the tissue. The heat conducted to the surrounding tissue is capable of coagulating blood vessels up to 0.5 mm in diameter. Vessels larger than this may also be sealed with minimal damage by applying laser energy directly to the end of the severed vessel after it has been clamped. The beam has also been demonstrated to seal lymphatic vessels spontaneously, raising the possibility that the CO_2 laser may have a unique advantage in dealing with lesions that have a potential for lymphatic spread[31].

The CO_2 laser can also be used as a tool for the vaporization and ablation of multiple superficial lesions. Here the beam is applied to the tissue in a defocused, large spot size (2 to 5 mm) at low power, resulting in irradiances in the range of 150 to 500 watts/cm^2. When the laser is used in this manner, the final result is a broad but shallow zone of tissue ablation. Condylomata acuminata, plantar and hand warts are very effectively treated by CO_2 laser vaporization. The laser is of particular value in difficult-to-reach lesions or when surface contour must be preserved with minimal damage. Furthermore, high local temperatures presumably sterilize and eliminate the virus from the treatment field. Other skin lesions such as papillomatosis, rhinophyma, skin tags, solar keratosis, seborrheic keratosis, xanthelasma and pyogenic granuloma are easily and rapidly vaporized with the CO_2 laser[32].

Skin tumors, both benign and malignant, have been treated with a variety of lasers. The CO_2 laser does have the potential for sealing blood vessels and lymphatics permitting almost a "no touch" tumor removal to be performed. It is possible that the cosmetic results in the treated areas will be superior to those obtained by other methods but conclusive long term studies have not been conducted. Photodynamic therapy (PDT) using red light and photosensitizing dyes has also been successfully used to treat widespread and recurrent head and neck cutaneous malignancies thus precluding the need for multiple, possibly deforming excisions[33].

The unique properties of lasers create an enormous potential for specific therapy of variety of skin diseases. In the next few years, it may be possible to define the tissue reflectance characteristics of specific lesions with the spectrophotometer and then choose a wavelength for laser phototherapy that will maximize lesion absorption and destruction while minimizing adjacent damage. Further laser light-tissue interactions, such as shock waves, remain to be explored and understood. Modification of current lasers and innovative advances with biomedical laser instrumentation may eventually allow us to match optimally the laser and the treatment procedure, with the lesion.

REFERENCES

1. S. H. Barsky, S. Rosen, D.E. Geer and J. M. Noe, "The nature and evolution of port wine stains: a computer assisted study", J. Invest. Dermatol., 74:154-157 (1980)

2. J. B. Mulliken and A. E. Young, "Vascular Birthmatks - Hemangiomas and Malformations", Philadelphia, PA: W. B. Saunders Company (1988)

3. L. Goldman, "Laser treatment of extensive mixed cavernous and port wine stains", Arch. Dermatol., 113:504-505 (1977)

4. D. B. Apfelberg, M. R. Maser and H. Lash, "Argon laser management of cutaneous vascular abnormalities", West. J. Med., 124:99-101 (1976)

5. B. Cosman, "Experience in the argon laser therapy of port wine stains", Plast. Reconstr. Surg., 65:119-129 (1980)

6. J. M. Noe, S. H. Barsky, D. E. Geer and S. Rosen, "Port wine stains and the response to argon laser therapy: successful treatment and the predictive role of color, age and biopsy", Plast. Reconstr. Surg., 65:130-136 (1980)

7. J. L. Finley, S. H. Barsky and D. E. Geer, "Healing of port wine stains after argon laser therapy", Arch. Dermatol., 117: 486-489 (1981)

8. J. L. Finley, K. A. Arndt, J. M. Noe and S. Rosen, "Argon laser port wine stain interaction: immediate effect", Arch. Dermatol., 120:613-619 (1984)

9. J. M. C. van Gemert and J. P. C. Henning, "A model approach to laser coagulation of dermal vascular lesions", Arch. Dermatol. Res., 270:429-439 (1981)

10. R. R. Anderson and J. A. Parrish, "The optical properties of skin", in J. D. Regan, J. A. Parrish (eds), The Science of Photomedicine, New York, NY: Plenum Press, pp. 147-194 (1982)

11. O. T. Tan, S. Murray and A. K. Kurban, "Action spectrum of vascular specific injury using pulsed irradiation", I. Invest. Dermatol. 92:868-871 (1989)

12. R. R. Anderson and J. A. Parrish, "The optics of human skin", I. Invest. Dermatol., 77:13-19 (1981)

13. R. R. Anderson and J. A. Parrish, "Selective photothermolysis: precise microsurgery by selective absorption of pulsed radiation", Science, 220:524-527 (1983)

14. J. G. Morelli, O. T. Tan, J. M. Garden, Y. Seki, J. Boll, M. Carney, R. R. Anderson, H. Furumoto and J. A. Parrish, "Tunable dye laser (577 nm) treatment of portwine stains", Laser Surg. Med., 6:94-99 (1986)

15. O. T. Tan and T. J. Stafford, "Treatment of portwine stain at 577 nm: clinical results", Med. Instrum., 21:218-221 (1987)

16. J. M. Garden, O. T. Tan and J. A. Parrish, "The pulsed dye laser: its use at 577 nm wavelength", J. Dermatol. Surg. Oncol., 13:134-138 (1987)

17. O. T. Tan, K. Sherwood and B. A. Gilchrest, "Treatment of children with port-wine stains using the flashlamp-pulsed tunable dye laser" New. Eng. J. Med., 320:416-421 (1989)

18. L. L. Polla, O. T. Tan, J. M. Garden and J. A. Parrish, "Tunable dye laser for the treatment of cutaneous vascular ectasia", Dermatologica, 174:11-17 (1987)

19. A. Orenstein and J. S. Nelson, "Treatment of facial vascular lesions with a 100 m spot 577 nm pulsed continuous wave dye laser", Ann. Plast. Surg., 23:310-316 (1989)

20. D. B. Apfelberg, R. A. Greene, M. R. Maser, H. Lash, J. L. Rivers, and D. R. Lans, "Results of argon laser exposure of capillary hemangiomas of infancy - preliminary report", Plast. Reconstr. Surg., 67:188-193 (1981)

21. L. W. Hobby, "Further evaluation of the potential of the argon laser in the treatment of strawberry hemangiomas", Plast. Reconstr. Surg., 71:481-489 (1983)

22. B. M. Achauer and V. M. Vander Kam, "Argon laser treatment of strawberry hemangioma of infancy", West. J. Med., 143:628-632 (1985)

23. H. Rosenfeld and R. Sherman, "Treatment of cutaneous and deep vascular lesions with the Nd:YAG laser", Laser Surg. Med., 6:20-23 (1986)

24. J. A. Dixon, R. K. Davis and J. L. Gilbertson, "Laser photocoagulation of vascular malformations of the tongue", Laryngoscope, 96:537-541 (1986)

25. D. B. Apfelberg, M. R. Maser and H. Lash, "Argon laser treatment of decorative tattoos", Br. J. Plast. Surg., 32:141-144 (1979)

26. P. L. Bailin, J. L. Ratz and H. L. Levine, "Removal of tattoos by CO laser", J. Dermatol. Surg. Oncol., 6:997-1001 (1980)

27. D. B. Apfelberg, M. R. Maser, H. Lash, D. N. White and J. T. Flores, "Comparison of argon and carbon dioxide lader treatment of decorative tattoos: a preliminary report", Ann. Plast. Surg., 14:7-16 (1985)

28. W. H. Reid, P. J. McLeod, A. Ritchie and M. Ferguson-Pell, "Q-switched ruby laser treatment of black tattoos", Br. J. Plast. Surg., 36:455-459 (1983)

29. D. R. Laub, R. B. Yules, M. Arras, D. E. Murray, L. Crowley and R. A. Chase, "Preliminary histopathological observation of Q-switched ruby laser radiation on dermal tattoo pigment in man", J. Surg. Res., 5:220-224 (1968)

30. R. R. Anderson, personal communication, 1990

31. P. L. Bailin, "Lasers in dermatology: principles and clinical applications", Prof. Dermatol., 21:1-8 (1987)

32. P. L. Bailin, J. L. Ratz and R. G. Wheeland, "Laser therapy of skin: a review of principles and applications", Dermatol. Clin., 5:259-285 (1987)

33. T. J. Dougherty, "Photodynamic therapy (PDT) of malignant tumors", Crit. Rev. Oncol. Hematol., 2:83-116. (1984)

FIRST EXPERIENCE WITH A LASER CLAMP IN NEUROSURGERY

V. A. Fasano(*), W. Cecchetti(**), S. M. Peirone(***),
R. Urciuoli(*), G. F. Lombard(*), M. Fontanella(*) and
L. Sansalvadore(*)

The laser clamp described in this paper was invented by the senior Author (V. A. F.) and was designed and built by one of the authors (W. C.).

The principal goal of a neurosurgeon is to remove the lesion while preserving the brain's functions; this can be accomplished only by very selective surgery. Neurosurgeons already have two instruments to gain this selectivity: the microscope and the bipolar forceps. The latter is particularly useful for tissue, but its effects on the vessels cannot be perfectly controlled; the forceps operates on the vessel wall, and because its effect is not easily controlled, in some cases a perforation of the wall cannot be avoided[1].

In 1980 in Chicago we presented the first results on the use of Nd-YAG and Argon lasers in vascular surgery of the brain; in this work, surgery was restricted to vessels of a small diameter and with a low flow[2]. These restrictions arose not only from the radiation sources, but also from the instruments, which did not permit total irradiation of the wall and did not stop the flow, within which a considerable part of the energy was dissipated.

We have therefore studied the possibility of developing an instrument enabling us to irradiate the entire wall while at the same time stopping the flow and modifying the power and the time of exposure, in order to obtain either effects on the wall (shrinking of the vessel with a preservation of the flow) or an endoluminal effect (i.e. an occlusion of the vessels)[1].

To obtain a more selective effect, we are also considering the feasibility of building a dedicated two wavelength Nd-YAG laser (1064, 1320 nm).

The clamp can be utilized with different laser sources. We used a Nd-YAG cw laser built by Messerschmitt-Boelkow-Blohm (W. Germany)[3] featuring continuously adjustable power (0-60 W), a closed circuit integrated cooling system, and a He-Ne pilot laser of 5 mW for targeting the invisible 1064 nm radiation.

(*)   Istituto di Neurochirurgia, Università di Torino, Italy
(**)   Dipartimento di Chimica Fisica, Univ. di Venezia, Italy
(***) Istituto di Anatomia Veterinaria, Univ. di Torino, Italy

The output beam of the Nd-YAG cw laser is split into two optical fibers[4] [5] fixed to the arms of a 20 cm long stainless-steel bayonet type clamp. The fiber tips are shaped in such a way that the output of the laser radiation is orthogonal to the fiber axis[4], and they can work in contact with water or physiological liquid, with a laser power of several watts[6]. The clamp is provided with an irrigation system to avoid adherence to the tissue walls and to cool the fiber tips. The pilot He-Ne laser illuminating the tips provides the continuous check of the transmission system working conditions. It is necessary to increase or decrease the power and exposure time depending on the diameter of the target artery.

Preliminary experiments with a prototype applied to the dural and cerebral vessels have been performed. Microscopic studies show that the laser clamp, operating at a power of 10 W, produces a modification of the collagen with a shrinking of vasal lumen, without damage to the wall. The bipolar forceps, used at a power of 35 W, produces microhemorrhages inside the vasal endothelium and in the vascular wall.

REFERENCES

1. V. A. Fasano, S. M. Peirone, R. M. Ponzio, M. M. Lanotte and A. Merighi, "Lasers Surgery and Medicine", 6, 308 (1986)
2. V. A. Fasano, "Advanced Intraoperative Technologies in Neurosurgery", Springer-Verlag Wien, New York (1980)
3. E. F. Downing, P. W. Ascher, L. J. Cerullo, C. R. Neblett, J. H. Robertson and J. M. Tew et al., "Laser in Neurosurgery", Springer-Verlag, Wien-New York (1989)
4. V. Russo, G. C. Righini, S. Sottini and S. Trigari, "Fiber Optics", Proc. SPIE Fibre Optics '85, 522, 166 (1985)
5. W. Cecchetti, O. Curcuruto, E. Ghezzo, R. Polloni and P. Traldi, "Applicazioni laser alla spettrometria di massa", Conf. Proceedings, SIF, Bologna, 21, 345 (1989)
6. W. Cecchetti, "Terminazione di fibre ottiche con deviazione ortogonale della radiazione di uscita", Dom. Dep. Brev. N.84120 A/90, 04/04/90.

EXCIMER LASER ANGIOPLASTY IN HUMAN ARTERY DISEASE

R. Viligiardi*,
V. Gallucci**,
R. Pini***, R. Salimbeni*** and S. Galiberti***

I. INTRODUCTION

The principal criteria guiding our activities have been as follows: a) the surgical method investigated must be simple and repeatable because we deal with a pathology which is chronic and evolutive; and, b) our purpose is to validate a surgical technique which could represent a potential improvement over the long-term results shown by other techniques.

We chose to use an excimer laser based system because of several potential advantages compared with other laser systems: i) a low histo-lesivity in the residual tissue; ii) the present availability of devices with emission characteristics suitable for this application; and, iii) the presence on the market of specially designed optical fibers for high power delivery of ultraviolet radiation.

Only the first of the aforementioned features is significantly related to long-term results, which derive primarily, in our opinion, from the diameter of the recanalized neolumen and from the traumatism of the procedure.

The possibility of creating neo-lumina in completely occlused vessels with a recanalized diameter only sufficient for a balloon catheter subsequent treatment represents a widening of the patient treatability index, but at the same time, does not improve the typical follow-up results of the balloon technique. Our efforts have hence been directed to recanalizing neo-lumina of suitable diameters exclusively through a photoablative process. This objective has been pursued by means of composite multifiber catheters with an overall diameter similar to the original diameter of the vessel. In this way the combination of the low tissue lesivity and the negligible mechanical damage induced on the wall can reduce the occurrence of the reconstructive phenomena which are usually the prelude to early re-occlusions, as is frequently observed with other techniques.

Our approach has been supported by the results obtained in a pre-clinical phase, which showed minimal wall damage in a relatively large recanal-

(*)   Policlinic of Careggi, Institute of General Surgery, Florence, Italy
(**)   Institute of Cardiovascular Surgery, University of Padua, Italy
(***) Institute of Quantum Electronics, CNR, Florence, Italy

ized diameter, and therefore in June 1989 we began a clinical trial. This activity was performed at the "Istituto di Chirurgia Cardiovascolare" of Padua University and in the iliac and femoral-popliteal department of the "Sezione Laser Angioplastica" of the Careggi (Florence) Polyclinic, General Surgery Division.

Experimental results and operational problems with excimer laser angioplasty in human arteries will be presented. These results involve clinical experience with the coronary region in association with an A.C. bypass procedure and in the iliac and femoral popliteal regions with a percutaneous approach.

II. MATERIALS AND METHODS

For the purposes of this research program an excimer system suitable for this medical application has been designed and constructed. Its main features are: 1) EMI compatibility for surgery room operation; 2) no connections are required except for the power cord; 3) no alignment procedure is needed for fast interchange of catheters; 4) average energy density of 3 J/cm^2 at 308 nm, out of a 3 m long catheter; 5) repetition rate variable in the 1-100 Hz range; and, 6) built-in push button gas refill procedure.

Different multifiber catheter geometries have been designed and tested. Typical configurations are 12×200 µm, 19×200 µm, 9×400 µm and 3×600 µm; the last one has been developed and is most frequently utilized.

For patients who underwent coronary angioplasty the technique we have employed is used with an A.C. by-pass procedure. This choice is justified by the need to operate under absolutely safe conditions, and to introduce a mixed method capable of treating patients with severe and multiple distal occlusions, reducing their number and eventually limiting treatment to proximal sites. The catheter introductory site was the arteriotomy required for the distal anastomosis of the by-pass.

In the iliac and femoral-popliteal regions, the procedure utilized has been the following:
- vessel access is gained under local anesthesia through a percutaneous technique or surgically, depending on the lesion characteristics: partially or completely occluded vessel, proximal or distal approaches;
- the progression of the catheter through the vessel was observed and controlled by means of intermittent radioscopy;
- in some cases endoscopy was also utilized.

III. CLINICAL EXPERIENCES

By means of the method described, a total number of 5 patients have been treated by laser coronary angioplasty in the period June-November '89. They are 4 male and 1 female, with an age ranging between 58 and 72 years, all suffering angina pectoris, CCS III and IV class. Three of them showed pathology in 2 coronary vessels and two in 3 vessels. A total of 7 stenoses have been treated by laser angioplasty: 4 anterior descending, 1 right, 1 circumflex and 1 diagonal. 4 were concentric and 3 eccentric, 2 were calcified.

The intraoperative evaluation of the results has been: good progression inside all the examined segments, with a reasonable advancement rate also in the total occlusions, except for the heavily calcified ones. Coronariographic control has been performed at 7 days from the operation with selective opaqueness of both the native coronaric net and the grafts. In 2 cases good results were obtained for this procedure: stenosis reduction in a right vessel from 75% to 20%, and in a diagonal from 75% to 10%. 2 patients exhibited a dissection of the anterior descending in the laser application site, one case with intramural hematoma and complete obstruction of the lumen. In one patient we observed a thrombotic occlusion of the circumflex and 2 in the anterior descending, always preceding the stenosis. All grafts were pervious and operative.

In the iliac and femoral-popliteal region 16 patents have been treated with an overall number of 5 and 21 segments respectively. Iliac lesions were total and extended in 4 cases and distally segmented in 1 case. Lesions located in the femoral-popliteal tract extended in 3 cases from the origin of the femoral artery through to 2/3 of its length; in the other 18 lesions they extended from the III medium to the popliteal one.

Iliac segments have been treated in the surgery room, under local anaesthesia, to allow surgery with reduced risk from possible perforation. All patients underwent angiographic examination 30 days following the operation. The results are as follows:
- 2 untreatable patients (because of heavily calcified lesions);
- 1 dissection;
- 1 popliteal perforation;
- 1 perforation at the III inferior of the femoral artery;
- 16 positive results.

In the iliac segments, except for one with a limited distal lesion, balloon angioplasty was provided to obtain a sufficient neo-lumen. Because the catheter is insufficiently flexible, its path follows the direction of minimum resistance, which makes it difficult to center it on the target. Heavily calcified plaques turn out to be untreatable, or carry a high hazard of perforation.

IV. CONCLUSIONS

The reasonable percentage of successes obtained up to now with a limited statistical base, is promising for the general validity of our premises. For this reason we consider it important and useful to pursue the development of an exclusively laser angioplasty, utilizing improved emission laser systems and catheters with larger emitting areas.

V. ACKNOWLEDGEMENTS

The authors wish to thank the "Progetto Finalizzato Tecnologie Biomediche e Sanitarie" of the Italian National Research Council for partial financial support.

REFERENCES

1. W. Grundfest, I.F. Litvack, T. Goldenberg, T. Sherman, L. Morgenstern, R. Carroll, J. Fishbein, J. Forrester, J. Margitan, S. McDermid, T. J. Pacala, J. M. Rider and J. B. Laudenslager, "Pulsed ultraviolet lasers and their potential for safe angioplasty", Amer. J. Surg., 150:220-226 (1985)
2. R. Pini, R. Salimbeni and M. Vannini, "Optical fiber transmission of high power excimer laser radiation", Appl. Optics, 26:4185-4189 (1987)
3. R. Viligiardi, G. Thiene, A. Angelini, V. Gallucci, S. Galiberti, R. Pini, R. Salimbeni and M. Vannini, "Multifiber catheter test to ap-

proach an A.C. by-pass associated with excimer laser angioplasty procedure", Proceed. 2nd Int. Symp. Lasers in Cardiovascular Diseases, Wien, Heart and Vessels, 4, p. 61 (1988).

THE EFFICIENCY OF LASER PANRETINAL PHOTOCOAGULATION FOR DIABETES

M. L. Wolbarsht(*) and M. B. Landers III(**)

(*) Department of Psychology, Duke University
Durham, North Carolina, USA
(**) Department of Ophthalmology, University of California
Sacramento, California, USA

I. INTRODUCTION

The proven efficacy of panretinal laser photocoagulation in controlling many varieties of proliferative retinal diseases is astonishing considering the lack of agreement on the physiological mechanisms underlying this effect. However, a review of the evidence supports a model in which the effects (and effectiveness) of photocoagulation depend on the interaction between the choroidal and retinal circulations. To a large degree, this interaction is controlled by the metabolism of the photoreceptor layers. The data available also suggest that in diabetes (and possibly other retinal disorders), altered states of metabolism influence retinal vascular oxygen tension and pH which in turn act together to induce the characteristic vascular pathology.

Under normal circumstances, the adult retina receives oxygen from two sources. The inner retina, from that portion closest to the vitreous up to the outer plexiform layer, is usually supplied by the retinal circulation. In contrast to this, the avascular outer retina, which includes the photoreceptors and the pigment epithelium, receives oxygen by diffusion from the choroidal circulation.

The retinal circulation is sensitive to changes in oxygen and pH. Hypoxia and hypercapnia induce an autoregulatory vasodilatation, while hyperoxia and hypocapnia cause constriction[1]. The choroidal circulation which supplies the outer half of the retina has a large flow rate and does not autoregulate significantly. Indeed, normal choroidal flow rate keeps the chroidal tissue oxygen tension close to arterial oxygen levels[2+4]. In fact, the majority of the choroidal O_2 comes from the O_2 dissolved in the blood plasma and lipid membranes of the blood cells rather than from the hemoglobin bound state. To understand the effect of photocoagulation on the retina, it is necessary to examine several extreme physiological and pathological conditions. One of these extremes is the retinopathy of prematurity (ROP), formerly termed retrolental fibroplasia (RLF), in which there is marked constriction, and even suppression of the retinal circulation is the earliest stage. An allied condition in adults is retinitis pigmentosa (RP) which also produces retinal vasoconstriction that increases as the condition progresses. These types of situations furnish a starting place for understanding the way photocoagulation is effective as a therapeutic measure for the proliferative retinopathies as a group.

Diabetes is a systemic disease, but as will be shown, it leads to a particular pathological vascular complication in the kidneys, brain, heart, muscles, and retina. The precursor alteration in the metabolic state in the retina is especially suitable to be controlled by panretinal photocoagulation. Based on these considerations, a model is proposed extending previous work dealing with the immediate and long-term effects of supplemental oxygen in the retina[5,6] to cover the altered metabolic state in diabetes. This model shows how laser photocoagulation and vitrectomy can halt the development of the vessel pathology related to neovascularization in general and specifically in diabetes.

Panretinal laser photocoagulation therapy effectively inhibits proliferative retinopathies such as diabetic retinopathy and the retinopathy of prematurity (ROP) by controlling the oxygen and pH levels in the inner retina. This control is effected through laser irradiation heating or photochemically altering the retinal pigment epithelium both of which result in destruction of the rods and cones. The removal of the large quantity of mitochondria and other active metabolic tissue associated with the rods and cones allows the choroidal oxygen to enter the inner retina in large quantities while at the same time lowering metabolic by-product levels from the receptor layer. Autoregulation of retinal O_2 and pH levels results in marked constriction of the retinal vasculature. This acts to protect the vasculature from deleterious effects of the chronic vasodilatation characteristic of diabetes and many other retinopathies. In diabetes, this vasoconstriction is a return to more normal blood flow and possibly also reflects a higher and more normal tissue pH. A model is presented which links the diabetic patho-physiology of the retinal capillaries with the characteristicly high glucose levels in the blood. Comparisons with metabolic changes in the retinopathy of prematurity and retinitis pigmentosa indicate the universality of the factors leading to vascular proliferation and atrophy. An analysis of the interplay between the retinal O_2 and pH levels is given to show how the total retinal metabolism must be considered in analyzing any long-term effects of the removal of photoreceptors by laser phototherapy.

II. EFFECTS OF BLOOD OXYGEN LEVELS ON RETINAL VESSEL DEVELOPMENT

Under normal circumstances, the adult retina vessels are static. Their pattern is a result of the total development. In the fetal retina, the vascular distribution differs from that in the adult retina. At first, there are no retinal vessels, and the metabolic demands of the inner retina are totally met by the choroidal circulation. As the photo-receptors mature and increase their rate of metabolism, the amount of oxygen which can diffuse past the inner retina is reduced. Consequently, the oxygen tension in the inner portion of the retina drops, and the retinal vessels dilate and progressively grow and mature toward the periphery[5,8], possibly led by the spindle cells. This development of the retinal vascular tree is not complete until term[9]. Presumably, the dilatation and growth of the retinal vasculature permits only a slight decrease in the oxygen tension of the inner retina as the retina matures. All of the data suggests that retinal vessels can continue to grow only when conditions exist which produce and maintain vasodilatation.

Neonatal retinal vessels autoregulate in response to tissue oxygen and pH levels as in the adult[1]. It is commonly observed that supplemental oxygen therapy immediately constricts retinal vessels of premature neonates of all birth weights. This is to be expected as the first normal breath suddenly exposes the neonatal retina to a highly elevated ambient oxygen

level, which is followed by a constriction of the retinal vessel[10,11].
Also, it has been demonstrated[12] that retinal vessels obliterated by inspiration of pure oxygen reopen when retinal detachment develops. Presumably, the increased diffusion distance from the choroid to the retinal circulation during a detachment lowers the inner retinal oxygen tension and induces an autoregulatory dilatation of the constricted vessels, although this dilatation may also be due to a lowered pH from the accumulation of metabolites (CO_2, lactic acid, etc.) from the inner retina or diffusing from the outer retina.

The O_2 tension in the choroidal tissue is very close to the arterial level as a (~ 90 mm Hg) result of the high flow rate of the blood in the very large choroidal capillaries. This is shown by the slight differences between the arterial and venous hemoglobin saturation levels in the choroid where the high choroidal O_2 levels are maintained almost entirely by O_2 dissolved in the circulating blood plasma rather than being derived from hemoglobin bound oxygen. For example, hyperbaric ambient O_2 levels raise the O_2 level in the choroid and inner retina when the retinal circulation is clamped off almost as much as in the arterial blood, although other somatic tissues have much smaller increases[13,14]. Indeed, it has been speculated that the high flow level in the choroid is rather for tissue heating that for oxygenation[15].

Vasodilatation is, at the very least, a part of the process that induces vessel growth. In all retinal vascular pathology, neovascularization is preceded by a phase of chronic vasodilatation[16]. More importantly, for the present discussion, vessel growth and maturation as well as neovascularization cease when vasoconstriction is produced no matter by what means. Perhaps, this stimulus to growth is similar to that already documented in the effects of tension upon the induction of many types of growth and cell division[17,18]. In any case, all these observations indicate that vasodilatation is a necessary if not a major part of the initiation of neovascularization[19]. The sequence of events in ROP clearly illustrate this dependence of growth on vasodilatation. The immediate constriction of the retinal vasculature following birth which is particularly accentuated in prematurity is ample demonstrative of this[9,10,12]. In ROP, neovascularization begins when vasodilatation occurs, usually at the end of oxygen therapy.

Another stimulus for vasodilatation in the neonatal retina may come from the accumulation of metabolic by-products during exposure to high levels of oxygen. The abrupt rise in choroidal oxygen tension at birth which may be accentuated by supplemental oxygen therapy allows a greater portion of the premature neonate's inner retina to receive its oxygen from the choroid. However, this high O_2 level does not increase the ability of the choroidal circulation to remove metabolic products. Thus, as the retinal metabolism increases with maturation of retinal neurons, the concentration of lactic acid and carbon dioxide rise in those portions of the retina not perfused by the as yet poorly developed retinal vascular bed. The potential vasodilatory stimulus produced by these accumulated metabolic products is balanced against the vasoconstrictive tendency induced by increased oxygen levels to produce the operational vasculature. However, upon cessation of oxygen therapy, the accumulated lactic acid and CO_2 certainly induce an exaggerated vasodilatation. The vasodilatation may in turn cause additional neovascularization. In any case, the initial action of the elevated O_2 levels in the retina associated with birth or supplemental O_2 is the constriction of retinal vasculature and relative or absolute inhibitiuon of further development.

As a companion to the ROP syndrome which shows the effect of increased oxygen tension in the developing retinal vasculature, it is instructive to

examine the pathology in these syndromes which mimic photocoagulation to
show the effect upon the adult structures. In retinitis pigmentosa (RP),
there is a loss of photoreceptors similar to the effects of photocoagulation. Retinitis pigmentosa (RP) covers a large group of hereditary degenerative disorders of the retina in which a functional loss of vision develops, initially in the periphery followed by increasing night blindness and
progressive constriction of the visual field. Overall, the retinal vessels
become progressively attenuated with marked optic nerve atrophy. In some
ways, the clinical picture resembles that encountered in chronic vitamin A
deficiency which also greatly diminishes photoreceptor function. There are
many reasons for considering RP similar to the hereditary retinal distrophy
in the RCS strain of dystrophic rats[20]. In these rats, exposure to light
produces degeneration of the retina within a few days. The time course is
much slower if the rats are kept in nearly total darkness. Histological
examination of the retina shows an accumulation of rod outer segments in
the early stages of degeneration in the retina. In the normal rat, the rod
outer segments continue to grow during life, but the length of the outer
segment is kept constant by phagocytosis in the pigment epithelium. The
balance between rod outer segment growth and phagocytosis is disturbed in
the rat with RCS inherited retinal dystrophy. In these rats, the rod outer
segment growth rate relative to the rate of phagocytosis is increased, and
a large amount of rod outer material accumulates as retinal debris[20,21].
As mature cone outer segments do not normally continue to grow rapidly, the
failure of the balance between growth and phagocytosis does not yield a
large cone contribution to the accumulation of retinal debris. Following
the accumulation of rod outer segment debris, and possibly as a result, the
outer neural layers of the retina degenerate.

In humans with RP, rod function is severely affected first, accompanied by less severe manifestation to the cones[22]. In the later stages of
RP, even the surviving cones (those in the fovea) show some signs of abnormality in their function[23]. The early signs of RP resemble in their
loss of photoreceptors the effects of minimal photocoagulation and the late
stages of RP, in which most of the photoreceptors with their high rate of
metabolism have disappeared almost completely, a more widespread panretinal
application. The loss of photoreceptor metabolic capabilities drops the
oxygen consumption in the outer retina to a very low value and allows the
oxygen from the choroid to diffuse freely to the inner retina and supple-
ment that available from the retinal vasculature which results in the vasoconstriction already noted. As has been presented so far, ROP has as its
initial symptom profound vasoconstriction leading to the development of
large areas of poorly perfused retina. The ROP at this stage is comparable
with the capillary constrictive phase of diabetic retinopathy following
extensive photocoagulation. At this point, no proliferation pathology can
develop.

The vasodilatation that begins the diabetic and other proliferative
retinopathies would, thus, be inhibited by some retinal pathology similar
to RP. For the present discussion, it is important to note that all proliferative retinopathies share one important symptom, vasodilatation, usually at a very early stage, but in any case, this symptom always immediately precedes neovascularization. In a similar fashion to RP, raising the
oxygen levels in the inner retina is, at least, part of the cause of the
vasoconstriction which inhibits both normal and abnormal vascular growth.

III. DIABETES AND PHOTOCOAGULATION

Diabetic retinopathy has an extended time course showing quickly progressive features interspersed with many periods of remission or slow progression. However, the late stages follow each other quickly. The patho-

logical signs are mostly in the near-central and mid-peripheral retina, often including the macula and optic disc. The initial sign is vasodilatation accompanied by leakage[24÷26], and exudates of various kinds. The vasodilatation after a variable delay is followed by microaneurysms and capillary dropout in the same area. Retinal hemorrhages and neovascularization can accompany this state but are usually most markedly seen later in the time course. More serious pathology can also be found: hemorrhage into the vitreous cavity, neovascularization from the retinal vessels expanding onto the surface of the vitreous face followed by retinal traction and detachment, and even iris neovascularization which in its final stage can spread into the angle giving an intractable glaucoma and loss of the eye.

Examination of a diabetic retina shows that capillary dropout is followed by neovascularization confined to the venous side of the retinal circulation. The dilated veins in the venous side of circulation are also the site of endothelial proliferation. Even the capillary microaneurysms seem to be in a large part due to endothelial proliferation. A similar sequence of pathology is also found in polycytaemia, leukemia, myelomatosis, circinate retinopathy, central vein thrombosis, macroglobulinaemia, and Eales' disease as well as retinopathy, CO poisoning, and as an early stage in radiation damage of the retina.

Although the initial vasodilatation in diabetes has often been ascribed to retinal hypoxia, many data do not fit this hypothesis. Normal subjects made hyperglycemic have both vasodilatation and a seemingly higher than normal O_2 time tension in the retina[27]. This vasodilatation at a time when the available retinal oxygen level is normal or has been slightly increased is strange. One speculation has been that increased glucose would require increased oxygen to metabolize it[19]. However, a more complete analysis of the biochemical situation indicates that higher than normal glucose levels in tissue should not increase oxygen consumption but rather decrease it mostly due to the lack of capacity of the enzyme systems involved in aerobic metabolism to cope with the increased levels of the oxygen and glucose. That is, additional glucose suppresses ordinary glycolysis and encourages an aerobic glycolysis through the Crabtree effect[28,29] with a resultant lactic acid build up.

The lactic acid increase described above, indeed, is possibly the most important factor in retinal vasodilatation, vascular pathology, and the increase in lactic acid production from additional glucose would be most marked in the photoreceptor layer where the largest portion of the retinal metabolism occurs and where the higher levels of glucose diffusing from the choroid will have their largest effect. However, although the photoreceptor layer receives all of its oxygen and glucose by diffusion from the choroid, its metabolic waste products must certainly diffuse equally in both directions. Thus, the additional lactic acid produced in the photoreceptor layer may have no pathological effect on the choroidal vessels due to dilution by the large volume and rapidity of blood turnover in the choroidal circulation, but it would certainly produce vasodilatation in the meager retinal circulation in addition to that already present from the higher levels of lactic acid produced by the metabolism in the inner retinal tissue adjacent to the retinal vessels. As is discussed below, the lactic acid also has other possibly more important effects.

One of the puzzling elements of the development of diabetic retinopathy is the widespread damage in the capillaries, microaneurysms, accumulation of exudates, and failure or dropout prior to vasoproliferation. This development over a long period of time may be connected with the dilatation of the vessels, but other disorders which have similar vasodilatation do not usually have as drastic later damage in the vasculature. There is a similarity of the sequence of capillary failure in diabetes to that in the so called reperfusion injury. In reperfusion injury, capillary damage follows a variety of ischemic tissue situations including organ transplants[30]. However, interruption to the blood supply may not be in itself the key agent. For example, the resemblance of reperfusion injury to the vascular pathology in muscles following heavy exercise where there is increased blood flow without any proceeding ischemic phase has been noted[31]. Even though the time course for the changes in diabetes is much longer - months and years - as compared with hours and days in the reperfusion situation, the sequence of events and the development of the pathology are quite striking.

The capillary damage in reperfusion injury is associated with a higher than usual concentrations of oxygen derived free radicals, usually superoxide, O_2^-. These free radicals are produced in response to a marked hyperoxia resulting from a pH induced Bohr shift in blood within the capillaries following ischemia or other metabolic stress[32]. Table I has some O_2 levels that result from various pH values in the blood. All values in mm Hg assuming the pO_2 of the entering arterial blood at 95 mm Hg. These values are calculated using the relation of:

$$\log pO_2 = (\text{Bohr coef.})(7.4 - pH).$$

Although the highest Bohr coefficient given above, .466, is probably closest to the proper one for ischemic tissue, the other may apply when the initial O_2 saturation of the hemoglobin is lower. A higher initial pO_2 of entering blood give a higher Bohr coefficient. From Table 1, 7.4 is the normal value, 6.4 is the value in exercising muscles which as mentioned above are a capillary pathology similar to that in reperfusion injury[31]. An increased glucose level produce vasodilatation and also hyperoxia[27] as in the very earliest stage of diabetic retinopathy. Thus, the same mechanism of free radical generation that acts acutely in reperfusion injury may act chronically in diabetes to injure the retinal capillary endothelium resulting in leakage and capillary dropout. As discussed above, the hyperoxia in diabetic tissue comes from the Crabtree effect combined with a Bohr shift and can be expected to produce O_2 damaging levels within the capillaries.

The capillary tissue O_2 levels in diabetes calculated from measurements of oxygenated/deoxygenated hemoglobin saturation differences may not accurately reflect the tissue PO_2 due the Bohr shift, the change in the hemoglobin O_2 affinity as a function of pH. That is, in an acid environment, capillary PO_2 may be much higher than would be calculated from the hemoglobin-oxyhemoglobin ratio when a normal pH level is assumed. This

Table 1. pH induced shifts in pO_2 for various Bohr coefficients

pH	.325	.375	.425	.466
7.4	95	95	95	95
7.3	102.4	103.6	104.8	105.8
7.2	110.3	112.9	114.2	117.7
7.1	118.9	123.1	125.2	131.1
7.0	128.2	134.2	137.3	145.9
6.9	138.1	146.3	150.6	162.5
6.8	148.8	159.5	165.1	180.9
6.7	160.4	173.9	188.5	201.3
6.6	172.9	189.5	207.8	224.1
6.5	186.3	206.5	229.2	249.5
6.4	200.8	225.3.	252.8	277.8

high tissue level of O_2 in the capillaries would in turn produce a higher level of superoxide O_2^- which would lead to endothelial damage and capillary fragility and leakage. This fragility will in turn cause capillary drop-out. The capillary dropout itself will act to enlarge the affected area, as dilatation and hyperoxia will now occur in the surrounding tissue. It should be noted at this time that raising the tissue O_2 level will not change the capillary O_2 level very markedly. This is due to the buffering action of the hemoglobin. That is, at a given pH, the hemoglobin will absorb O_2 if the tissue PO_2 rises due to exogenous oxygen. On the other hand, changes in the pH of the hemoglobin environment will change the PO_2.

The pathology of damage in the dilated vessels in diabetic retinopathy follows a somewhat complex course. An examination of physiological conditions in the retina, as well as the clinical course of the pathology, suggests a model that includes changes in the PO_2 and pH in the tissue as well as the capillaries.

For example, retinal vasodilatation is induced by lactic acid production in the photoreceptors when it diffuses from the photoreceptor layer into the retinal vessels. Although vasodilatation itself seems to initiate vasoproliferation, this may be only when acute and exaggerated such as occurs in the ROP, late stage diabetic retinopathy and some other types of proliferative retinopathies. Although all of these have some lesser degree of chronic vasodilatation than in diabetic retinopathy, they do not have as one of their phases the extensive leakage and exudates that accompany the chronic dilatation in diabetic retinopathy. Perhaps this is the key to the difference. Along with the chronic vasodilatation maintained by the lactic acid goes the hyperoxia in the capillaries induced by the Bohr shift of hemoglobin in the capillaries. The hyperoxia must in turn be accompanied by much higher levels than usual of oxygen derived free radicals. These free radicals have very destructive effects on the capillary endothelium and produce leakage and fragility[33] as in reperfusion injury[34]. That the free radicals are produced in other cases of tissue acidosis is well documented in other similar uses by the specific effect of superoxide dismutase (SOD) in controlling or preventing reperfusion injury. Perhaps in diabetic retinopathy, it is these oxygen generated free radicals that are responsible for the leakage of capillaries and ultimately their blockage. This may be termed the first stage of diabetic retinopathy. The second stage begins with capillary dropout which initiates the vicious cycle of much greater vasodilatation followed by vasoproliferation. Within the second stage, the capillary dropout may itself induce a widening ring of capillary dropout in part fueled by lactic acid build up in the non-perfused area. It can be seen that photocoagulation has a dual role. By destroying photoreceptors, the inner retina pH is raised possibly to nearly normal levels, and the oxygen is raised above that normally found there. Both of these together act to reduce the vasodilatation. The extra oxygen from the choroid does not, however, raise the oxygen level in the capillaries to the point of producing superoxide, as the buffering action of the hemoglobin only allows a slight increase to persist.

Diabetes is a systemic disorder, and the vascular complications occur in many sites in addition to the retinal vessels, notably the glomerular vessels in the kidney, the vessels in the tissue surrounding peripheral neurons, and those in the heart tissue. In all of those, the chronic reperfusion injury model of vascular injury could apply as well as it does in the retina. That is, all of those tissues, including the retina, have normal glucose metabolism in diabetes, as they are noninsulin dependent. Thus, these tissues will all have pronounced Crabtree effects with lactic acid accumulation when the glucose level is higher than normal. This model, thus, accounts for the rather puzzling pattern of the occurrence of vascular complications in diabetes, that they are confined to the non-in-

sulin dependent parts of the body. Also, in the kidney, only the glomerular tissue is involved, as this is only part of the kidney to receive
fresh, highly oxygenate blood and, thus, only in this region is the hemoglobin susceptible to a marked Bohr shift with a correspondingly excessive
superoxide generation.

Although the destructive effect of the oxygen derived free radicals is
an important factor in capillary drop out, other factors connected to vasodilatation are as important and cannot be neglected. The hyperemia with
its pronounced vasodilatation is certainly a major stress upon the integrity of the capillary wall. The faster flow and increased turbulence, associated with dilatation, can induce endothelial cell damage by shear
stress [35÷38]. Also, the changes in blood rheology during diabetes [39] partially due to the higher levels of circulating fibrin, may lead to increased
clot formation in the contorted small capillaries characteristic of vasodilatation. These factors all act to place additional stress on the capillaries.

The cycle of capillary dropout once initiated, is independent of the
original cause, the chronically increased glucose levels characteristic of
all forms of diabetes. The terminal stages occur rapidly after perhaps
many years of slow deterioration of the capillary bed due to the chronic
vasodilatation and variable but offer highly elevated level of oxygen derived free radicals.

The beneficial effects of vitrectomy, which only raises the inner
retina O_2 level and also lowers the pH and, thus, produces vasoconstriction
indicate that vasoconstriction itself will minimize pathology in the capillaries. Certainly, following vitrectomy, capillary leakage is markedly reduced, capillary dropout almost stops, and no further neovascularization
develops in the retina.

The efficacy of pan-retinal photocoagulation in halting the progression of diabetic retinopathy has led to its almost universal application.
The interplay between the changes in the oxygen distribution in the retina
induced by photocoagulation which causes destruction of the rods and cones
and allows the choroidal oxygen to diffuse more freely into the outer retina to cause vasoconstriction which is the presently accepted basis of the
therapy [5,40]. However, as the theory has been presented so far, it is
obvious that, in fact, photocoagulation is specific not only for allevi-
ating the symptoms but for removing part of the cause that is, the cones
and rods whose metabolism is operating under the abnormal condition of a
high glucose level to produce the exaggerated lactic acid levels which in
turn gives the initial dilatation of the retinal vessels. This dilatation
is accompanied by an increased O_2 tension in these vessels due to the Bohr
shift. The excessive O_2 tension produces an excess of superoxide which
injures the capillary endothelium and certainly exaggerates the effects of
the vasodilatation itself to lead to the rest of the observed pathology.
Thus, the removal of this highly metabolic tissue in the rods and cones by
a conversion into scar or glial cells is a major benefit of photocoagulation in diabetes.

The above analysis suggests that light or minimal photocoagulation is
all that is necessary and that, indeed, it is probably the optimum type.
It is not necessary to destroy more of the retina than the photoreceptor
layer, and certainly any injury to the vessels in the superficial choroid
should be avoided, as it will both interfere with the removal of the remaining lactic acid, and it lowers the O_2 tension in the inner retina.
Thus, any wavelength of laser radiation that would preferentially penetrate
past the pigment epithelium into the choroid should not be used. These
considerations, of course, indicate that the optimum wavelength(s) for

photocoagulation would be in the blue or bluegreen region of the spectrum where melanin has the highest absorption, and damage to the choroidal vessels can be minimized[41]. Along these lines, it may be best to use the blue or ultraviolet end of the spectrum for photocoagulation which encourage photochemical rather than thermal processes of photoreceptor damage[42]. A near ultraviolet wavelength (perhaps the argon laser line around 350 nm) would be ideal, but this would require by-passing the crystalline lens, possibly by an intraocular or pars plana delivery system[43].

IV. ALTERNATIVES TO PHOTOCOAGULATION AND VITRECTOMY

As the vascular difficulties in diabetes seem to be a chronic form of reperfusion injury or exercise injury, it is tempting to consider that the various biochemical therapies that are effective in the acute syndromes might also be applied to the chronic disorders. Some examples are superoxide dismutase (SOD) which detoxifies O_2^- and Allopurinol which blocks the enzyme xanthine oxidase in the capillary endothelium responsible for converting O_2 into superoxide. Although it is unlikely that SOD can be infused on a chronic basis to prevent deleterious O_2^- activity, Allopurinol is a likely candidate. It is already administered on a chronic basis to treat gout, and it has very few and minor side effects. It is interesting to note that searches through several large diabetes clinics have failed to yield any adult on-set diabetic patients with vascular complications who also were treated for gout. As the incidence of diabetes is 5% and that of gout is 1%, at least 5% of the adult onset (Type II) diabetics should also have gout, particularly as purines are not especially banned from diabetic diets. It is tempting to speculate that Allopurinol blocks the vascular complications of diabetes all over the body, and, thus, although the gout patients may have adult onset diabetes, it is never diagnosed as the diabetic vascular complications which usually lead to diagnosis of this form of diabetes do not occur in a virulent form. The probable biochemical basis of the vascular involvement in diabetes begins with the failure in maintaining normal glucose levels in the blood at all times which causes low tissue pH on a chronic basis.

V. OTHER PROLIFERATIVE RETINOPATHIES

The previous discussion covered diabetic retinopathy, but the proliferative retinopathies previously mentioned can also be thought of in the same way, although the etiology of these do not necessarily include an excess of lactic acid and low pH causing vasodilatation. Nevertheless, they all have some pathology which causes an early retinal vasodilatation. This vasodilatation should be counteracted by some (but not too much) additional oxygen from the choroidal circulation following laser photocoagulation to cause vasoconstriction of the retinal vessels[5].

VI. RUBEOSIS IRIDIS

Rubeosis iridis, in which the vessels may grow into the angle to cause neovascular glaucoma, is a frequent accompaniment of late stages of diabetic retinopathy it often follows a vitrectomy, particularly if this procedure requires removal of the lens, converting the eye into a single chamber which connects the fluid bathing the iris to the retina in a convective way. Normally, the inner surface of the retina at the vitreous junction has an oxygen tension far below that in the anterior chamber. Experimental data shows that the O_2 tension in the anterior chamber following lensectomy is closer to the retinal level, thus placing the iris in a low O_2 tension environment as compared with its normal one[44]. The resulting iris vasodilatation can lead to neovascularization. The situation may be complicated by the lactic acid diffusing out of the retina. Certainly, the lower lactic acid level in the retina will help inhibit O_2^- production in the retinal capillaries, but its effect on the iris vessels is unknown. It is also possible that the metabolism of the iris will be altered by the high glucose levels characteristic of diabetes sufficiently to add, as in the retina, some superoxide free radical damage. Rubeosis iridis and neovascularization may even accompany a vitrectomy and lensectomy when the retinal metabolism is normal, thus indicating that vasodilatation itself is a major contributor to the observed vascular pathology. However, as the later stages of diabetes have an overall large hypoxic area of the retina due to capillary dropout, the vitreous fluid in contact with the retina would have a much lower oxygen tension than normal, and, thereby, in a single chamber eye, a very low O_2 tension might appear around the iris. Panretinal photocoagulation alleviates, if it does not completely prevent, this whole problem by raising the oxygen tension of the inner retina, and, thereby, indirectly the oxygen tension around the iris. In any case, the present theory suggests that an important part of the therapy of any proliferative retinopathy is photocoagulation of the lightest kind possible consistent with the destruction of the necessary amount of the rods and cones to bring about vasoconstriction.

VII. CONCLUSION

Many different types of proliferative retinopathy induced by various types of initial disorders have a common pathology in their mid and terminal stages. Thus, proper therapy is devoted toward elimination of the initial cause as well as alleviation of the proliferative processes. Vasodilatation appears to be a consistant feature in all stages of diabetic retinopathy, the mid and late stages it seems very dependent upon capillary dropout, whereas the primary vasodilatation may derive from quite different causes. The efficacy of photocoagulation as a therapy for all stages seems to derive from decreasing the metabolism in the photoreceptor layer sufficiently to result in vasoconstriction of the retinal vessels. A model has been proposed to show how diabetes, by altering the metabolism in the photoreceptor layer, causes the initial destruction of the retinal capillary bed followed by the visible signs of diabetic retinopathy. Photocoagulation, thus, is even more appropriate for this particular syndrome than previously had been thought, as it not only reduces potentially destructive vasodilatation but also removes the metabolic cause of the free radical induced destruction of the capillary endothelium which is the initial step in capillary drop-out. A review of the present data indicates that the best type of pan-retinal photocoagulation is a very light type affecting the photoreceptors only with a minimal amount of damage to other parts of retina and the vessels in the choroid. The possible use of photochemical types of destruction of the photoreceptor as a therapeutic modality is attractive, but it is certainly too speculative to use until more detailed investigations have been completed. However, the basic therapeutic approach of choice may be to prevent the initial vasculature involvement by keeping the tissue pH normal or by blocking the generation of superoxide with Allopurinol or similar medication.

VIII. ACKNOWLEDGMENTS

Many of our colleagues have contributed criterium, suggestions, and we thank them for their help: Professors John Marshall, Irwin Fridowich, and Einar Stefansson; Doctor Henry Wagner; and Professor Doctor Berger of Düsseldorf.

VII. REFERENCES

1. J. B. Hickam and R. Frayser, "Studies on the Retinal Circulation in Man. Observations on Vessel Diameter, Arteriovenous Oxygen Difference and Mean Circulation Time", Circulation, 33:302-316 (1966)
2. S. S. Elgin, "Arteriovenous Oxygen Difference Across the Uveal Tract of the Dog Eye", Invest. Ophthalmol. 3:417-426 (1964)
3. A. Alm and A. Bill, "Blood Flow and Oxygen Extraction in the Cat Uvea at Normal and High Intraocular Pressures", Act. Physiol. Scand. 80:19-28 (1970)
4. A. Bill, "Ocular Circulation", in: "Adler's Physiology of the Eye", (R. A. Moses, ed.), C. V. Mosby Co., St. Louis (1975)
5. M. L. Wolbarsht and M. B. Landers, III. "The Rationale of Photocoagulation Therapy for Proliferative Diabetic Retinopathy: A Review and Model", Ophthal. Surg., 11:235-245 (1980)
6. M. L. Wolbarsht, G. S. George, J. Kylstra, M. B. Landers III, and W. A. Shearin Jr., "Speculation on Carbon Dioxide and Retrolental Fibroplasia", Pediatrics, 71:859-860, (1983a)
7. M. L. Wolbarsht, G. S. George, W. A. Shearin Jr., J. A. Kylstra, and M. B. Landers III, "A New Look at an Old Disease", Ophthal. Surg., 14:919-924 (1983b)
8. J. J. Weiter, R. Zuckerman, and C. L. Schepens, "A Model for the Pathogenesis of Retrolental Fibroplasia Based on the Metabolic Control of Blood Vessel Development", Ophthalmic Surg. 13, 12:1013-1017 (1982)
9. N. Ashton, B. Ward, and G. Serpell, "Role of Oxygen in the Genesis of Retrolental Fibroplasia", Br. J. Ophthal., 38:433-440 (1953)
10. D. Bracher, "Changes in Peripapillary Tortuosity of the Central Retinal Arteries in Newborns. A Phenomena Whose Underlying Mechanisms Need Clarification", A. V. Graefe's Arch. Ophthal., 218:211-217 (1982)
11. R. Fahrni, J. Thalmann, J. Weber, G. V. Muralt, D. Sidiropoulos, and D. Bracher, "Central Retinal Arteries in the Full-Term Newborn: Decrease in Width and Tortuosity During Uneventful Adaptation", Helv. Paediat. Acta, 36:107-121 (1981)
12. N. Ashton and C. Cook, "Direct Observation of Oxygen on Developing Vessels", Preliminary Report, Br. J. Ophthal. 38:433-440 (1954)
13. M. B. Landers III, "Retinal Oxygenation Via the Choroidal Circulation", Trans. Amer. Ophthalmol. Soc., 76:528-556, (1978)
14. M. L.Wolbarsht, E. Stefansson, and M. B. Landers III, "Retinal Oxygenation from the Choroid in Hyperoxia", Exper. Biol. Environ. Sens. Aspects, 47:42-52 (1987)
15. L. M. Parver, C. R. Auker, and D. O. Carpenter, "Choroidal Circulation as a Heat Dissipative Mechanism in the Eye", Amer. J. Ophthalmol. 89:641-646 (1980)
16. E. Stefansson, M. B. Landers III, and M. L. Wolbarsht, "Oxygenation and Vasodilatation in Relation to Diabetic and Other Proliferative Retinopathies", Ophtal. Surg. 14:209-226 (1983)
17. A. G. Curtis and G. M. Seehar, "The Control of Cell Division by Tension or Diffusion", Nature, 274:52-53 (1978)
18. J. Folkman and A. Moscona, "Role of Cell Shape in Growth", Nature, 273:345-349 (1978)
19. M. L. Wolbarsht, M. B. Landers III, and E. Stefansson, "Vasodilatation and the Etiology of Diabetic Retinopathy: A New Model", Ophthal. Surg. 12:104-107 (1981)
20. L. Feeney, "The Phagolysosomal System of the Pigment Epithelium. A Key to Retinal Disease", Arch. Ophthal. 12:635 (1971)
21. W. K. Noell, V. S. Walker, B. S. Kang, and S. Berman, "Retinal Damage by Light in Rats", Invest. Ophthal. 5:450-473 (1966)

22. E. L. Berson and J. Howard, "Temporal Aspects of the Electroretinogram in Sector Retinitis Pigmentosa", Arch. Ophthal. 86:653-665 (1971)

23. H. Kolb and P. Gouras, "Electron Microscopic Observations of Human Retinitis Pigmentosa, Dominantly Inherited", Invest. Ophthal., 13:487 (1974)

24. J. G. Cunha-Vaz, "Studies on the Permeability of the Blood Retinal Barrier, Breakdown of the Blood Retinal Barrier by Circulation Disturbances", Br. J. Ophthalmol. 50:505-516 (1966)

25. J. G. Cunha-Vaz, J. R. F. Abreu, A. J. Campos, G. M. Figo, "Early Breakdown of the Blood Retinal Barrier in Diabetes", Br. J. Ophthalmol. 59:649-656 (1975)

26. J. G. Cunha-Vaz, "The Blood-Ocular Barriers", Surb. Ophthalmol. 23:279-296 (1979)

27. J. S. Reed, J. T. Ernest, T. K. Goldstick, "Hyperglycemia and the Retinal Circulation in Man", Invest. Ophthalmol. Vis. Sci. 19 (Suppl.):168 (1980)

28. P. C. Brazy, G. Gullans, L. J. Mandel, and V. W. Dennis, "Metabolic Requirements for Inorganic Phosphate by the Rabbit Proximal Tubule: Evidence for a Crabtree Effect", J. Clin. Invest. 70:53-62 (1982)

29. H. G. Crabtree, "Observations on the Carbohydrate Metabolism of Tumours", Biochem. J. 23:536-545 (1929)

30. L. Huang, C. Privalle, D. Serafin, and B. Klitzman, "Increased Survival of Skin Flaps by Scavengers of Superoxide Radical", Fed. Amer. Soc. Experm. Biol. J., 1:129-132 (1987)

31. K. J. A. Davies, A. T. Quintanilha, G. A. Brooks, and L. Packer, "Free Radicals and Tissue Damage Produced by Exercise", Biochem. Biophys. Res. Comm. 107:1198-1205 (1982)

32. M. L. Wolbarsht and I. Fridovich, "Hypothesis: Hyperoxia During Reperfusion is a Factor in Reperfusion Injury", Free Radical Biol. Med. 6:61-62 (1989)

33. U. Fuch, W. Tinius, S. Gonschorek, and V. J. Scheidt, "Gesteigerte Kapillare Vulnerabilität beider Diabetischen Retinopathie", Klin. Monat. f. Augenheilkunde, 192:234-236 (1988)

34. D. A. Parks, G. B. Bulkley, D. N. Granger, S. R. Hamilton, and J. M. McCord, "Ischemic Injury in the Cat Small Intestine: Role of Superoxide Radicals", Gastroenteroly, 82:9-15 (1982)

35. P. F. Davies, A. Remuzzi, E. J. Gordon, C. F. Dewey, Jr., and M. A. Gimbrone, Jr., "Turbulent Fluid Shear Stress Induces Vascular Endothelial Cell Turnover In Vitro", Proc. Natl. Acad. Sci. (U.S.A.) 83:2114-2117 (1986)

36. C. F. Dewey, Jr., S. R. Bussolari, M. A. Gimbrone, Jr., and P. F. Davies, "The Dynamic Response of Vascular Endothelial Cells to Fluid Shear Stress", J. Biochem. Engrg. 103:177-185 (1981)

37. C. F. Dewey, Jr., "Effects of Fluid Flow on Living Vascular Cells", J. Biochem. Engrg. 106:31-35 (1984)

38. D. Mathews, P. La Sala, and Su. Chien, "Blood Rheology and Oxygen Transport", IEEE Engineering in Medicine and Biology Magazine, 15-18 (1986)

39. H. L. Little, "The Role of Abnormal Hemorrheodynamics in the Pathosenses of Diabetic Retinopathy", Trans. Amer. Ophthalmol. Soc., 74:573-636 (1976)

40. M. L. Wolbarsht, M. B. Landers III, and L. Rand, "Modification of Retinal Vascularization by Interaction Between Retinal and Choroidal Circulation", Invest. Ophthal. Vis. Sci. 12 Suppl., 224 (1978)

41. M. L. Wolbarsht and M. B. Landers III, "Some Considerations for Choosing the Wavelength Appropriate for Laser Photocoagulation of the Retina", pp. 11-19, In: Laser Treatment and Photocoagulation of the Eye", ed. by R. Birngruber and V. P. Gabel, Docum. Ophthal. Proc. Series 36, Dr. W. Junk, The Hague (1984a)

42. W. T. Ham, H. A. Mueller, J. J. Ruffolo, P. Guerry, III and R. K. Guerry, "Action Spectrum for Retinal Injury from Near Ultraviolet Radiation to the Aphakik Monkey", Amer. J. Ophthalmol. 93:299-306 (1982)

43. M. L. Wolbarsht and M. B. Landers III, "Endophotocoagulation With Near Ultraviolet Radiation", pp. 289-293, In: "Laser Treatment and Photocoagulation of the Eye", R. Birngruber and V. P. Gabel, eds., Docum. Ophthal. Proc. Series 36, Dr. W. Junk, The Hague (1984b)

44. E. Stefansson, L. M. Cobo, D. Robinson, M. L. Wolbarsht, and M. B. Landers III, "Anterior Chamber Oxygen Tension Following Lens Extraction", Invest. Ophthal. Vis. Sci. 25(3) Suppl., 254 (1984)

LOW POWER LASER THERAPY

EFFECTS OF VISIBLE LASER RADIATION ON CULTURED CELLS

T. I. Karu

Laser Technology Center
USSR Academy of Sciences
Troitzk, Moscow Region, USSR

I. INTRODUCTION

A developing therapeutic role for laser phototherapy in treating patients with skin diseases[1] has led to interest in the effects of visible light on cultured cells. The need to examine the action of different visible radiation wavelengths upon cellular cultures arises partially from the knowledge that disorders which respond to laser phototherapy, such as indolent wounds and trophic ulcers[2,3] may be associated with increased proliferation of cells surrounding the injuries[4].

To investigate the mechanism of wound healing by laser light, we looked for a model system which would more closely resemble tissue cell population (as compared with cultures of, e. g., microorganisms), while still permitting the simplicity of in vitro manipulation. Usually mammalian cellular cultures in an actively proliferating state (exponentially growing cultures) are used as models in such experiments.

In this type of culture a large number of cells is traversing the cell cycle. Actually, the cellular populations in vitro are heterogeneous, containing a number of subpopulations. In terms of the diversity of cellular subpopulations, cultured mammalian cell populations in the plateau phase of growth are close to cellular populations in vitro[5,6]. Plateau phase cellular cultures as model systems have been proved to be useful for studies of cellular effects of chemotherapeutic drugs and γ-radiation[7,8]. In our experiments we used both types of models (exponentially growing and plateau phase cells). The effects of laser irradiation on colony forming ability, DNA and RNA synthesis and progression through the cell cycle will be described.

Literature available to us contained no data about the action of low power laser light on the proliferative activity of plateau phase cells. In experiments with exponentially growing populations using radiation from various visible light lasers, the irradiation has been found to act in a stimulating, indifferent or even inhibiting manner on the proliferation of cellular cultures. These data are reviewed in detail in Reference 9.

Some experiments performed to explain the quantitative effects of monochromatic visible light on various microorganisms and cellular cultures, as well as the possible mechanisms have been discussed earlier in the International School of Quantum Electronics Course on "Laser Photobiology and Photomedicine"[10]. Some data may also be found in References 11 and 12.

II. CELLULAR CYCLE

Cell proliferation encompasses the overall dynamics of a cell in passing from its initial resting state through the entire division procedure, including its interaction with the remaining cell population[13]. The chain of events which a cell passes through while proceeding towards division has been termed the cell cycle. A typical cell cycle can be represented schematically as shown in Figure 1. It consists of a mitotic phase (M), a G_2 phase, a DNA synthetic phase (S), and a G_1 phase prior to next division. A cell can leave the cycle either during the G_1 or the G_2 period, entering into G_0 (resting or quiescent) state[13]. Replicative synthesis of DNA occurs only during the S period of cell cycle, reaching a maximum approximately midway through it[14], and is easily monitored by administering a radioactive precursor of its synthesis, thymidine. Synthesis of RNA occurs during G_1, S and G_2, but undergoes a two-fold increase in rate during the first half of the S phase [15,16] and can be monitored by means of a radioactive precursor of its synthesis, uridine (Figure 1).

In our experiments, He-La cells were cultivated as monolayers in scintillation vials with a bottom diameter of 24 mm, in 2 ml of nutrient medium (199 synthetic medium, supplemented with 10% calf serum and 100 units/ml kanamycine or lincomycine). Under these conditions the growth curve of the population resembles a typical curve of mammalian cells growth in vitro . The curve presents an initial phase in which cell number does not increase (lag phase), followed by an exponential region (log phase), and finally by a plateau phase. The laser irradiation experiments were performed when the number of cells in the vial was from 2.5 to $6 \pm 0.2 \times 10^5$ (exponentially growing population) or $1.0 \pm 0.3 \times 10^6$ cells (plateau-phase population). The labeling index (I_s) of the log-phase population was 19.1 ± 3.0 %, and that of plateau-phase culture was 5.0 ± 0.8 %. The mitotic indices (I_m) of the log-phase and plateau-phase populations were 1.1 ± 0.1 % and 0.1%, respectively.

III. VIABILITY AND CLONIGENITY OF PLATEAU-PHASE HeLa CELLS AFTER IRRADIATION WITH He-Ne LASER

To investigate the viability and clonigenity of HeLa cells, the plateau-phase cultures were irradiated with a He-Ne laser and replated at various intervals after the irradiation[17]. Using this technique, not all subpopulations will be attached and start to divide after the replating. Figure 2 illustrates the diversity of subpopulations of a plateau phase cell culture [6], and shows the clonigenic cells starting to divide after trypsinization and replating.

In the first series of experiments we studied the growth curves and, in the second series, the clonigenity of cells by the Puck technique after the irradiation of plateau-phase HeLa cells with a He-Ne laser.

A He-Ne laser (model LG-52/1, λ = 632,9 nm) was used for irradiation (I = 5 W/m^2, t = 20 sec, D = 100 J/m^2). This dose has been found to be optimal to stimulate synthesis of DNA[17-20] : a short focus positive lens was used to expand a beam from a light source to a diameter of 24 mm, equal to that of the monolayer of cells. The cells were irradiated in darkness through the bottom of the flask. During irradiation the flask was attached to a special support. At various intervals (from 5 to 240 min) after the irradiation the monolayer was trypsinized and the cells were replated into fresh nutrient medium.

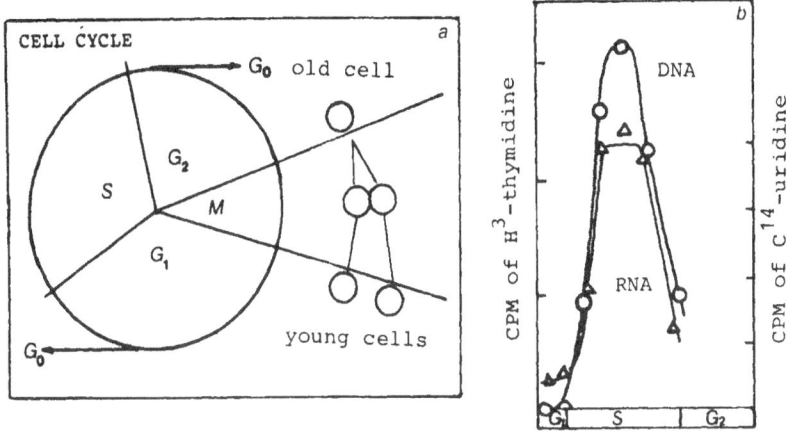

Fig. 1. Schematic diagram: (a) of the cell cycle of a mammalian cell, and (b) the synthesis of nucleic acids in the cell cycle.

The laser growth stimulation during the exponential phase of growth was observed to last 6-7 days when the interval between irradiation (λ = 632,8 nm, Δ = 100 J/m^2) and plating was 30 min and more (Figure 3). Worthy of notice is the difference in the shape of growth curves between the irradiated and non-irradiated cultures. In the control, beginning from the 8th day there appears the plateau-phase of growth and the number of cells in the flasks does not change till the 15th day. As distinct from the control, in the irradiated cultures where the interval between the

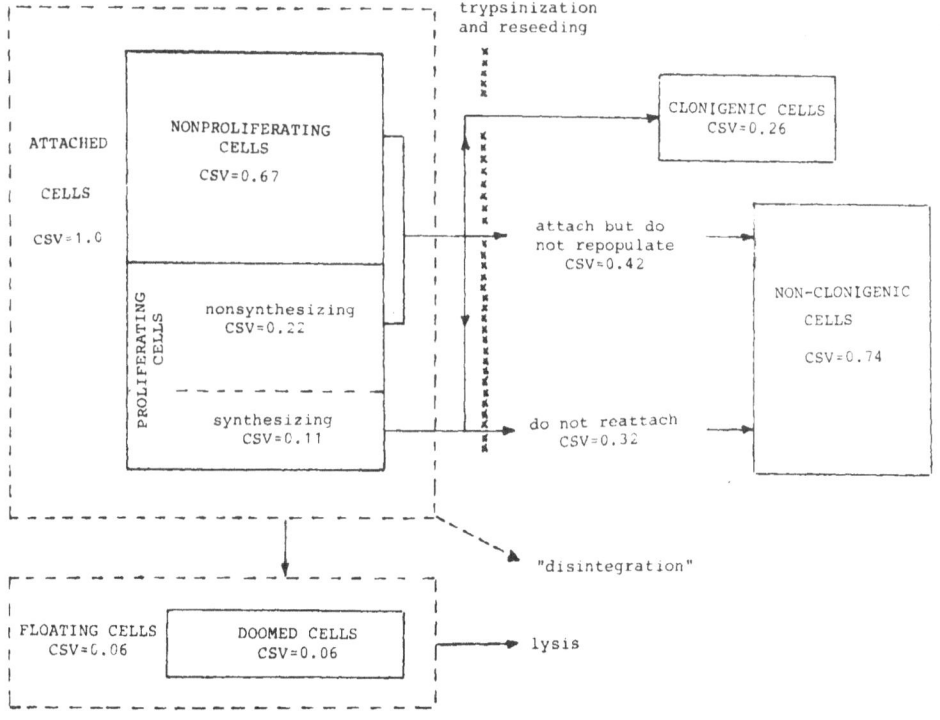

Fig. 2. Scheme of compartments of a plateau-phase population of Chinese hamster cells[6].

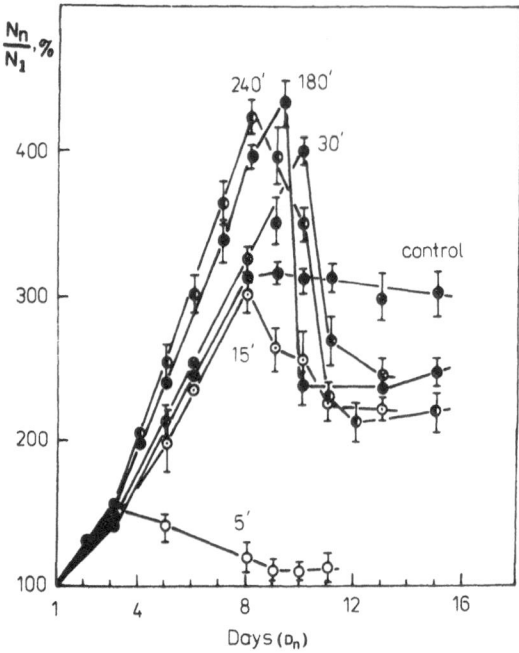

Fig. 3. Subcultivation growth curves of HeLa cells following irradiation
of plateau-phase cultures with a He-Ne laser with 100 J/m² dose
and inoculation of the irradiated cells into fresh nutrient
medium 5, 15, 30, 180 or 240 min after irradiation. The ordinate
gives the ratio of the number of cells on day n (N_n) to the
number at the end of the first day (N_1), and the abscissa gives
the days (D_n) after plating.

irradiation and plating ranged between 30 and 240 min, the number of cells
decreases sharply at the end of the long phase of growth (8th-9th day), and
the plateau occurs below that in the control and practically at one and the
same level for all groups, except for the culture for which the interval
between irradiation and plating was short (5 min). In the latter case, the
culture stopped growing altogether in three days (Figure 3).

The differences in growth kinetics between the control and test groups
of cells can apparently be explained as follows. As an effect of the radiation, in the plateau-phase cells there occur changes which become mani-
fested upon subcultivation. On the one hand, these changes are due to the
acceleration of proliferation with subsequent loss by some members of the
population of the ability to enter the plateau phase of growth. Such a
phenomenon was observed when irradiating plateau-phase HeLa cells with
small doses of γ-radiation (0.1 Gy)[21]. On the other hand, the sharp cell
growth retardation in the case of 5 min interval between irradiation and
plating is evidently explained by the fact that the action of He-Ne laser
radiation is in this case a sensitizing factor aggravating the damage to
the cells due to subcultivation.

We also studied the influence of irradiation on the number of clonigenic cells as well as on the clone size distribution, by the Puck technique[22]. The plateau phase He-La cells were irradiated with a HeNe laser
at a dose of 100 J/m², trypsinized 180 min after irradiation, resuspended
in Hanks solution and counted with a hemocytometer[11]. Suitable delutions
were made and the same number (100) of cells was inoculated into all vials,
both control and test.

The number of clones as well as the diameters of clones were counted after 14 days. The percentage of clonigenic cells in the nonirradiated culture was 45.2 ± 0.4 %, and this number increased after irradiation, reaching 50.4 ± 0.5, 58.3 ± 0.4 and 54.5 ± 0.9 % at irradiation doses 10, 10^2 and 10^3 J/m^2, respectively. This increase of the plating efficiency may be due to increased attachment of cells caused by irradiation. The histograms of the frequency distribution of clone sizes are presented in Figure 4. The clones were counted in four groups: clones with diameter 0.25 mm and less (abortive clones), from 0.25 to 0.35 mm (small clones), from 0.35 to 0.5 mm (middle clones), and from 0.5 to 0.8 mm (big clones).

As seen in Figure 4, the per cent of abortive clones practically does not change after irradiation as compared with the non-exposed control. The per cent of small clones decreased and the number of medium and large clones increased in a dose-dependent manner. The size of the large clones did not increase. Therefore the stimulative effect of He-Ne laser irradiation is most noticeable on the proliferative activity of the slowly-growing subpopulations (the slowly dividing cells yield small and medium clones). After irradiation the clone-size distribution becomes more homogeneous. For example, when irradiating with a dose of 100 J/m^2, the percentage of small, medium and large clones is almost equal (near 30%). In the control experiment the distribution was 42:36:11 %.

From this group of experiments it is possible to conclude that irradiation with a He-Ne laser stimulates the proliferation of He-La cells under

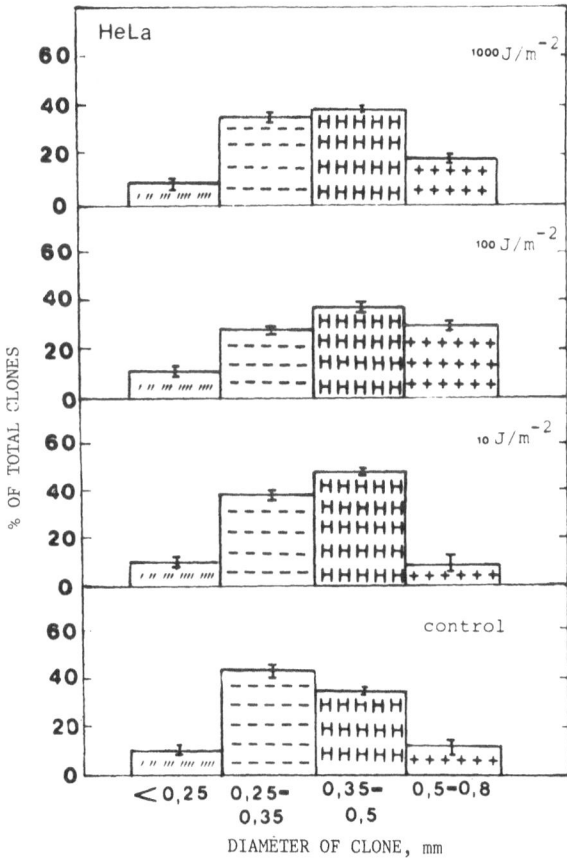

Fig. 4. Histograms of clone-size distribution of cell population at 14 days after the irradiation of He-La cells with a He-Ne laser (D = 10, 100 or 1000 J/m^2).

our experimental conditions (irradiating the plateau phase cells, with post-irradiational replating).

IV. CHANGES IN DNA AND RNA SYNTHESIS RATE AFTER THE IRRADIATION

Replicative DNA synthesis occurs in the S phase of the cellular cycle (Figure 1), and synthesis of RNA occurs through G_1, S and G_2 with a maximum in the S phase[15]. The synthesis rates can be monitored by means of radioactive precursors, ^{3}H-rhymidine and ^{14}C-uridine respectively. These two parameters were used to characterize the quantitative laws of laser action (dependence on the wavelength, dose, intensity and irradiation regime)[4,17÷20,22,24].

Reparative DNA synthesis occurs during the entire cell cycle. DNA, RNA and proteins do not absorb visible light[25], and the energy of visible light photons is too small to cause damage to these molecules by molecular bonds (as in the case of γ-radiation, for example). For these reasons we believe that in our case we are dealing with replicative DNA synthesis.

IV.1. Exponentially growing He-La cells

Dose dependence. The dose dependence of the DNA synthesis stimulation effect in proliferating HeLa cells was determined at various wavelengths and it was found that the effect was different in the red and blue-ultraviolet parts of the spectrum. In the red and far red regions the optimal dose was 100 J/m^2 (Figure 5) and it was relatively insensitive to the wavelength. In the blue-ultraviolet region the dose causing maximal effect was lower by about an order of magnitude (see λ = 404 nm as an example in Figure 5). The dose dependence of the RNA synthesis stimulation effect in proliferating HeLa cells is shown in Figure 6.

Role of light intensity. In next series of experiments, the influence of light intensity on the DNA synthesis stimulation effect was studied. I

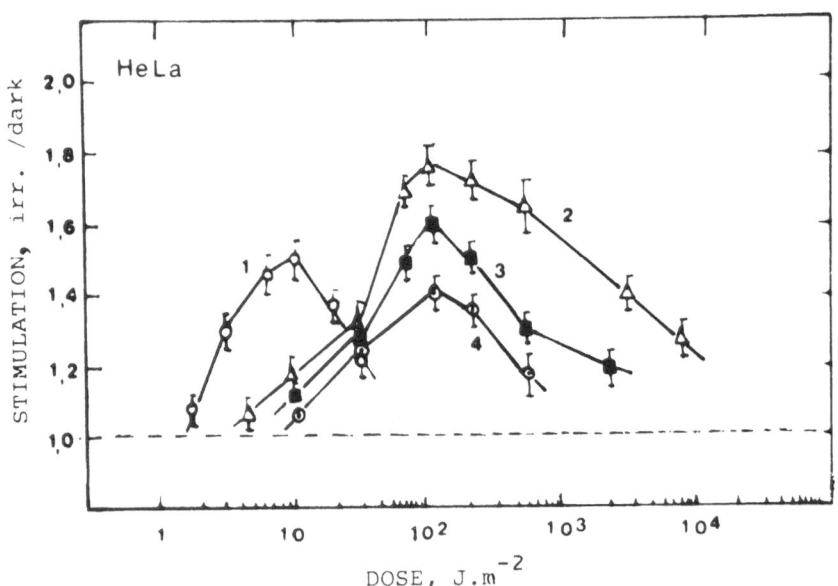

Fig. 5. The effect of cw light on stimulation of DNA synthesis in exponentially growing HeLa cells (1 - λ = 404 nm; 2 - λ = 760 nm; 3 - λ = 620 nm; 4 - λ = 680 nm) 1.5 h after irradiation.

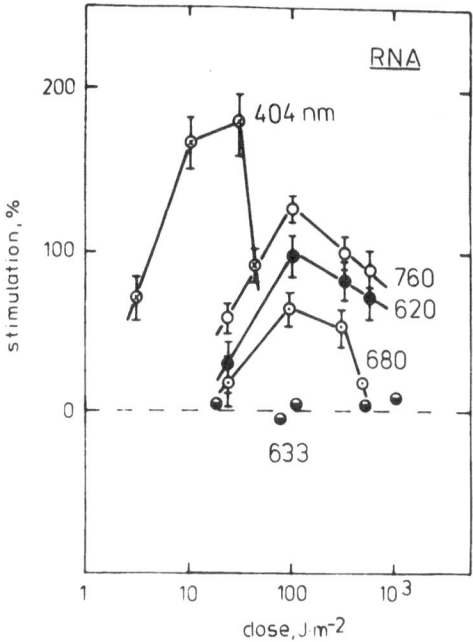

Fig. 6. The effect of cw light irradiation dose on stimulation of DNA
synthesis in exponentially growing HeLa cells (⊗ - λ = 404 nm;
O - λ = 760 nm; ● - λ = 620 nm; ◐ - λ = 680 nm; ◑ - λ = 630 nm)
1.5 h after irradiation.

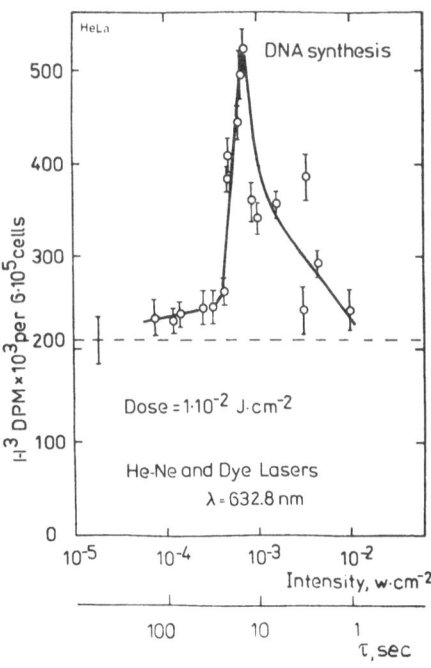

Fig. 7. The influence od cw light intensity at λ = 633 nm
(D = 10 J/m^2 = const.) on the rate of DNA synthesis
in exponentially growing HeLa cells (H^3 - thymidine
incorporation was measured 1.5 h after irradiation.

experiments with cw light at λ = 633 nm the radiation dose D = I x τ received by the cells was changed by varying the intensity of light I and the duration or irradiation τ. A possible influence of the duration or of the radiation intensity on the stimulation effect in the case of a fixed dose is quite possible. This measurement was verified by using He–Ne laser radiation (I_{max} = 10 W/m^2) as well as radiation from a cw dye laser pumped by an argon laser (λ_{out} = 633 nm, I_{max} = 80 W/m^2). The intensity of light was attenuated with calibrated neutral density filters. The radiation dose received by the cells was constant and amounted to 100 J/m^2, which was optimal for the rate of DNA synthesis in a culture of proliferating HeLa cells (Figure 5).

The results of measurements of the influence of the intensity of light on the rate of DNA synthesis are presented in Figure 7. It is clear from this figure that the DNA synthesis stimulation was very sensitive to the irradiation time and to the intensity of light when the light intensity was 8–10 W/m^2 or when the irradiation time was 10–12 sec. It decreased very rapidly with an increase in the irradiation time or a reduction in the light intensity, being observed clearly during an irradiation time from 2 to 20 sec in our experimental conditions.

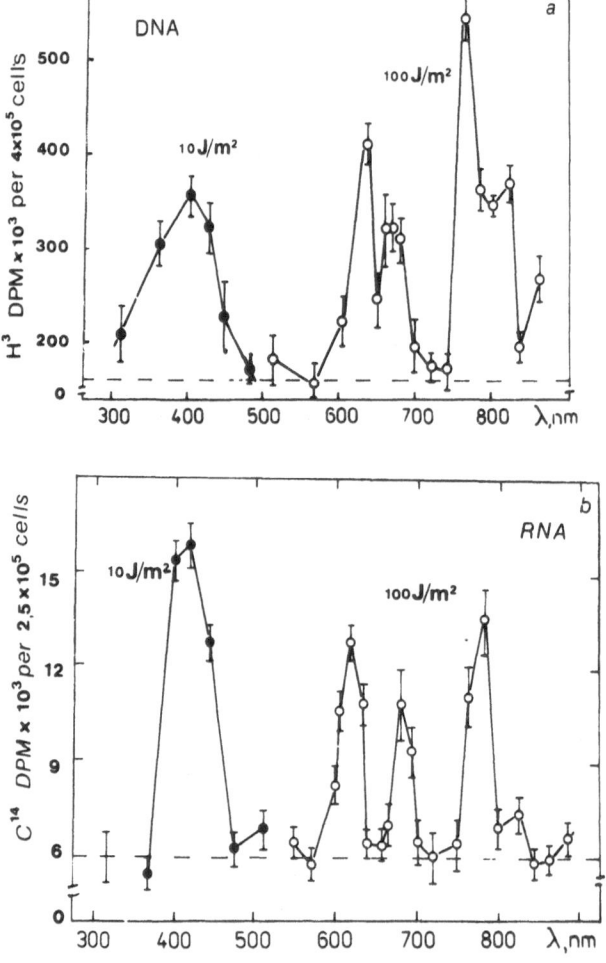

Fig. 8. Action spectra of visible monochromatic light on
(a) DNA and, (b) RNA synthesis in exponentially growing
HeLa cells, measured 1.5 h after irradiation.

Dependence on wavelength. Figure 8 shows the results obtained from measuring the action spectra of light, on the nucleic acid synthesis rate 1.5 h after the irradiation of proliferating HeLa cells. From these figures (Figures 8a and 8b) it follows that the synthesis rate of DNA and RNA increases in some spectral intervals. Stimulation of DNA synthesis can be observed in a wavelength range of about 320 to 450, 600 to 650, 660 to 720, 740 to 840 nm with maxima nearby 400, 630, 680 and 760 nm. In the infrared region there is a structure in the form of a peak near 820 nm. The action spectrum of RNA synthesis is very close to the corresponding action spectrum of light on DNA synthesis. The maxima in the stimulation of RNA synthesis correspond approximately to the wavelengths of 400, 615, 680, 780 and 820 nm. In the range from 450 to 580 nm, we have not observed statistically valid changes in the DNA and RNA synthesis rate under these experimental conditions. Such an effect might be detectable when other parameters (irradiation time, intensity) are used. It is quite possible that intensity dependences are not similar for all spectral regions.

IV.2. Plateau-phase HeLa cells

As concerning the plateau phase HeLa cells, the irradiation causes changes in DNA and RNA synthesis beginning 2 to 2.5 h after the irradiation (Figure 9). The maximum effect is observed 4 to 6 h after irradiation. A comparison of results for DNA and RNA synthesis (Figures 9 and 10) gives grounds for assuming that the changes in the synthesis rate of DNA during the first 3 h after irradiation occur faster than for RNA. This conclusion, however, needs further investigation. The results of the measurements show that the stimulation of DNA and RNA synthesis is maximum at irradiation doses 100 J/m^2, practically the same as for proliferating cells.

The action spectra of DNA and RNA synthesis stimulation in plateau-phase HeLa cells were measured for a dose 100 J/m^2 in the range of wavelengths from 580 to 880 nm (Figure 11). These action spectra obtained are very similar to the ones observed earlier for proliferating HeLa cells (Figure 8).

The basic difference in the reaction of cells of the plateau culture as compared with that of exponentially growing cells is that their stimulation after irradiation occurs several hours later than in the case of proliferating cells. This result can be explained by the deceleration of many metabolic processes in the plateau-phase cells.

V. CHANGES IN CELL CYCLE PARAMETERS AFTER IRRADIATION

From experiments performed with radioactive precursors, it was not clear whether the increased incorporation of H^3-thymidine, was connected with the enhancement of DNA synthesis in the S-phase cells, with changes in the proliferation kinetics of population, e.g. shortening of the duration of the G_1 phase of the cell cycle of cells of the proliferative pool, or with an increase in the proliferative pool. To answer these questions we conducted the following autoradiographical experiments[4].

To determine the changes in the number of S-phase cells and M-phase cells after irradiation, the cultures were pulse-labeled at various times after the irradiation and fixed. The results of these experiments with exponentially growing cultures indicate (Figure 12) that the number of DNA-synthesizing cells increases during first 3-4 h after irradiation and then diminishes to the control level. On the basis of this curve, one can suppose that the increased number of S-phase cells originates from a part of the G_1-phase population that is ready to pass into the S-phase.

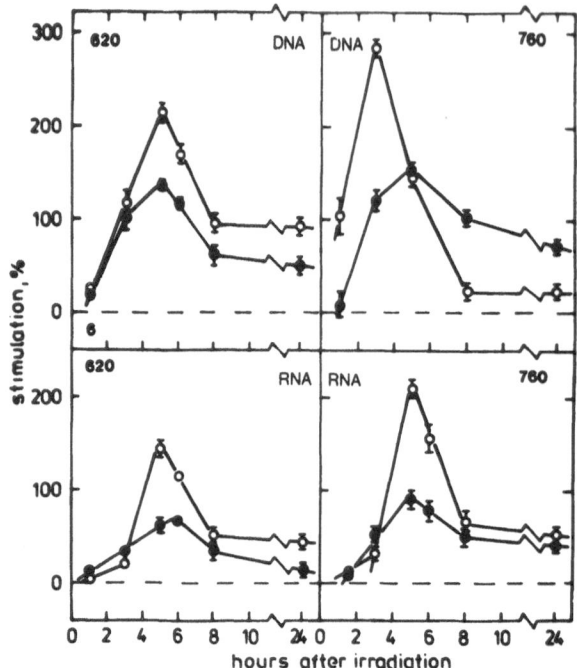

Fig. 9. The influence of irradiation with red (λ = 620 nm) and far red (λ = 760 nm) light on DNA or RNA synthesis in plateau-phase HeLa cells. Doses were (-o-o-o-) 100 J/m^2 or (-•-•-•-) 800 J/m^2.

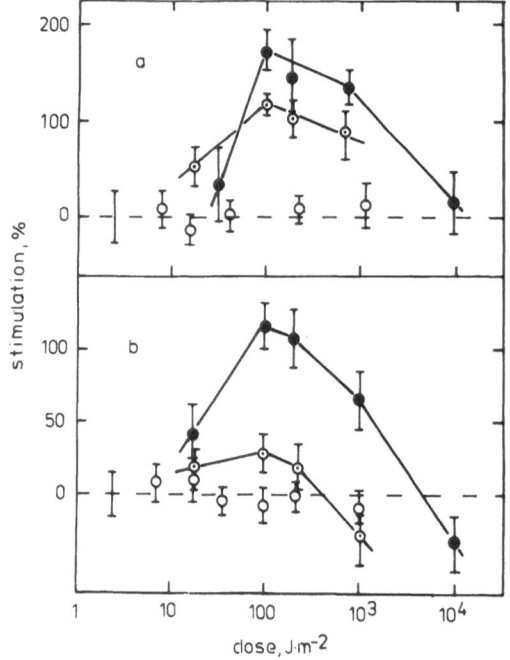

Fig. 10. The dependence of (a) DNA and, (b) RNA synthesis stimulation by an irradiation dose of 100 J/m^2, for times 1.5 h (-o-o-o-), 3.0 h (-⊙-⊙-⊙-) and 6 h (-•-•-•-) after irradiation. Plateau-phase HeLa cells, irradiated with red light at λ = 620 nm.

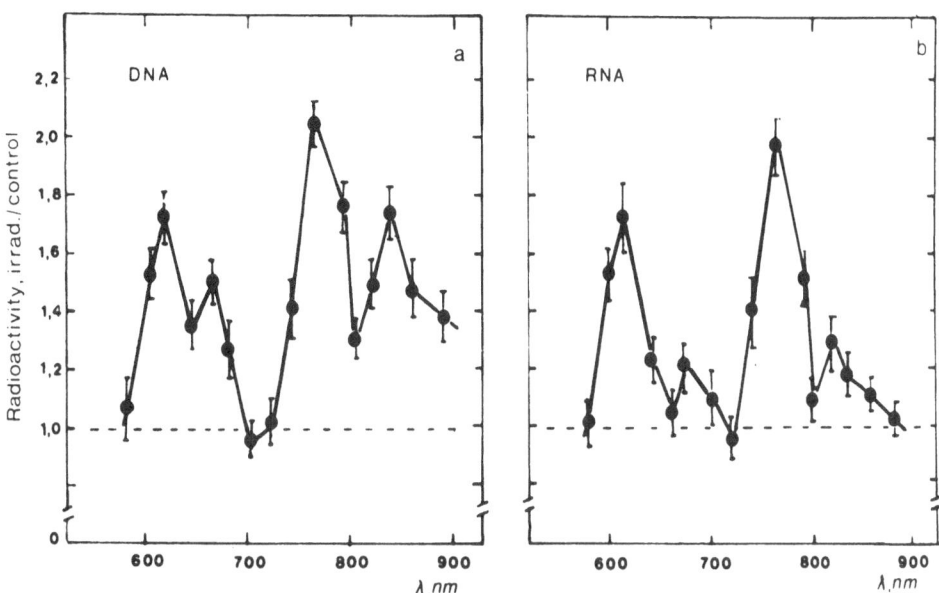

Fig. 11. Action spectra of visible monochromatic light on (a) DNA and, (b) RNA synthesis in plateau-phase HeLa cells for an irradiation dose of 100 J/m^2, measured 4.0 h after irradiation.

In case of plateau-phase cells (Figure 12), the increase of I_s above the baseline level only starts 3-4 h after irradiation, and is probably connected with $G_1 \to S$ transition of the cells of the proliferating fraction of the population, because 6 h is not sufficient time for a $G_0 \to S$ transition.

The stimulative effect of the irradiation on the progression of G_1-phase cells into S-phase was confirmed in autoradiographic experiments with continuous labeling (Figure 13). When a cell population, immediately after the irradiation, is treated with H^3-thymidine and continuously incubated with it for hours, the fraction initially labeled represents the cells in S-phase at the moment of irradiation, while the subsequent increase in the I_s reflects the flow of the cells from G_1 into S during the interval studied. It is obvious from Figure 13 that the percent of labeled cells increases after the irradiation of both log-phase and plateau-phase populations.

To answer the question, does the irradiation influence the rate of DNA synthesis in S-phase cells, the number of grains was counted on the labeled nuclei. As seen in Figure 12b, the average grain count in the nuclei of cells increases after the irradiation, being detectable above the control level 3-6 h after the irradiation. The average grain count does not allow us to determine how the individual cells are influenced by irradiation. In Figure 14, we show the distributions of grain number per nucleus after the irradiation. The behavior of the cumulative curves indicates that the number of cells with higher grain counts increases following irradiation.

The percentage of cells in mitosis (I_m) does not change during the first few hours after irradiation (Table 1). The only significant difference from control level was noticed 6 h after the irradiation.

On the basis of these data it is possible to conclude that the enhanced incorporation of H^3-thymidine into DNA found earlier could be due to intensification of DNA synthesis in S-phase cells, as well as due to an

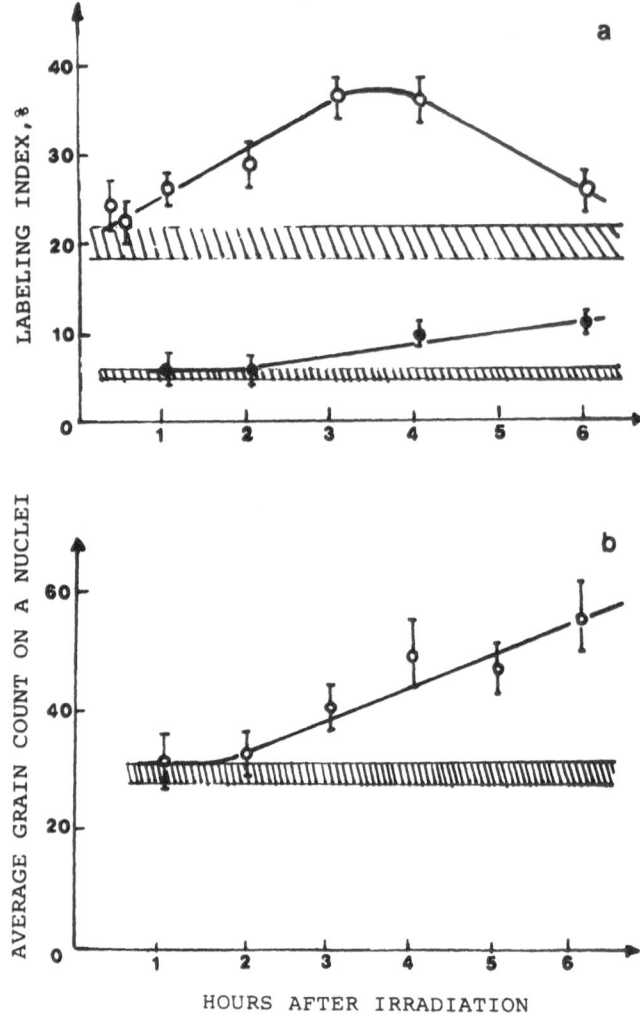

Fig. 12. Changes in: (a) labeling index and, (b) average grain count in
(-o-o-o-) exponentially growing and (-●-●-●-) plateau-phase
HeLa cells, pulse labeled with H^3-thymidine at various times
after irradiation with a He-Ne laser at a dose of 100 J/m^2.

enhanced $G_1 \rightarrow S$ transition for a part of the population. Different kinetics of both these processes after irradiation makes it possible to suggest
that in the first hours after the irradiation the increase of H -thymidine
is caused mainly by an increased number of S-phase cells, but at longer
times after irradiation (e.g. 6 h in our experiments) by enhanced DNA
synthesis in S-phase cells.

Table 1. Changes in mitotic index of exponentially growing HeLa culture
at various times after irradiation (D = 100 J/m^2, λ = 632.8 nm)

Time after irradiation, h	I_M, %
1.0	1.1 ± 0.1
2.0	1.3 ± 0.1
3.0	1.2 ± 0.1
6.0	2.2 ± 0.1

VI. INFLUENCE OF DICHROMATIC IRRADIATION

The experiments described above showed that the effects were cau
relatively narrow-band radiation (±7 nm). Therefore it is of interes
investigate the stimulation effect when the radiation spectrum is wid
beyond the biologically active spectral intervals, going in the limit
white light.

A monolayer of cells at the bottom of a scintillation flask was
radiated with stimulating red light (λ = 633 ± 7 nm) through the bott
the flask and wide band radiation was directed to the cells through t
open neck of the flask (arriving from the opposite side (Figure 15).
comparison was made of the effects of simultaneous irradiation of cel
with red and wide band light with the effects of separate irradiation
The red light dose was optimal for DNA synthesis (100 J/m^2) and the d
of wide band radiation was selected to be about 4 times greater. Thi
ratio of doses corresponds approximately to the fraction of the red l
(600-650 nm, 750-840 nm) in natural solar radiation to which the vari
biological objects have become adapted in the course of their evoluti

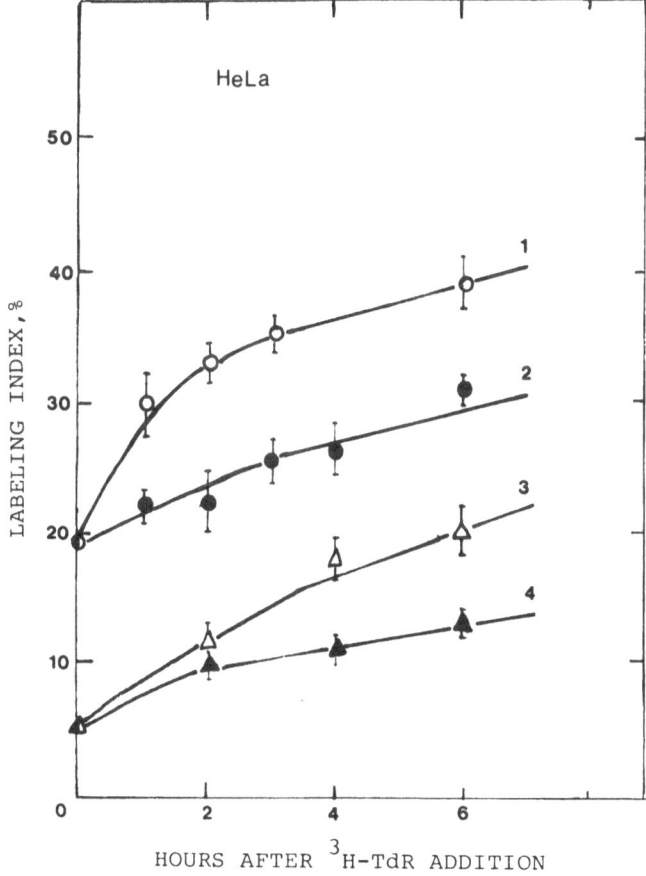

Fig. 13. The variation of the percentage of labeled cells during coi
tinuous labeling with H^3-thymidine in: (1, 2) the exponentially growing and, (3, 4) the plateau-phase. HeLa cells
(2, 4), without irradiation or, (3, 4) after irradiation
with He-Ne laser with a dose of 100 J/m^2.

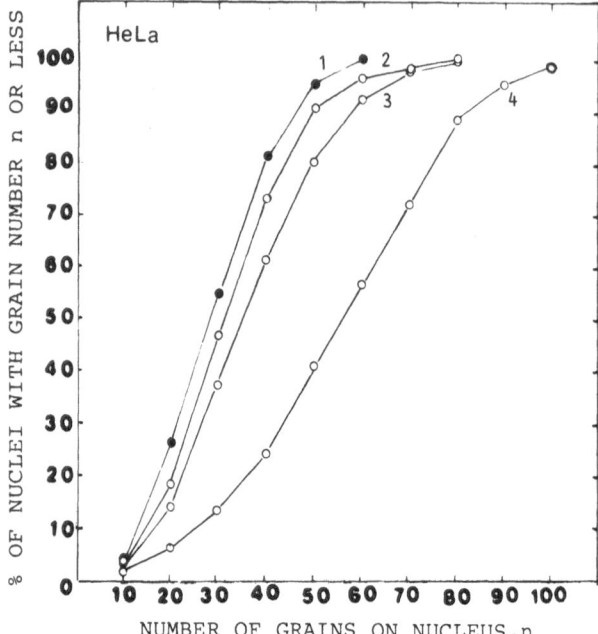

Fig. 14. Autoradiographically measured distribution of silver grains
per nucleus of exponentially growing HeLa cells in: (1) control
culture and, (2, 4) a culture irradiated with a He-Ne laser
(100 J/m^2), for (2) 15 min, (3) 3 h, (4) 6 h after irradiation.

Figure 15 shows the results obtained. Irradiation of cells with just
wide band light, including the red interval, in these experimental conditions resulted in practically no statistically significant variations in
DNA synthesis as compared with the nonirradiated control.

The disappearance of the DNA stimulation effect (caused by irradiation
with light 633 ± 7 nm) when irradiating simultaneously with the narrow-band
and the wide-band light from the blue-yellow region can be explained qualitatively as follows.

Light of the investigated range of wavelengths is not absorbed directly by DNA. Therefore, it is natural to assume that light is first absorbed
by some other molecule or molecules M. The photoproducts formed as a
result of reactions of an excited molecule M which acts as a light acceptor
(dissociation and isomerization products, complexes with other molecules,
etc.) should clearly influence the metabolic processes in cells.

We shall assume that light in the blue-yellow region is absorbed more
effectively by the photoproducts of M than by the initial molecule M itself. Consequently, following the process $M \rightarrow M_{prod}$, as a result of the
subsequent photoreactions, the concentration of the biologically active molecules M_{prod} decreases. The different efficiencies of the two processes
in the spectral range under consideration may be due to differences between
the absorption of the M and M_{prod} molecules or due to the spectral dependences of the quantum efficiencies (yields) of the photoreactions involved.

We may assume that a reduction in the rate of DNA synthesis as a result of dichromatic irradiation is not due to the wide-band nature of light
but due to the specification of those wavelengths which are within the
investigated spectral range.

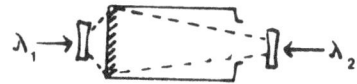

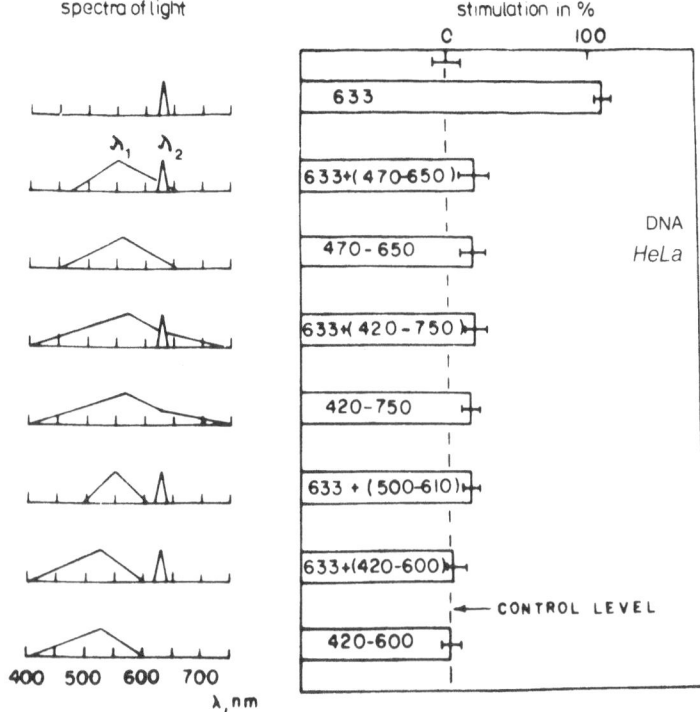

Fig. 15. Effects of broadening the spectrum of the irradiating light in simultaneous dichromatic irradiation with red (λ_2) and wide-band (λ_1) light, on DNA synthesis in exponentially growing HeLa cells. Simplified spectra of the wide-band visible light (λ_1) and of the narrow-band stimulating red light (λ_2) are shown on the left. The corresponding DNA synthesis rates are shown on the right. The dashed line is the control level, taken as 0%.

Thus, it may be stated that low-intensity visible laser radiation causes clearly observable changes in the biochemical processes in a cell involving biomolecules which do not directly absorb laser radiation. Monochromatic light is needed but the spectral band can be 50-150 nm. This bandwidth is more or less 10^5 times wider than the spectral bandwidth of laser radiation (e.g. $\delta\lambda$ of He-Ne laser 10^{-3} nm). That is why the correlation between the stimulative effect of light and its monochromaticity in the range $\Delta\lambda \gg \delta\lambda$ laser is not negligible.

Since there are several maxima in the action spectra of light-nucleic acids synthesis response (Figures 8 and 11) in our next experiments with dichromatic irradiation, we used those wavelengths that had been found to give maximal responses. Dichromatic irradiation experiments can answer the question of whether the hypothesized photoacceptor molecules are photoreversible pigments. In this way also, it is possible to determine those radiation wavelengths in the visible light spectrum which made the stimulative effect of the red component of the spectrum disappear (Figure 15).

HeLa cells in the exponential growth phase were cultivated and irradiated [17],[19],[20]. The rate of DNA and RNA synthesis was estimated 1.5 h after irradiation by a radiometric technique[24].

In the first series of experiments[24], HeLa cells in the form of a monolayer culture on the bottom of a scintillation vial were simultaneously irradiated with red light (λ = 632.8 nm) through the open neck of the flask and a variable wavelength monochromatic light (λ_{add}) through the bottom of the flask from the opposite side (Figure 16). The red light irradiation dose was always kept fixed at 100 J/m^2. The irradiation dose for the variable wavelength light was taken at 100 J/m^2 in the range (600-800) nm and 10 or 25 J/m^2 in the range (400-570) nm. Figure 16 presents the action spectra of dichromatic irradiation for the synthesis of DNA and RNA. In the red and far red regions of the spectrum (λ_{add} = 600-820 nm), the maximum at 620 nm remained the same as in the case of monochromatic irradiation (Figure 8) whereas that at 760 nm disappeared. With the irradiation dose for the variable wavelength light λ_{add} set at 10 J/m^2, the maximum in the blue region of the spectrum, which was observed at 400 nm in Figure 8, now occurred at 450 nm, i.e. it shifted into a region of longer wavelengths. With the irradiation dose for λ_{add} increased to 25 J/m^2, the maximum in the blue region vanished and a new maximum appeared in the green region (λ = 550-570 nm). The synthesis of both DNA and RNA was observed to be inhibited in the green region when the irradiation dose for λ_{add} was set at 10 J/m^2.

When the cells were subjected to consecutive dichromatic irradiation in the sequence λ_{add} + 633 nm with a 60 min interval between the successive irradiation events (Figure 17), these effects vanished and the respective action spectra for the synthesis of DNA and RNA became similar to those obtained earlier (Figure 8) for the wavelengths equal to λ_{add} in this spectral region. This means that only the first irradiation event proved effective, and, by the onset of the second irradiation event with λ = 633 nm, the system had changed into another state from which no further change was possible.

VI.1. Action of red and far red light upon HeLa cells

It is clear from results of the above experiments that the time inter-

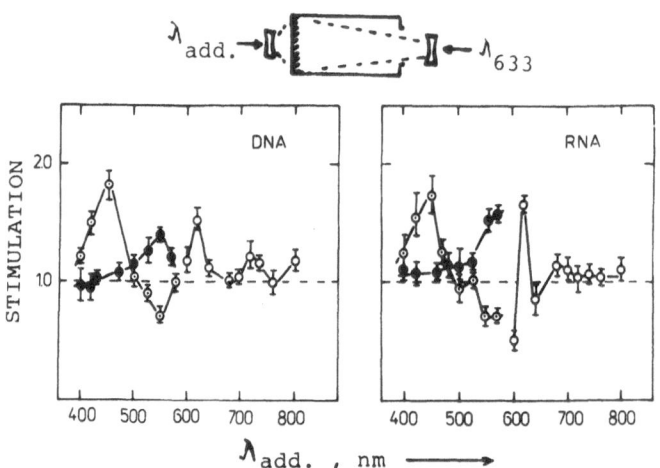

Fig. 16. Action spectra of concurrent dichromatic irradiation with λ = 632.8 nm and λ_{add} on the synthesis of DNA and RNA in exponentially growing HeLa cells, measured 1.5 h after irradiation. $D_{632.8}$ = 100 J/m^2; $D\lambda_{add}$ = 100 J/m^2 (-o-o-o-); 25 J/m^2 (-•-•-•-); 10 J/m^2 (-⊙-⊙-⊙-).

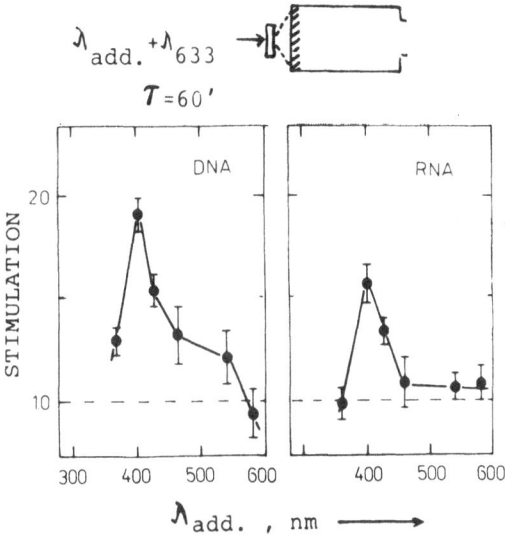

Fig. 17. Action spectra of consecutive dichromatic irradiation, in the
sequence λ_{add} + 632.8 nm, on the synthesis of DNA and RNA in
exponentially growing HeLa cells, 1.5 h after the last irradiation event. $D_{632.8}$ = 100 J/m^2; $D\lambda_{add}$ = 10 J/m^2.

val between the successive irradiation events in dichromatic irradiation
plays an important part. We performed consecutive dichromatic irradiation
with λ = 760 nm and 633 nm, varying the time interval between the irradiation events over a broad range (from 1 sec to 2 h) (Figure 18). The irradiation doses were always taken at 100 J/m^2, which corresponded to the
maximum stimulative effects for these wavelengths (Figure 5). Recall that
the concurrent dichromatic irradiation with λ = 633 nm and 760 nm caused no
changes in the rate of synthesis of DNA and RNA (Figure 16).

Figure 18 shows the rate of synthesis of DNA and RNA as a function of
the time interval between the successive irradiation events. With short
(110 sec) time intervals, the rate of DNA synthesis does not deviate from
the control level. As the interval grows longer, the rate of DNA synthesis
changes, the sense of the effect depending on the sequence of dichromatic
irradiation wavelengths. Irradiation first with the far red light (λ = 760
nm) and then with the red light (λ = 633 nm) stimulated DNA synthesis,
whereas that in the reverse order (633 + 760 nm) inhibited it. These effects reach their maxima when the time interval between the successive irradiation events becomes 1 to 3 min and becomes progressively less pronounced with further increases in the interval. It should be noted that
the effects in their maxima are not equal in magnitude: stimulation amounts
to 60%, while inhibition is only 20%.

Variations in the rate of RNA synthesis caused by consecutive dichromatic irradiation do not follow the pattern described above for DNA. RNA
synthesis is observed to be stimulated with either of the two irradiation
wave-length sequences (760 + 633 nm or 633 + 760 nm), but the maximum of
the stimulative effect for the sequence 633 + 760 nm is observed to practically vanish (Figure 16). This fact provides every reason to search
for some possible connection between these two wavelengths.

In the case of consecutive dichromatic irradiation with λ = 760 nm and
633 nm, variations in the rate of DNA synthesis (stimulation or inhibition
of the synthesis) depend on both the irradiation wavelength sequence and
the time interval between the successive irradiation events (Figure 18).

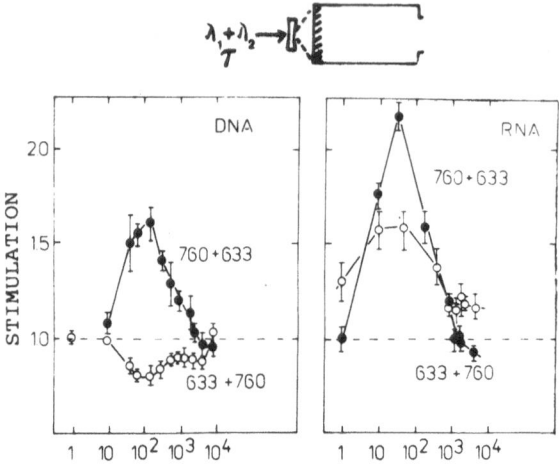

Fig. 18. Rate of DNA and RNA synthesis (1.5 h after last irradiation
event) of exponentially growing HeLa cells, irradiated
(D = 100 J/m^2) consecutively with λ = 760 nm and λ = 632.8 nm,
as a function of the irradiation wavelength sequence and the
time interval between the successive irradiation events.

These data point to the possibility that the photoacceptor may exist in two
interconvertible forms having their absorption maxima located in the red
and far red regions of the spectrum.

A well known example of such a photochromatic photoreversible pigment
is phytochrome[26]. There are no published experimental data that would confirm the existence of a similar system in mammal cells. There are only
data on the antagonistic action of red and far red light on the process of
oxidizing phosphorylation in mitochondria[27], and on chromosome aberrations
in mammalian cells[28]. In experiments described in Reference 27, the photo-
reversal of stimulation and inhibition of ATP synthesis in isolated rat
live mitochondria was realized for two cycles.

VI.2. Action of blue and red light upon HeLa cells

The next series of experiments follows a method similar to the preceding one, the only difference being that instead of far red light we used
blue light with a wavelength of 404 nm (D = 10 J/m^2), which is close to one
of the maxima in the action spectrum (Figure 8). Recall that concurrent
irradiation with λ = 404 nm and 633 nm practically does not change the rate
of synthesis of DNA and RNA as compared with the control level (Figure 19).
In the case of consecutive dichromatic irradiation (Figure 19) with short
(1–10 sec) time intervals between the irradiation events, the rate of DNA
synthesis remains at the control level no matter what irradiation wavelength sequences are used. Increasing the time interval between the irradiation events in the case of the sequence (404 + 633 nm) enhances the rate
of DNA synthesis, whereas the reverse irradiation wavelength sequence has
for all cases no effect on the synthesis. RNA synthesis is observed to be
stimulated with both irradiation wavelength sequences, the maximum of the
stimulative effect being higher in the case of the sequence (404 + 633) nm.
This effect already appears at short (1–10 sec) time intervals between the
irradiation events. As the interval is increased (up to 2 h in our experiments) the effect reaches its maximum and then remains practically unchanged.

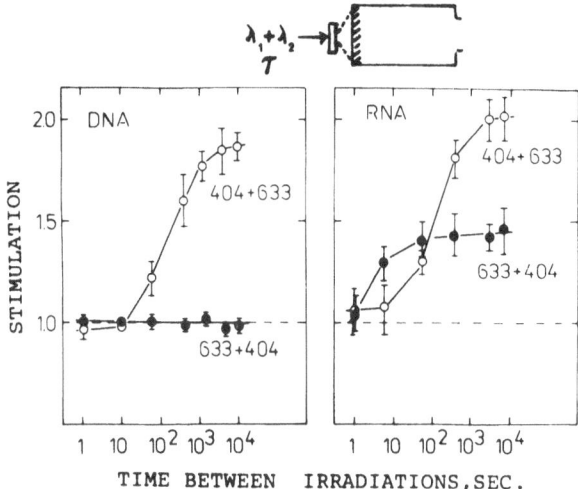

Fig. 19. Stimulation of DNA and RNA synthesis (1.5 h after last irradiation event) of exponentially growing HeLa cells, irradiated
consecutively with λ = 404 nm and λ = 633 nm (D_{633} = 100 J/m^2,
D_{404} = 10 J/m^2), as a function of the irradiation wavelength
sequence and the time interval between the successive irradiation events.

In our experiments, consecutive dichromatic irradiation with red and
blue light with different time intervals between the successive irradiation events and in different wavelength sequences has different effects on
the synthesis of macromolecules. Irradiation in the sequence 633 + 404 nm
has either no effect at all (in the case of DNA) or a considerably weaker
effect (in the of RNA, Figure 19) as compared with that in the sequence
404 + 633 nm. Irradiation in this latter sequence (404 + 633 nm) is observed to stimulate the synthesis of both DNA and RNA, irrespective of the
time interval between the irradiation events. Our data agree with the data
on the consecutive dichromatic irradiation of a culture of human embryonic
skin-muscular fibroblasts[29]. No changes have been noticed in the mitotic
activity of the cells in the case of irradiation in the sequence
633 + 441 nm, whereas, in the case of irradiation in the sequence
441 + 633 nm, the mitotic index was observed to increase by 40 to 60 %
within a 5 min interval between the successive irradiation events. Non-
additive stimulation was also noticed in cases in which one of the two
irradiation components has a inhibitive effect when used individually.

In our experimental conditions, we failed to observe in the action
spectra for the rate of synthesis of DNA and RNA any statistically significant deviations from the control level in the green region of the spectrum
(Figure 8). A blue maximum was found at around 404 nm, which was approximately 10 times as sensitive as the red and far red maxima (maximum effects
at irradiation doses 10 and 100 J/m^2, respectively - see Figure 5). Concurrent dichromatic irradiation was found to affect the action spectra: the
blue maximum shifted from 404 nm to 450 nm and changes were observed in the
green region of the spectrum (Figure 16). These changes depend on the irradiation dose of the blue or green light (the red-light irradiation dose
was kept fixed at 100 J/m^2). Thus, the light of these wavelengths is
"biologically active". For these effects to occur, it is necessary that
the irradiation doses should be low and the irradiation times short.

Here we are dealing with some photoreceptor(s) which can initiate
changes in the final photoresponse. One may assume that concurrent dichromatic irradiation (Figure 16) shifts the equilibrium of the electron trans-

fer chain (respiratory chain[30]) in such a way that the action spectrum reflects the other redox state of a chromophore absorbing at λ = 450 nm as well as of that chromophore absorbing in the green region (probably cytochromes). At 450 nm, the oxidized form of flavin features an absorption maximum[31]. The one electron reduction of oxidized flavins leads to the formation of either an anionic form of semiquinone having its absorption maxima located at 380, 400 and 490 nm (with a continuation up to 650 nm) or neutral semiquinone which absorbs light within almost the whole of the visible light region (up to 750 nm). The two-electron reduction of flavin molecule results in almost complete disappearance of the absorption band at 450 nm. It is quite possible that some flavoprotein is the primary photoacceptor for which we search. Flavoproteins play a key part in redox regulation of mitochondrial activity. Due to their ability to form a stable intermediate semiquinone form, they serve as connecting links between one and two electron reduction-oxidation systems in the respiratory chain.

The picture here is complicated by the fact that hemoproteins to which the cytochromes belong also absorb light in the blue (λ = 400 nm) and green regions of the spectrum[25]. Therefore changes in the final macroeffect may occur, as in fact is shown by the sensitivity of nucleic acid synthesis to the irradiation dose of λ_{add} (Figures 16 and 17). It can be expected that for other ratios between the irradiation doses of the red light and λ , the action spectra will be different.

Thus, proceeding from the data on dichromatic irradiation, flavoproteins can be thought of as a photosensitive regulatory system in which the chromophores are $flavin_{reduced} \rightarrow flavin_{semiquinone} \rightarrow flavin_{oxidized}$. There is information in the literature showing that irradiation with visible light at λ = 400 nm alters the reduction-oxidation state of flavins. It has also been demonstrated that there is a functional relationship between the degree of oxidation of flavin coenzymes and the ATP synthesis activity of mitochondria in vivo[32]. Regulation is thought to occur through changes in the spatial conformation of flavoproteins, which is determined by the reduction-oxidation state of the chromophore[33]. Problems connected with possible primary photoacceptors and light signal transduction chains in cells (e.g. from respiratory chain to DNA) are discussed in detail in the References 12 and 30.

VII. STIMULATION AND INHIBITION

In experiments described above, the positive (stimulating) effects of irradiation were described. However, there also exists a tremendous amount of data describing inhibitory, and even lethal, effects of light (especially blue and fluorescent light) on various types of cells. Table 2 shows some light-growth responses of mammalian cell cultures[9]. An analysis of this data suggests that the dose and the intensity of the light used determine the sense stimulating versus inhibitory of the end macroeffect. The stimulative doses are 3-4 orders lower than inhibitory ones.

To verify the importance of light dose on the sense of the final effect, two series of experiments with HeLa cells were performed.

In the first series of experiments, we studied the dose-response dependences for lethal effect of different visible light wavelengths. HeLa cells were cultivated as usual and taken for the experiment 72 h after plating (during the log-phase culture). The culture was trypsinized, and a suspension of 2×10^5 cells in 1 ml of Hank's solution was prepared for irradiation. The irradiation was performed in a special quartz cuvette with volume 90 μl or 250 μl (S_{irrad} = 6.38×10^{-2} and 2.12×10^{-1} cm^2, respectively). For irradiation, a Rhodamine S dye laser pumped with cw Ar laser (λ = 578

Table 2. Action of various bands of light on proliferation activity of mammalian cellular cultures.

Wavelength nm	Culture	Stimulation	Inhibition
		Dose, J/m^2	Dose, J/m^2
400	HU-274 WI-38	10^4	10^5-10^6
441.6 632.8 694.3 741	Human embryonic skin fibroblasts	10^3-10^4	10^5-10^6
Cool white fluorescent lamps	Human embryonic diploid lung fibroblasts	Dose not shown. Exposition every day during 150 days for 2 h: cells came through 70 divisions; 40 h:60 div.; 6 h:53 div.; dark control: 53 div.	Dose not shown. The cells exposed constantly, died within 2-3 days
405	Human lympho-blastoid cells	–	10^5
630-633	Chinese hamster fibroblasts	10^3	–
632.8	L	7.5	–
632.8	Human embryonic foreskin fibro-blasts	10^2	–
632.8	HeLa	10^2	–
546-579	HeLa	Dose not shown, dose rate 10-50 W/cm^2	dose not shown, dose rate 100-300 W/cm^2
694.3	Human skin fibroblasts	$(1-4) \times 10^3$	–
694.3	Human melanoma in culture	$(2.4-48) \times 10^6$	–
694.3	Mouse fibro-blasts L929	10^5	–
694.3	Human epidermal and mouse lung cells in culture	–	No damage till 10 J/m^2

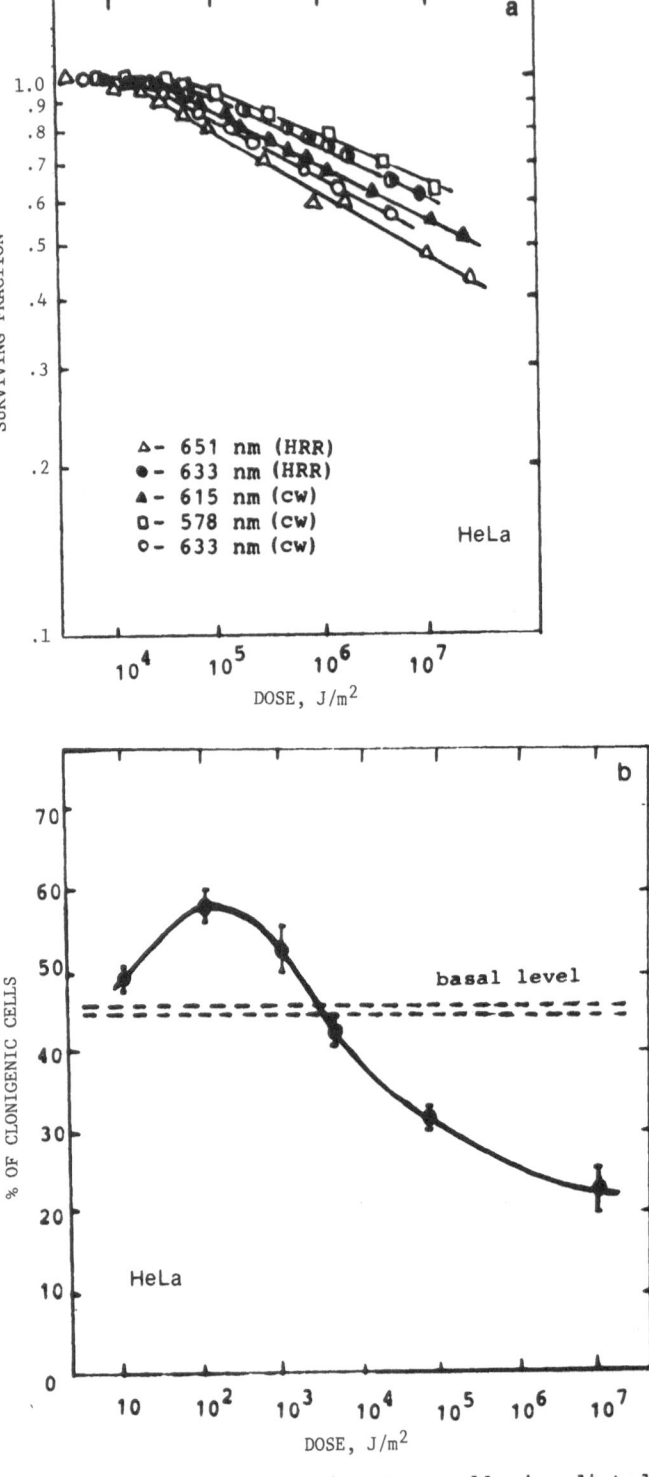

Fig. 20. (a) Survival of proliferating HeLa cells irradiated with a
dye laser at different wavelengths and with different irradi
ation modes (cw or repetitively pulsed light) and, (b) change
of number of clonigenic cells as a function of HeNe laser radiation dose (plateau-phase HeLa cells were irradiated as a
monolayer, and their clonigenic ability was evaluated by the
Puck technique after replating).

or 615 nm), and Oxazin 170 dye laser pumped by copper vapor laser (λ = 633 651 and 670 nm, high repetition rate nanosecond pulses) were used. Just after the irradiation, cell viability was evaluated by trypan blue exclusive test. The viability of cells in nonexposed samples was 89 ± 3 %.

As seen in Figure 20a, the dose-effect dependences are very similar for all wavelengths used in our experiments. There are no lethal effects till approximately 10^4 J/m^2, beyond which increasing doses give an increasing lethal effect.

In the type of test just described it is impossible to investigate the positive and negative effects in one experiment. For that reason we performed the next series of experiments using the type of experiment described at the beginning of this chapter. Plateau-phase HeLa cells were irradiated with a He-Ne laser in different doses, trypsinized 3 h after irradiation and replated. The number of viable cells was determined by counting the number of clones.

Figure 20b shows the results obtained. As seen in Figure 20b, there is a rather abrupt switch from a positive to a negative effect upon increasing the light dose. A similar type of curve was obtained earlier when investigating the influence of near IR light at λ = 890 nm on the growth of E. coli[34], and at λ = 904 nm on ATP production in Saccharomyces calbergensis[35].

On the other hand, the negative effect seems not to increase monotonically with increasing irradiation dose. Irradiation of Saccharomycodes ludwigii with He-Ne laser did not cause growth inhibition more than 20%, even when the dose was increased by 2 orders of magnitude[36][37].

In explaining the positive (stimulating) effects of visible light, the respiratory chain components were proposed to be the primary photoacceptors[12,19,30]. Also, in the case of negative effects (which inhibit cellular metabolism or are lethal), the respiratory chain components have been shown to be primary photoacceptors[38÷44]. For example, it has been demonstrated[43] that cytochromes c and c_1 failed to act as hydrogen acceptors following 10 pulses of 1 mW/cm^2 of green laser light at 530 nm, and cytochrome a/a_3 showed a similar response when irradiated at λ = 609.6 or 601.3 nm.

These data illustrate the principle that laser wavelengths which are appropriately matched to the absorption characteristics of target molecules can not only stimulate but also selectively inhibit specific molecular components in cells. The lack of wavelength specificity in our experiments can be probably explained by absorption and resulting changes in various molecules in the respiratory chain, the final results being overall inhibition of the electron transfer chain.

REFERENCES

1. J. Parrish, "Photomedicine: a status report", Photochem. Photobiol., 49 S, 649 (1989)
2. E. Mester, "Über die stimulierende Wirkung der Laserstrahlung auf die Wendheilung", In: Der Laser, K. Dinstl and P.L. Fischer, Eds., Berlin-Heidelberg-New York: Springer, pp. 109-119 (1981)
3. N. F. Gamaleya, "Laser biomedical research in the USSR", in: Laser Application in Medicine and Biology, M. L. Wohlbarsht, Ed., New York, London: Plenum, 3:1-175 (1977)
4. T. I. Karu, L. V. Pyatibrat and G. S. Kalendo, "Biostimulation of HeLa cells by low-intensity visible light. Y.Stimulation of cell

proliferation *in vitro* by He-Ne laser irradiation", Il Nuovo Cimento, D, 9:1485-1494 (1987)

5. A. H. W. Nias, "Clone size analysis: a parameter in the study of cel. population kinetics", Cell. Tissue Kinet., 1:153-165 (1968)

6. F. Mauro, B. Falpo, G. Briganti, R. Elli and G. Zupi, "Effects of antineoplastic drugs on plateau-phase cultures of mammalian cells. I. Description of the plateau-phase system", J. Natl. Cancer Institute, 52:705-713 (1974)

7. M. Hahn and J. B. Little, "Plateau-phase cultures of mammalian cells: An *in vitro* model for human cancer", In "Current Topics in Rad. Res.", M. Ebert, Ed., Amsterdam-London: North-Holland, 8:39-83 (1972)

8. O. I. Epifanova, I. N. Smolenskaya and V. A. Polunovsky, "Responses of proliferating and non-proliferating Chinese hamster cells to cyto-toxic agents", Br. J. Cancer, 37:377-385 (1978)

9. T. I. Karu, "Effects of visible radiation on cultured cells", Photo-chem. Photobiol., 52, N6 (1990)

10. T. I. Karu and V. S. Letokhov, "Biological action of low-intensity monochromatic light in the visible range", In: Laser Photobiology and Photomedicine, S. Martellucci and A. N. Chester, Eds., New York-London Plenum Press, pp. 57-66 (1985)

11. T. I. Karu, "Photobiological fundamentals of low-power laser therapy", IEEE J. Quant. Electr., QE-23, 1703-1717 (1987)

12. T. I. Karu, "Photobiology of Low-Power Laser Therapy", Chur, London, New York: Harwood Academic Publ. (1989)

13. R. Baserga, "Multiplication and Division in Mammalian Cells", New York, Basel: Marcel Dekker Inc. (1979)

14. J. C. Schaer, L. Ramsier and R. Schindler, "Studies on the division cycle of mammalian cells. III. Incorporation of labeled precursors into DNA of synchronously dividing cells in culture", Exp. Cell. Res., 65:17-22 (1971)

15. S. E. Pfeiffer and L. A. Tolmach, "RNA synthesis in synchronously growing populations of HeLa S cells. I. Rate of total RNA synthesis and its relationship to DNA synthesis", J. Cell Physiol., 71:77-94 (1968)

16. S. D. Kazmin, "Biochemistry of Mitotic Cycle of Tumour Celles", Kiev: Naukova Dumka (in Russian) (1984)

17. T. I. Karu, G. S. Kalendo, V. V. Lobko and L. V. Pyatibrat, "Kinetics of tumour HeLa cell growth under subcultivation after irradiation by low intensity red light at the stationary growth phase". Eksperimentalnaya Oncologiya, 6, N 1:60-63 (1984)

18. T. I. Karu, G. S. Kalendo, V. S. Letokhov and V. V. Lobko, "Biostimu-lation of He-La cells by low intensity visible light", Il Nuovo Cimento D, 1:828-840 (1982)

19. T. I. Karu, G. S. Kalendo, V. S. Letokhov and V. V. Lobko, "Biostimu-lation of He-La cells by low intensity visible light. II. Stimula-tion of DNA and RNA synthesis in a wide spectral range". Il Nuovo Cimento D, 3:309-318 (1984)

20. T. I. Karu, G. S. Kalendo, V. S. Letokhov and V. V. Lobko, "Biostimu-lation of He-La cells by low-intensity visible light. III. Stimu-lation of nucleic acid synthesis in plateau phase cells", Il Nuovo Cimento D, 3:319-325 (1984)

21. G. S. Kalendo, "Early Responses of Cells to Ionizing Radiation and Their Role in Radioprotection and Sensibilization", Moscow: Energoizdat (in Russian)(1982)

22. T. T. Puck, P. I. Marcus and S. J. Cieciura, "Clonal growth of mammalian cells in vitro", J. Exp. Medicine, 103: 272-283 (1956)

23. T. I. Karu, G. S. Kalendo and V.S. Letokhov, "Control of RNA synthesis rate in tumour cells HeLa by action of a low-intensity visible light of a copper laser", Lettere al Nuovo Cimento, 32:55-59 (1981)

24. T. I. Karu, V. S. Letokhov and V. V. Lobko, "Biostimulation of HeLa cells by low-intensity visible light. IV. Dichromatic irradiation", Il Nuovo Cimento D, 5:483-496 (1985)

25. P. H. Kirschenbaum, ed. "Atlas of Protein Spectra in the Ultraviolet and Visible Regions", New York, Washington, London: IFI/Plenum Press (1972)

26. W. Rüdiger and H. Sheer, "Chromophores in photomorphogenesis", In "Photomorphogenesis", W. Shropshire, Jr. and M. Mohr Eds., Berlin, Heidelberg, New York, Tokyo: Springer, p. 119-183 (1983)

27. S. A. Gordon and K. Surrey, "Red and far red action on oxydative phosphorylation", Radiat. Res., 12:325-339 (1960)

28. S. A. Gordon, A. N. Stroud and C. H. Chen, "The induction of chromosomal aberrations in pig kidney cells by far red light", Rad. Res., 45:274-287 (1971)

29. N. A. Bogush, V. A. Mostovnikov, A. T. Pikulev and I. V. Khokholov, "Effect of biostimulation increases by combined action of blue and red laser light", Dokl. Akad. Nauk Belorusskoi SSR (Proc. Beloroussian Acad. Sci.), 26:951-954 (1982)

30. T. I. Karu, "Fundamentals of low-power laser photomedicine", in "Laser Science and Technology", A. N. Chester, V. S. Letokhov and S. Martellucci Eds., New York, London, Plenum Press, p. 217-232 (1988)

31. A. White, P. Handler, E. Smith, R. Hill and I. Lehman, Eds., "Principles of Biochemistry", New York, McCraw Book Co. (1978)

32. E. S. Vishnevskaya, T. A. Lozinova, O. N. Brzevskaya, O. S. Nedelina and L. P. Kayushin, "Effect of visible light on ATP-synthetase function of mitochondria", Biofyzica, 29:637-639 (1984)

33. O. S. Nedelina, O. N. Brzevskaya and L. P. Kayushin, "Redox regulation in ATA synthesis", Biophyzika, 30:119-191 (1985)

34. V. P. Zharov, T. I. Karu, Yu. P. Litnikov and O. A. Thiphlova, "Biological effect of radiation of a semiconductor laser in near infrared region", Kvantovaya Electronika, 14:2135-2136 (1987)

35. U. Warnke and W. H. Weber, "Influence of light on cellular respiration", in "Electromagnetic Bioinformation", F. A. Popp Ed., München, Wien, Baltimore: Urban and Schwarzenberg (1987)

36. G. E. Fedoseyeva, T. I. Karu, T. S. Laypunova, N. A. Pomoshnikova and M. N. Meissel, "The activation of yeast metabolism with He-Ne laser radiation. II. Activity of enzymes of oxidative and phosphorous metabolism", Laser Life Sci., 2:147-154 (1986)

37. G. E. Fedoseyeva, T. I. Karu, T. S. Laypunova, N. A. Pomoshnikova and M. N. Meissel, "The activation of yeast metabolism with He-Ne laser radiation. I. Protein synthesis in various cultures", Laser Life Sci., 2:137-146 (1988)

38. L. N. Edmunds, "Blue light photoreception in the inhibition and synchronization of growth and transport in the yeast Saccharomyces", in "Blue Light Syndrome", M. Senger, Ed., Berlin, Heidelberg, New York: Springer, p. 584-596 (1980)

39. B. L. Epel, "Inhibition of growth and respiration by visible and near visible light", in Photophysiology, A. L. Giese Ed., New York, London, Academic Press, 8:209-229 (1965)

40. B. L. Epel and W. L. Butler, "Cytochrome a : destruction by light", Science, 166:621-622 (1969)

41. J. Frederick, "Effects de differentes longueurs d'onde spectre visible sur des cellules vivantes cultivées in vitro", C. R. Soc. Biol., 148:16781682 (1954)

42. D. E. Rounds and R. S. Olson, "The effect of intense visible light on cellular respiration", Life Sci., 6:359-366 (1967)

43. D. E. Rounds, R. S. Olson and F. M. Johnson, "The effect of the laser on cellular respiration", Z. Zellforsch., 87:193-198 (1968)

44. R. B. Webb and M. S. Brown, "Sensitivity of strains of E. coli differing in repair capability to far UV, near UV and visible radiations", Photochem. Photobiol., 36:425-492 (1976).

MEDICAL APPLICATIONS OF LOW POWER LASERS IN CHINA

Fu-Shou Yang

Department of Laser Medicine
Shanghai Seamen's Hospital
Shanghai, People's Republic of China

I. INTRODUCTION

The clinical applications of low power laser irradiation (power output < 200 mW, power density < 50-100 mW/cm^2) have been widely explored in China. These applications have principally included: wound healing, photodynamic therapy (PDT), tissue welding, biostimulation, medical diagnosis and acupuncture[1]. According to imperfect statistics during the most recent ten years, more than a million laser treatment sessions have been given in China. This paper will introduce the clinical and functional mechanism of laser acupuncture, laser tumor immunization and, medical diagnosis of laser fluorescence for tumors (except diagnosis of HPD fluorescence). The prospect for low power lasers in medical and surgical applications will be discussed at the same time.

II. LASER ACUPUNCTURE

II.1. Functional Mechanism

II.1.i. The laser acupuncture therapy which is combined with the theory of traditional Chinese medical acupuncture has a biological effect consistent with traditional Chinese medical theory, therefore its characteristics are different from those of general biostimulation[2÷4]. Experiments have shown the following (Figure 1):

(a) The sensitive points stimulated by laser beam irradiation on acupoints have fixed location and repeatibility, mostly corresponding to the traditional acupoints.
(b) The line relating with the sensitive point basically corresponds to the classic main and collateral channels circular line, comparatively shallow, that can be hindered with a very small pressure (about 2-3 mm Hg); when the pressure is removed, and it is irradiated again with the laser beam, the stimulation will reappear.
(c) The sensitive conduction is two-directional and has the phenomenon of relay sensitive conduction.
(d) The sensitive main channel line and the diseased internal organs are related; for instance, the gastric main channel of the patient with gastric disease is particularly sensitive, the gall main channel of the gall bladder diseased patient is also sensitive, and so forth (Figure 2).

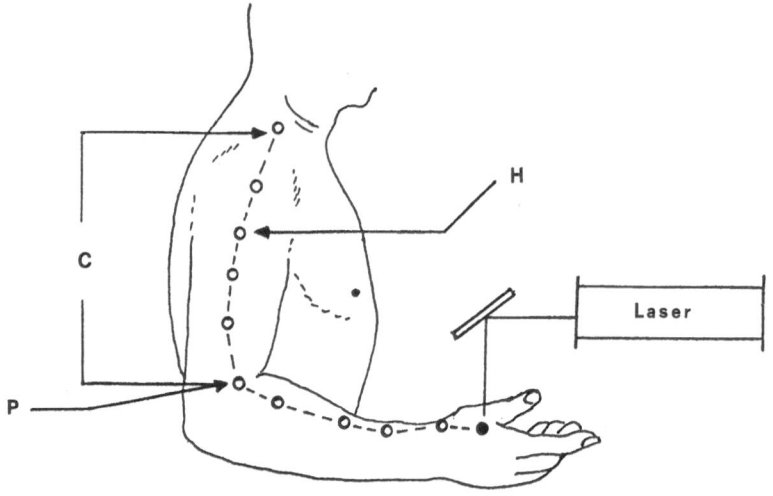

Fig. 1. Relation of occult conduction channels and laser beam
irradiation on acupoint of "L.I. Channel". --- = occult
conduction channels, o = sensitive, ● = acupoint "HEGU",
c = hysteresis conduction, P = adding pressure,
H = magnetic field may cause the interference of occult
conduction channels.

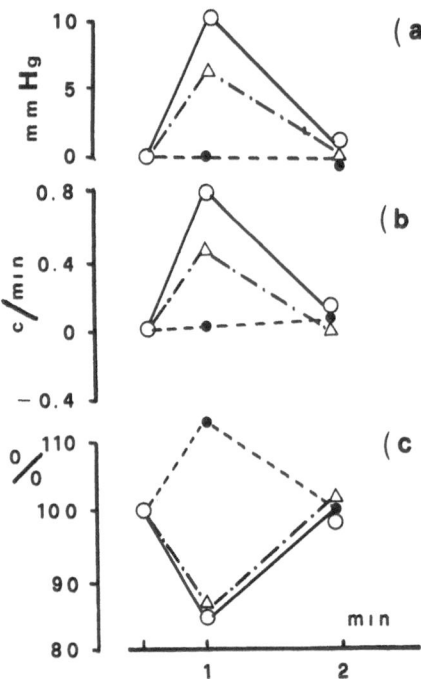

Fig. 2. Biological effect on normal and ill persons caused by
laser acupuncture on different acupoints. N = 30 (gastric
lesion 10); a) blood pressure; b) gastric peristalsis;
c) gastric nerve discharge.
o = gastric illness, "ZUSANLI" (L. I. channel);
△ = normal gastric system, "ZUSANLI" (L. I. channel); and,
● = normal kidney, "YINDU" (K. channel).

116

(e) The occult conduction can be disturbed by a magnetic field; the mag-
     netic field can stop, reverse, weaken or re-direct the conduction.
     The removal of the magnetic field can cause the original conduction to
     recover (this is called "the change of the main channel by magnetism";
     it is one of the objective arguments that demonstrate the magnetic
     character of the occult conduction).

II.1.ii. Laser acupuncture can promote a defensive immunofunction in the
          human organism

     The promotion of defensive immunofunction, which correspoonds to the
function of "helping the right and eliminating the wrong" in traditional
Chinese medicine, is one of the principles of laser acupuncture and traditional acupuncture theory. Experiments have shown that laser acupuncture can regulate the defensive immunofunctions of organisms; it has antiinflammatory, antipyretic and healing effects, it regulates the balance of
the immune network, improves the health, and provides the patients with
perfect recovery from the relevant illness. Experiments have shown that
after laser beam irradiation, the relevant acupoints in the organisms may
increase the PHA and SK-SD positive value reaction, significantly increase
"T" lymphocytes in blood, phage percentage and phagocytic index of neutro-
philic leucocytes, and increase electrophoretic gamma-globulin and albumin
in serum (Figure 3).

     In clinics, we have treated 120 cases of asthma with laser acupunc-
ture therapy. According to traditional Chinese medicine, we chosed
"TIANTU", "SHANZHONG", "DINGCHUAN", etc. as acupoints, and we obtained a
significant therapeutic effect. Before and after the therapy, we measured
the immune index and we found that the body fluid immune level was noticeable increased (Table 1).

II.1.iii. Regulation of function by laser acupuncture on the human
           organism

     Through numerous experiments with human and animal subjects, we have
shown that the mechanism of regulation of function by laser acupuncture and
traditional acupuncture takes place through the autonomic nerve system -
medium - blood vessel reaction system. Experiments have revealed an effect
of regulation on circulation, digestion and endocrin systems etc.

A) Regulation of function on brain blood flow

     If the relevant acupoints of patients with brain blood vessel or non

Table 1. Body fluid immune performance variation of asthma patients
          during laser treatment.

120 cases	IgG (mg/ml)		IgA (mg/ml)		IgM (mg/ml)	
	before	after	before	after	before	after
patients (average value)	16.9	18.4	3.7	3.65	1.06	1.38
normal value comparison	7.0	18.0	0.9	4.5	0.6	2.5
	(average value = 12.0 mg/ml)		(average value = 2.00 mg/ml)		(average value = 1.00 mg/ml)	

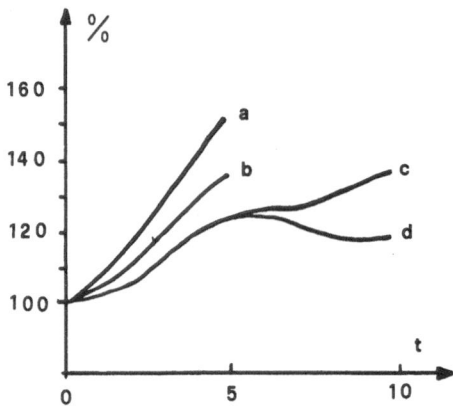

Fig. 3. He-Ne laser acupuncture on acupoint "ZUSANLI" of the
         domestic rabbit caused cell immune response. CO_2 laser
         acupuncture on the acupoint "BIYU" of the deer caused
         variation of "T" cell value and white cell phagocytosis.
         a) "T" cells; and, b) phagocytosis % (W.C.). He-Ne laser
         acupuncture: c) PHA; and, d) SK-SD.

brain blood vessel diseases (BAIHUI, YINTANG, HEGO, ZUSANLI, and TAIZHONG,
etc.) are irradiated by laser, after treatment, the circulation of brain
blood flow is regulated; thereby, the blood flow is improved, and patients
gain the benefit of recovery from illness (Table 2).

B) The effect of laser acupoint irradiation on gall bladder function

     Figure 4 summarizes the experimental results: (i) Ultrasonic waves
were used to detect the volume of gall bladder and to measure its size
electronically; (ii) After the gall bladder acupoints were irradiated for
ten minutes by laser, measurements were commenced; (iii) After one-half
hour of irradiation, we displayed the ultrasonic image and measured again.
The experiment showed that irradiation of the gall bladder acupoints with a
laser produced shrinkage of the gall bladder.

C) Laser irradiation on acupoints can improve the regulation of the
    endocrine system

     After laser acupoint irradiation of "INDOCRINE" on the ear, limb frac-
ture "QUCHI", we need ^{32}P as a tracer to measure the dynamic variation of

Table 2. Analysis of brain blood circulation before and after laser
          acupuncture

	Wave amplitude	Time of elevation	Circulation speed in vessel	Elevating angle	Main peak angle	Blood vessel resistance index
Cases	44	44	44	44	44	44
P value	< 0.01	< 0.01	< 0.05	= 0.01	> 0.05	> 0.05
Result after laser therapy	prominent elevating	prominent decreasing	prominent quickening	prominent sharpening	non-prominent sharpening	non-prominent decreasing

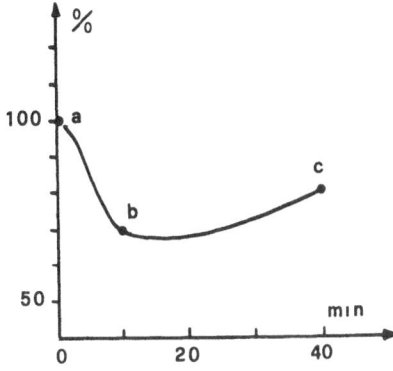

Fig. 4.   Function effect on "Gall-Bladder"after He-Ne laser
          acupoint irradiation, a) v = 72205.266 mm^3;
          b) $\overline{v}$ = 54013.633 mm^3, $\downarrow$- 25.19%; and,
          c) $\overline{v}$ = 57697.766 mm^3, $\downarrow$- 20.09%.

Calcium absorption from the illside and healthy side of a body with a
fractured limb.

   As mentioned above: all the clinical practices and related experimental research proved that laser acupuncture provides functional regulation
of the whole human body (Figure 5). When we accurately select acupoints
according to traditional Chinese medical theory, we obtain a strong therapeutic effect.

II.2.   CLINICAL APPLICATION

   Although He-Ne lasers are most commonly used in laser acupuncture, CO_2
lasers, Ar$^+$ lasers, N_2 lasers, etc. are also used[5]. Different lasers are
selected according to the different diseases to be treated.

II.2.i. Instruments and methods

A. The output of the He-Ne laser is 25 mW, which is conducted through an

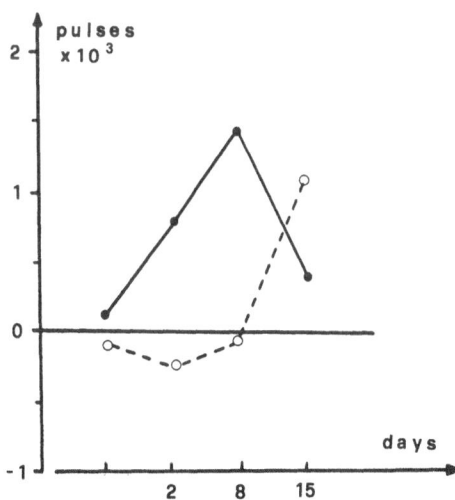

Fig. 5.   Comparison of sick and normal side of body by isotope
          ^{32}P metabolism.
          ---- = normal side; —— = sick side.

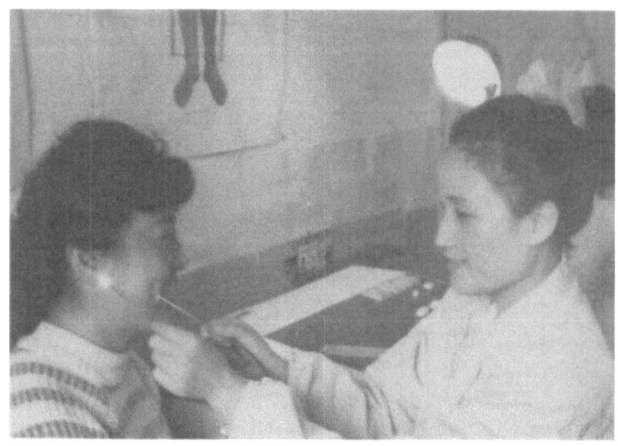

Fig. 6. He-Ne laser (laser acupuncture) clinical application.

optical fiber to irradiate the acupoints (Figure 6). Each acupoint is irradiated 3-5 minutes; every day one choose 2-5 acupoints, and one set contains 10-15 treatments.

B. The rules of acupoint selection are as follows: ordinarily according to
traditional Chinese medicine there are four rules for selection: (a) The
specific characteristics of acupoint treatment: each acupoint has its particular curative function. For instance, the acupoint of "HYPNOSIS" is
used to treat hypnosis, the acupoint of "DINGCHUAN" is used to treat wheez
ing (asthma), etc. (b) The double direction of acupuncture treatment. For
instance, the acupoint "TIANSHU" is used to stop diarrhea but also to pass
stools, the acupoint "NEIGUAN" is used to treat not only tachycardia to ge
bradycardia but also bradycardia to get tachycardia. (c) The character of
integration of acupoint treatment; from the 12 regular channels, we determine the acupoints for the relative channel. (d) The vicinity of acupoint
treatment. For example, the acupoints in the vicinity of ears, eyes, mouth
nose, etc. are used for curing the diseases of these organs.

C. Clinical therapy with laser acupuncture obtains the most effective
results in the following 30 diseases (Table 3).

For example, let us consider acupuncture used in the treatment of
hypertension. We treated 108 patients for hypertension, among them stage
I-II 80 cases, stage III accompanied by cardiopathy 29 cases. Duration of
sickness: 1-3 years, 45 cases; 3-5 years, 30 cases; 5-10 years, 15 cases;
10-17 years, 19 cases.

A 6328 Å He-Ne laser with output power 6 mW (31,8 mW/cm^2) was applied
conducting the laser beam through an optical fiber, irradiating on the acupoint "RENYING" on both sides of the neck. According to the conditions of
the disease, the addition of acupoints "QUCHI", "TAICHUNG", etc. was necessary. Every acupoint was treated for 3-5 minutes, once per day, for 1 set
of 10 times. Before and after irradiation, hypertension was measured. For
25 cases, the blood pressure relaxed after 1 radiation, for 16 cases the
blood pressure relaxed after 2 radiations. At least 4 radiations were necessary, and 30 radiations were the maximum. From analyzing the therapeutic effects on 109 hypertension patients, a satisfactory result was obtained. The best results were for the stage I-II patients. The most remote therapeutic effect attained was 94.4 % and no untoward reaction was
noted.

120

Table 3. Reference table of the acupoints used in treating diseases by laser acupuncture.

No	Diseases	Acupoints
1	Asthma	Dingchuan, Feishu, Tientu, Shouzong
2	Infantile pneumonia	Feishu, Dazhui
3	Cough	Chize, Kongzui, Lieque
4	Baby's diarrhea	Tianshu, Shenjue, Zhixie
5	Hypertension	Renying, Fengchi, Yintang, Yongquan
6	Enuresis	Guanyuan, Zhongji, Sanyinjiao, Tianshu
7	Gastroptosia	Guanyuan, Quepen, Zhongwan, Zusanli
8	Arthritis of lower jaw	Xiaguan, Tinggong, and others
9	Headache	Fengchi, Zhongchong, Hegu
10	Myopia	Jingming, Qiuhou, Shuangxue
11	Infantile chronic enteritis	Zusanli, Daheng, Sanyinjiao
12	Insomnia	Baihui, Shenmen, Neiguan
13	Neurasthenia	Baihui, Neiguan, and others
14	Tinnitus	Baihui, Fengchi, Tinggong, Neiguan
15	Rhinitis	The bilateral sides of Yingxiang
16	Frequent micturition	Guanyuan, Zhongji, Shenmen, Sanyinjiao
17	Chronic prostatitis	Guanyuan, Qixue, Shenshu, Ciliao
18	Diabetes mellituria	Jimen, Zusanli, and others
19	Lumbago	Shenshu, Mingmen, and others
20	Paralysis, cramp, neuralgia of lower limbs, sciatica	Chengshan, Chengfu, Weizhong, Weiyang
21	Paralysis of limbs, ache of shoulder	Quepen, Jianzhongshu, Janjing, Yanyu, Shousanli, Qingling
22	Peripheral nervous paralysis	Xiaguan, Yangbai, Sibai, Jiache, Dicang, Hegu
23	Neurogenic dermatitis	Focus
24	Schizophrenia	Yamen
25	Chronic diarrhea	Tianshu, Zusanli, Pishu
26	Keratitis	Focus, Taiyang, and others
27	Scleritis	Focus, Taiyang
28	Mumps	Focus
29	Central retinitis	Focus, Taiyang, Yanming
30	Tympanitis	Focus

As another example, consider the therapeutic effect of 25 cases of bronchial asthma treated with laser acupuncture. According analysis and standard models, we diagnosed 25 cases of bronchial asthma. Among them were 14 male cases and 11 female cases. The minimum age 19, maximum age 65. 21 cases were the most serious. There were 5 cases complicated by pneumonectasis, 1 case complicated by pulmonary heart disease, and 1 case complicated by pneumothorax. A 6328 A He-Ne laser was used, with output power 25 mW (approx. 125 mW/cm^2). Through an optical fiber, the laser beam was used to irradiate acupoints: "TIANTU", "SHANZHONG", "DINGCHUAN", "FEISHU" (added). Every acupoint was irradiated 3-5 minutes, once every day, for 1 set of 10 times. Through clinical observation of those patients who have been treated by laser acupuncture, their asthma attack ordinarily ceased or relaxed, and did not re-occur. Asthmatic wheezing and moist rale disappeared after 5 irradiations. The total effective rate achieved was approximately 90%. After 1/2 year or 1$\frac{1}{2}$ years' following treatment, the results were as shown in Table 4.

Table 4. Analysis of the therapeutic effect of 25 cases of bronchial asthma after laser acupuncture irradiation.

therapeutic effect	non recur- rence	repeated recurrence after treatment			no follow-up inspection result
		1 year	$\frac{1}{2}$-1 year	$\frac{1}{2}$ year	
reinspection of 25 cases	19	2	2	1	1
%	76%	8%	8%	4%	4%

II.2.ii. Clinical applications of laser acupuncture anesthesia

Methods of laser acupuncture anesthesia:
a) Selection of acupoint: according to the internal organs channels theory of traditional Chinese medicine acupuncture and nervous segment distribution.
b) Irradiation: He-Ne laser, output power 6 mW (31.8 mW/cm^2), through the optical fiber, to irradiate the relevant acupoints. Neck surgery: "FUDE" two sides, "HEGU" (continuous irradiation during operation); Hernia and scrotum surgery: "WEIDAO", "HENGGU", "ZUSANLI" (ill side), ear acupoint "GAOUAN" (continuous irradiation during operation); Caesarean section surgery: "SANYINJIAO", "ZUSANLI", "WAIFEI" (acupoint on ear, continuous irradiation during operation) etc.

In recent years, laser acupuncture anesthesias has been applied in China during surgical operation. The subsidiary hospital of Sansi Medical College in China has utilized various laser acupuncture anesthesias for various surgical operations. About 9 categories of 104 cases were success-fully treated with an excellent success rate of 89.43%. The second People's Hospital in Ganshou has applied laser acupuncture anesthesias in the surgery, obstetric and gyniatrical departments, reporting good results in cases such as thyroid, caesarean section, subtotal gastrectomy and hernia repairing operation. Among the 29 cases, 25 cases experienced excellent results, i.e. 86.2 %. All these were 100% successful, the patient: had no sharp pain, and the results were satisfactory. The clinical units in Chentu, Beijing, Fukian of China used laser acupuncture in extracting teeth; they also obtained satisfactory results.

III. LASER IMMUNIZATION IN MALIGNANT TUMORS

III.1. The Laser Biological Effect and Its Action on Malignant Tumors

III.1.i. In recent years, by exploring numerous laser sources with different wavelengths, we have discovered that certain wave lengths may act as a non specific stimulatory source, by means of which immune performance of organisms might be stimulated, and direct or indirect suppression of malignant tumors may be achieved.

A) The N_2 laser is a di-atom ultraviolet 3371 Å laser. By clinical experience we discovered that it possessed the function of restraining the growth of human liver cancer cells (BEL-7404 series). Laboratory experiments in animals revealed that this type of laser was a non-specific stimulating source. It could excite the immune function of the mice, prolonging thus the lives of the mice and restraining the growth of tumors.

B) In combination with traditional Chinese acupuncture theory, the therapy of applying the N_2 laser beam to irradiate the related acupuncture points of the patients possesses a specific character. Clinical data (1976-1988) show that in these 12 years, we analysed the curative effect of 245 cases of late stage tumors, and we discovered that 72.1% of tumors had noticeably shrunk or regressed.

III.1.ii.   Biological effect on immune liver RNA by laser irradiation: experimental results

A) Immune liver RNA after UV laser (Figure 7) irradiation had an inhibiting rate of 100% for mice liver cancer (in vitro). This occurred as follows: immune liver RNA was introduced into liver cancer cell. When the immune liver RNA assisted in the processing of copy-record and metabolism, it caused the liver cancer cell to lose its growth and proliferation[6,7].

B) As a result of laser treatment, immune liver RNA exhibit RNase antiactivity in mice, and also strengthen the immunity of anti-cancer cells in mice liver, leading to the inhibiting rate attained, approx. 66% in vivo. The above experiments showed that exogenous laser immune liver RNA might pass through not only RNA template action, but also genome regulation of cancer cells, to change the genome performance of cancer, and to cause phenotypical reversion toward a definite direction, rendering liver cancer cells incapable of reproduction (Table 5).

C) The laser immune liver RNA combined with HPD(c) photo-irradiation may obviously improve the action of killing liver cancer cells[7]. Experiments showed: when ascitic type liver cancer cells were cultivated with 3250 Å RNA or 3371 Å RNA in vitro, only 1/50 dose of HPD(c) photo-irradiation was necessary for killing all the liver cancer cells, thus the strength of this synergistic killing effect was 50 times more than that of simple HPD(c). The Figure 8 shows the 125 IudR penetrating inhibition (percentage) as a function of HPD(c) dose, with HeNe laser at 5 minutes 5 mW/cm^2 irradiation.

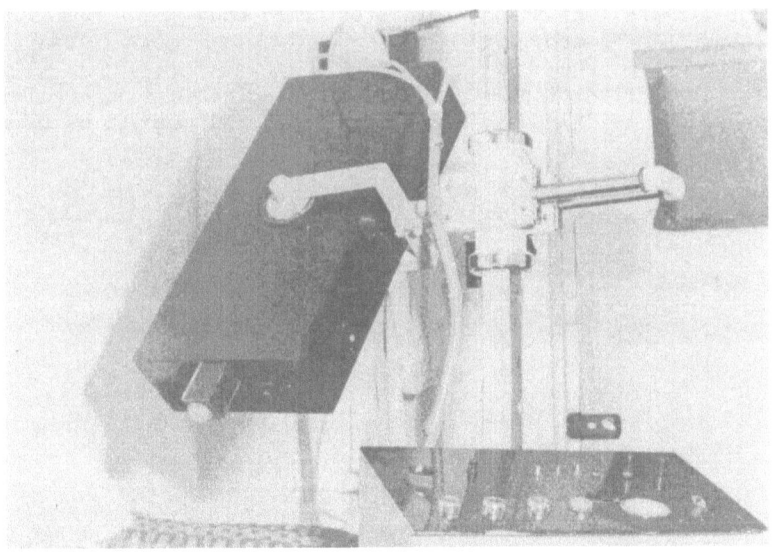

Fig. 7. N_2 Theralaser. 3371 Å wavelength. Output: 2 mJ
Pulse repetition frequency of 3 times/sec.

Table 5. The inhibiting action of immune RNA irradiated by laser, on the growth of ascitic type liver cells of C57DL series.

group	number of mice	number of cells	inhibition rate	condition
contrast	10	4.98×10^8	------	0.9% NaCl/mouse
3250 Å RNA	10	1.70×10^8	66%	0.5 mg
3371 Å RNA	10	1.76×10^8	65%	0.5 mg
4416 Å RNA	10	2.98×10^8	41%	0.5 mg
5145 Å RNA	10	3.60×10^8	28%	0.5 mg
P V S	10	4.56×10^8	8%	0.1 mg
PVS + immune RNA	10	2.83×10^8	44%	PVS 0.15 mg + immune RNA 0.5 mg
immune RNA	10	4.20×10^8	16%	0.5 mg

III.2. Clinical Applications of Laser Immunology [8,9]

III.2.i. Therapeutic process: a nitrogen laser of 3371 Å wavelength, powe: output 2 mJ and pulse repetition frequency of 3 times/sec was used. The selected tumor and the related acupoints were imaged and irradiated by laser through a quartz lens. In selecting acupoints, the following principles were observed:

A) In accordance with the dialectics of traditional Chinese medicine, when selecting acupoints, we should consider the patient's body as an organic whole. For example, the Strong Points of human body are: "ZUSANLI", "TACHUI", and "HEGU", etc.;

B) When following traditional Chinese medicine classics: non-recorded special acupoints as "PIGEN", "AIGEN" and "ZAISHENG" should be chosen;

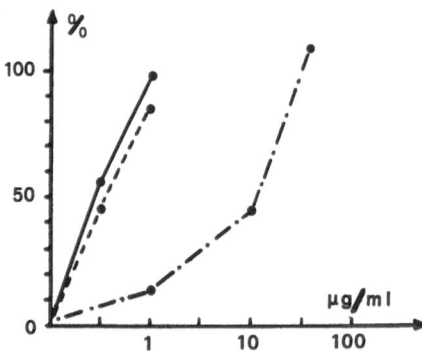

Fig. 8. HPD photosensitive effect on ascitic type liver cancer cells cultured with 3250 Å RNA and 3371 Å RNA. ——— = 3250 Å RNA + cells (cancer); ---- = 3371 Å RNA + cells (cancer); and, -·-·- = cancer cells.

C) According to one Chinese medicine writing on acupuncture theory, choose the acupoint along the nerve system: one should pick "TACHANSHU" in case of intestinal cancer and pick "FEISHU" in case of lung cancer. In addition to irradiation on the tumor, every patient should receive 5 minutes of laser irradiation on 3-5 acupoints, chosen in the abovementioned ways. For those patients whose tumors were located in body regions unreachable by the laser beam, kaser irradiation was only applied to acupoints. Each acupoint was subjected to one irradiation of 10 minutes and each set of therapy consisted of 10 irradiations. The total dose of irradiation received by every patient was no less than 50 Joules.

III.2.ii. Patients selected: from 1976 to 1988 we picked 245 patients with malignancies. All of them after ordinary and ineffective therapy had already abandoned their treatments. Cases such as gastric cancer had colon metastasis, and ovarian cancer had widely spread metastasis in womb, liver, rectum and bladder, etc. All cases were definitely diagnosed by pathologic examinations and research tests.

III.2.iii. Analysis of curative effects: the total improvement rate of 245 patients, with late stage tumors reached 72.1%. Most of them prolonged their life, and in 24 cases significant effects were noted. In following 1-8 years periodic check up on these patients, all were found to be alive and healthy. Further examinations discovered that tumors had noticeably shrunk, regressed or disappeared. Some of them had resumed work. In 85% of cases, they reported that their appetite had improved. In 50% of cases, they reported that they had better spirits and slept well. All appeared to have relief from pain, no longer suffered bloated feelings and had no bad effects.

IV. AUTOFLUORESCENCE AND EARLY CARCINOMA DIAGNOSIS

In recent years, following the continuous development of photomedicine, we utilized laser fluorescence as a detection technology for the diagnosis of disease, with special application to early stage tumor diagnosis. According to the spectra of autofluorescence of normal organ and tumor tissue, through the different intensity of fluorescence, we can discriminate the characteristics of lesions. This diagnostic procedure is simple and fast, and it is easy to use for the patient.

Now numerous reports[10,11] show that the N_2 laser, Xe^+ laser, etc. may be used as an exciting source; target area produces an autofluorescence spectrum after irradiation by the laser. This may be used in early stage diagnosis of cervical carcinoma, gastric cancer, etc. Its precision in diagnosis is superior than that of routine inspection (Figure 9).

IV.1. N₂ laser Used for Early Diagnosis of Cervical Carcinoma, Precancerous Lesions and Other Carcinomas

IV.1.i. The experiment: a 3371 Å, pulsed, 0÷50 pulses/sec repetition rate, tunable N_2 laser was used to irradiate the tissue of the cervix. The fluorescence spectrum was used to analyse the characteristics of an absorption peak of 4700 Å in normal tissue and a special characteristic tumor absorption peak of 4200 Å in cancerous tissue. By measuring this special peak, we could diagnose early carcinoma of the cervix and precancerous lesions.

IV.1.ii. Clinical applications: 112 cases of severe cervical erosion, doubtful or positive patients after Papanicolaous smear were examined, using a nitrogen laser to stimulate intrinsic fluorescence, for the purpose of detecting early cervical cancer and precancerous lesions. When compared with pathologic diagnosis, this method has a precision rate of

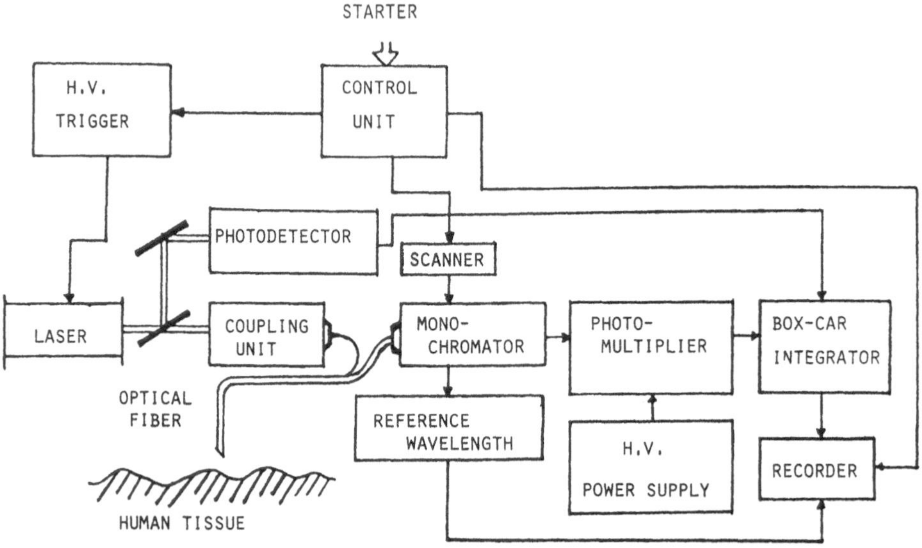

Fig. 9. Diagnostic instrument to analyze N_2 or Xe^+ laser fluorescence
spectrum.

100% for cervical cancer and 90.6 % for atypical cervical proliferation.
This method is safe, fast, and non-painful. It has shown no undesirable
side effects, and is accepted by patients. A certain scholar also used the
N_2 laser to stimulate the autofluorescence of an oral malignant tumor,
and discovered its intrinsic special characteristic peak of 565.1 nm.

IV.2. Xe^+ Laser Used To Diagnose Gastric Cancer

IV.2.i. Autofluorescence: the source of gastric cancer's self-induced particular characteristic peak of fluorescence may possibly be characteristic
porphyrin compound produced internally. Since cancer may accumulate such a
compound, we select a laser of maximum intensity which has its wavelength
near the porphyrin compound acting as the stimulating source. Thus it will
cause porphyrin to produce a stronger fluorescence effect. The inspection
of the particular characteristic fluorescence of porphyrin would be helpful
for diagnosing malignant tumors. A Xe^+ laser with wave length 365 nm,
width of pulse 0.5 µs, single pulse energy 400 µJ, and a repetition rate of
$0 \div 10$ Hz was used. Through an endoscope and optical fiber, the Xe^+ laser
was used for irradiating various internal gastric tissues; it produced the
particular characteristic autofluorescence spectral peaks of 630 nm and
690 nm.

IV.2.ii. Clinical applications: a gastrofiberscopic Xe^+ laser induced
autofluorescence spectra in various gastric diseases for testing 104 patients. These patients were diagnosed by endoscopy and pathologic biopsy,
including 28 gastric cancers (4 cardia cancers, 7 corpus ventricule cancers, 17 antrum cancers), 9 gastric ulcers, 46 chronic atrophic gastritis,
21 superficial gastritis. A precision rate of 80% was obtained when a
particular characteristic peak was compared with the pathologic diagnosis.
The results of this method show that under the endoscope, the Xe^+ laser
stimulated gastric mucosa to form autofluorescence spectra of a particular
type which characterizes gastric cancer. This method is fast, convenient,
safe and harmless. Compared with the HPD laser, it has no photosensitive
auxiliary function.

V. PROSPECTS OF LOW POWER LASERS FOR MEDICAL APPLICATIONS

In this chapter, we have introduced Chinese conditions for medical applications. We believe that China has a leading position in the field of clinical low power laser therapy. To open up broad prospects for applications, according to actual Chinese conditions (although we have done much in wound healing, PDT, tissue welding, biostimulation and therapeutic diagnosis), we consider laser acupuncture as the most significant developing direction. This is because China has traditional Chinese medical acupuncture theory and actual clinical experience, which were accumulated during a very long period. This is quite different from other countries. Based on this particular ancient theory in combination with laser techniques, we have obtained great success in clinics. Therefore we wish to research more about the relationship between ordinary acupuncture and that of laser treatment: why should the laser beam produce biostimulation according to traditional Chinese medical acupoints on the human body? The improvement and development of therapeutical techniques for laser acupuncture will be one part of the developing prospect for low power laser applications in China. At the same time, we ought to concern ourselves with low power laser immune techniques applied in tumor therapy, laser techniques used in early stage tumor diagnosis, etc. We consider it important to continue to develop low power laser applications in medicine.

REFERENCES

1. Fu-Shou Yang, "New frontiers in laser medicine", Published by the Anhui Institute of Optics & Fine Mechanics, Academia Sinica, Anhui, China (1987)
2. Zen-Dao Dien, "First report of stimulation of He-Ne laser in human occult conduction channels", ShanDong Medical College Bulletin, 1:73 (1979)
3. Wei-Leng Dang, "Observation of the effect of laser acupoint irradiation on gall bladder function by ultrasonic display", Shanghai Acupuncture Journal, 2 (1984)
4. Zan-Loh Wang, "The acupuncture function of He-Ne laser", Acupuncture Research, China, 11 (1983)
5. Ko-Ziang Zi, "Laser medicine", GanDong Education Press, China (1987)
6. Fu-Shou Yang, "Preliminary observation and approach of the laser biological effect of the immune liver RNA in the action of malignant tumor", in: "Laser Optoelectronics in Medicine", Wilhelm and Raphaela Waidelich, eds., Springer-Verlag (1988)
7. Fu-Shou Yang and Xi-Bing Wei, "The effect of laser irradiation on human liver carcinoma cells" (BEL-7404 series), Chinese Journal of Digestion, Vol.1, (2) (1981)
8. Fu-Shou Yang, Medical Laser Research & Clinic, 1:28 (1987)
9. Son-Ling Shu, "Laser Technic and Medical Application", The People Health Press, China (1989)
10. Ai-Hue Ding, "N_2 laser stimulating intrinsic fluorescence method used in diagnosis of precancerous lesions and early carcinoma of cervix", The 2nd Congress of Asianpacific Association for Laser Medicine & Surgery, Shanghai, China, Sept. 20-22 (1988)
11. Shu-Dong Xiao, "Autofluorescence excited with endoscopic Xe^+ laser in the diagnosis of gastric cancer", The 2nd Congress of Asianpacific Association for Laser Medicine & Surgery, Shanghai, China, Sept. 20-22 (1988).

THE USE OF LOW LEVEL LASER BIOSTIMULATION FOR THE TREATMENT

OF CHRONIC PAIN SYNDROMES

K. C. Moore

Laser Therapy Clinic
The Royal Oldham Hospital
Oldham, U.K.

I. INTRODUCTION

High powered lasers such as CO_2 and YAG which are used in surgery exert their effect by generating heat in the body tissues. At temperatures between 40-100 °C there is a progressive cellular change ranging through protein denaturation and coagulation to cellular disruption and destruction. Above 100 °C tissue vapourisation occurs and at even higher temperatures there is tissue carbonisation and burn off.

The low powered lasers do not have this photodestructive effect, temperature changes in irradiated tissue being less than 1°C. This non-thermal application of laser energy is referred to as photobioactivation and the laser systems used for therapy have variously been described as low power, low energy, low intensity, mid-laser, soft, cool and even cold lasers. However, it is the tissue effect that is important. Whilst the surgical lasers exert a high level of tissue reaction which is macroscopically evident, the therapeutic lasers do not produce a macroscopic tissue change and hence are deemed to cause a low level of tissue reaction. Their use in medicine will therefore be referred to as Low Level Laser Therapy (LLLT).

In 1969 the first report of the potentially beneficial effect of low powered lasers in wound healing was made by A. Mester of Budapest[1]. This initial finding has been supported by other researchers[2,3]. During the intervening years clinicians from a variety of medical disciplines have reported their use of lasers in plastic and reconstructive surgery[4], rheumatology[5,6], a wide variety of painful muscolo-skeletal conditions[7], and in post herpetic neuralgia[8]. The use of LLLT for pain management has been extensively reviewed in Reference 9. C. Shiroto and his colleagues have also reported their results in more than 3000 patients treated with LLLT for chronic pain[10].

Although the precise mechanism of the action of LLLT has yet to be fully explained, there is a wealth of clinical information supporting its application in medical practice. Current opinion suggests that there are two types of response to photobiostimulation. The first or primary response takes place in the irradiated tissues at cellular level and consists of increased microvasculature, improved lymphatic drainage, increased macrophage activity and increased fibroblast activity with improved collagen synthesis, all of which help to promote wound healing. The secondary response involves the release into the circulation of a number of chemical substances as a result of laser irradiation which produce a systemic effect. These chemicals include prostaglandins, enkephalins and endorphins which are most probably involved together with the local healing effect in the mechanism of pain relief.

The early low powered systems were based on the Helium-Neon (HeNe) laser with very low power outputs of the range 1-30 mW at a wavelength of 633 nm in the visible red spectrum. In 1980 the development of infra-red (IR) diode lasers greatly extended the therapeutic potential of LLLT. The laser medium for these semi-conductor diodes is either gallium arsenide (GaAs) or gallium aluminium arsenide (GaAlAs). The wavelength of this laser light is largely dependent on the strength of the electric current applied to the diode and may vary from 780 nm to 904 nm in the invisible near infra-red. Power outputs are usually of the range 10-60 mW. Most of the new systems are either pure diodes or a combination of HeNe and IR diode[11].

Many of these IR diode systems are designed to produce pulsed laser energy at a wide range of frequencies. Theory suggests that just as variations in wavelength and power output have differing tissue effects so do variations in pulse width and repetition rates. Thus for a more intense local effect short pulses (nanoseconds) with high repetition rates (1-5 kHz are utilised whilst for more widespread tissue effect longer pulses (milliseconds) and lower frequencies (less than 500 Hz) are employed. These variations in wavelength, power and output mode make it extremely difficult to accurately reproduce the same treatment parameters and irradiated tissue effect when using different diode laser systems and virtually impossible to monitor the power density delivered to target tissues during therapy.

II. CHRONIC PAIN

Because of subjective influences pain has always been difficult to quantify. Acute pain usually has an identifiable origin and is of limited duration with an expected outcome. By definition chronic pain implies a long standing condition present for months, years or even decades. Patients will usually have tried a wide variety of different therapies and have consulted a number of medical specialists. Eventually intractable pain that is unremitting and unrelieved will produce some degree of emotional overlay and a varying disturbance of affect. These subjective changes usually result in patients adopting a negative attitude to their pain which only serves to make treatment more difficult. Any therapeutic LLLT regime prescribed for patients with chronic pain must take account of these factors.

This chapter reviews more than 1000 treatments by LLLT of patients suffering from chronic pain. The classification or grouping of chronic pain patients into different categories tends to be personal to the Therapist and is only of real benefit in helping to identify more accurately those patients for whom a specific treatment regime is indicated. Thus different parameters such as aetiology, anatomical and systemic distribution or pain characteristics may be utilised. Classification by aetiology factors gives 4 main groupings:

(1) Post Traumatic Pain (PTP) - 23% Chronic pain subsequent to an identifiable incident such as road traffic accident, industrial injury, accidental fall or sports injury.
(2) Post Surgical Pain (PSP) - 17% Chronic pain following one or more operative procedures including orthopaedic, gynaecological, neurosurgical, thoracic and general surgery.

(3) Musculo-Skeletal Pain (MSP) - 20% Chronic pain involving the locomotor system but with no previous history of trauma or surgery. Mainly
associated with degenerative bone or joint disease.

(4) Post Herpetic Neuralgia (PHN) - 36% Neuralgic pain of greater than 6
months duration and showing little or no response to conventional
therapy.

These groupings can be further subdivided into their anatomical and
systemic distributions.:

Post Traumatic Pain

Cervical and upper thoracic spine	20%
Lumbo-sacral spine	40%
Upper limbs	12%
Lower limbs	28%

Post Surgical Pain

Abdominal surgery	38%
Neck and shoulder girdle surgery	22%
Lumbar spine surgery	22%
Upper and lower limb surgery	18%

Musculo-Skeletal Pain

Neck and shoulder girdle	42%
Lumbo-sacral	40%
Lower limb	18%

Post Herpetic Neuralgia

Trunk T3-L3	62.5%
Cephalic	22.5%
Limbs	15%

III. TREATMENT PROTOCOL

The laser used for treatment in all cases was a prototype Japanese
GaAlAs diode laser system (Japan Medical Laser Laboratory) producing a continuous wave output of 60 mW at a wavelength of 830 nm. There is evidence
that there is a "tissue window" between 820-840 nm at which wavelength
water absorption of GaAlAs laser light is at a minimum and hence maximum
penetration of tissues by the laser beam is achieved. All treatments were
delivered in the contact mode of application generating a power density at
point of tissue contact of 3 W/cm^2.

Initial therapy sessions were carried out on a twice weekly basis until a recognisable and sustained response was achieved. Thereafter LLLT
was continued weekly until either a satisfactory level of pain relief was
obtained or no further improvement was evident. The method of laser application was varied according to the origin and distribution of the pain.
Body surface areas of pain were treated using a 2 cm^2 grid method of application, the centre of each square being irradiated for 8-10 sec giving an
energy density of 24-30 J/cm^2 to each point. For spinal column and joint
pain trigger points were elicited by deep palpation and each joint was
treated for 16-20 sec (energy density 48-60 J/cm^2).

In the limbs, in addition to the above the neurovascular and lymphatic
pathways were also treated. Each therapy session lasted 20-30 min depending on the extent of the area to be treated. A record was kept of the
number of therapy sessions and the total treatment time for all patients.

Table 1. Treatment for Chronic Pain Syndromes

Analgesics	MUA
NSAIDs	Operation
Physiotherapy	TNS
Trigger Point	Osteopathy
Injection	Acupuncture
Epidural Steroids	Hypnotherapy

IV. PAIN ASSESSMENT

Patients were initially assessed for the severity and extent of their pain and for the effect it had on their daily living. Pain intensity was measured by patient self assessment on a linear analogue scale of 0-10. Where appropriate pain distribution was mapped out over the affected area and measured by body surface perimetry using an universal goniometer. Progress was monitored at regular intervals throughout therapy. Patients were periodically requested to complete a Quality of Life Questionnaire giving details of the level of their activity at work, in the home and socially; details of any sleep disturbance; and reviewing any concurrent medication. It was also found to be helpful to involve patients in an objective assessment of their pain by asking them to keep a daily diary recording the extent and duration of their pain. In this way small yet significant changes can be identified and patients may be more easily persuaded of a slow but gradual improvement in their condition. Similar assessments were carried out at follow up review every 6 weeks after the completion of therapy.

V. RESULTS OF THERAPY

All patients were taken from routine Pain Clinic referrals and had previously been treated unsuccessfully with most of the currently available therapies which are summarised in Table 1.

In view of the failure of previous therapy to achieve a reasonable degree of pain relief it could be argued that any improvement gained with LLLT was of significance. However, the placebo effect of a new therapy must be considered and in order to accommodate this factor a minimum improvement in symptoms of 20% was considered to be necessary before LLLT could be deemed responsible for patients' pain relief.

The PTP, PSP and MSP groups showed very similar characteristics in terms of age and duration of symptoms. The mean age for these 3 groups was 47 years (Range 19-69) and the mean duration of symptoms was 4 years (Range 4 months - 16 years) (Table 2). The age range for patients suffering from PHN was considerably higher (55-85) with a mean age of 71 years, 50% of patients being between 65-75 years of age. 45% of patients had suffered the pain of PHN for less than 1 year; 22% for between 1-2 years and 33% for more than 2 years. The duration of symptoms varied between 6 months and 11 years (mean 2.5 years).

The results of therapy for the PTP, PSP and MSP groups are presented in Table 3. The number of treatment sessions required varied greatly between 5-20. Similarly the response to LLLT showed an equally wide variation with some patients showing little or no response to therapy whilst others approached 100% pain relief. In general, therapy was discontinued after 6 sessions if the measured improvement was still below 20% i.e., placebo level. A detailed review of these cases has suggested that a more rigorous

Table 2. Chronic Pain Syndromes: PTP PSP MSP (all values are mean)

Site	Age (Yrs)	Duration (Yrs)	Treatments
Neck/shoulder	49	7.5	8.6
Low back	43	4.5	7.5
Upper limb	43	2	6.3
Lower limb	51	2.5	6
Abdomen	49	3.3	6.7

approach to patient selection could improve the prognosis. In particular, the exclusion from therapy of patients showing radiological evidence of a marked degree of bone deformity, destruction or disease will greatly increase the expected benefit of LLLT.

Some 40% of patients in these groups showed gross skeletal changes on X-ray. Co-incidentally these some patients showed little or no response to LLLT, the mean improvement after therapy being only 16%. The remaining 60% of patients had little or no radiological skeletal abnormality and these patients demonstrated a much greater response to LLLT achieving a mean improvement in symptoms of 70%.

Of the PHN sufferers, 15% showed little response to LLLT (mean improvement 20%). All these patients had more than 1 years history of pain, all were affected in the thoracic dermatomes, and all exhibited a major degree of cutaneous hyperaesthesia. This latter manifestation may be partly responsible for the poor response to LLLT. The remaining 85% of patients showed a better than 50% response to therapy. Overall the PHN group demonstrated a mean improvement of 71% (Range 10-95) with a mean number of treatment sessions of 8.6 (Range 5-22) and a mean total treatment time of 120 min (Range 50-231 min).

There was no appreciable difference in either the number of treatment sessions or the total treatment time when comparing results for groups with varying durations of PHN symptoms (Table 4). Long standing unrelieved PHN should, therefore, not be a contraindication for LLLT.

Table 3. Chronic Pain syndromes: PTP PSP MSP (all values are mean)

Site	X-ray	% Improvement
Neck/Shoulder	+	18
	−	67
Low back	+	20
	−	72
Upper limb	+	20
	−	67
Lower limb	+	13
	−	71
Abdomen	+	13
	−	72

Table 4. LLLT for PHN: Results (all values are mean)

Duration of PHN	Sessions (n)	Time (min)
Up to 1 year	8	114
1-2 years	8.5	109
More than 2 years	9	132

VI. FOLLOW UP

All patients who showed a greater than 20% improvement were followed up at 6 weekly intervals. The follow up period currently extends from 4 months (100%) to 1 year (42%). 8% of patients have died during follow up of unrelated causes. Pain relief achieved with LLLT has been maintained in 35% of patients since the end of therapy. In 42% there has been a continued improvement in monitored parameters during follow up of between 5-20%. In the remaining 22.5% there has been a recurrence of pain which has occurred at any time up to 9 months after the completion of treatment. In these cases a repeated course of LLLT has again provided good pain relief.

VII. DISCUSSION

There is still much to learn about the application of LLLT for chronic pain. Careful assessment of patients prior to therapy is obviously of importance if an accurate prognosis is to be given. Clinicians should pay particular attention to radiological findings as evidence of gross skeletal damage or disease makes the successful use of LLLT for associated chronic pain unlikely. Where pain is thought to arise mainly in soft tissues the prognosis is much more favourable. Nevertheless, the response of patients to LLLT can vary greatly. In this series the range of treatment sessions varied between 5-22. 75% of patients received a course of 10 treatment or less, whilst 25% required more than 10 sessions. Similarly the range of symptomatic improvement varied between 10%, which is merely a placebo response, to as high as 100%.

Detailed follow up of patients undergoing a new type of therapy is vital and frequently provides as much information as is gained by assessment during treatment. More than 40% of patients in this series showed a continuing improvement during follow up which suggests an ongoing effect of LLLT. 35% maintained their improvement in pain relief. However, 22.5% required further therapy because of a significant recurrence of pain. It is possible that in these cases LLLT had been discontinued at too early a stage. Further work is being undertaken to try to clarify these problems.

More than a century ago Peter Mere Latham (1789-1875) said "It would be a great thing to understand pain in all its meanings". When considering chronic pain this statement still rings true. In spite of a wide variety of available therapies a significant number of patients with chronic pain gain little or no relief. The introduction of LLLT would appear to give these sufferers a real chance of gaining pain relief. In addition current experience suggests that LLLT is not associated with any adverse reactions or troublesome side effects. What is now required from the experienced therapists are guidelines for therapy which will help establish LLLT as a proven therapeutic modality.

REFERENCES

1. A. Mester, "Experimentation on the interaction between infra-red laser and wound healing", Z. Exp. Chirurgie, 2:94 (1969)

2. M. Dyson and S. Young, "Effects of laser therapy on wound contraction" In G. Galleti (Ed) "Laser" Monduzzi Editore, Bologna, p. 215 (1985)

3. A. F. Mester and A. Mester, "Wound healing", Laser Therapy, 1:7-15 (1989)

4. T. Ohshiro, "Objective evaluation of the diode laser in the revitalisation of failing grafts and flaps", In: T. Ohshiro and R. G. Cal-Calderhead "Low Level Laser Therapy: A Practical Introduction", John Wiley & Sons, Chichester, pp. 86-91 (1988)

5. J. B. Walker, L. K. Akhanjee, M. M. Cooney, J. Goldstein, S. Tamayoshi and F. Segal-Gidan, "Laser therapy for pain of rheumatoid arthriritis", Clin. J. Pain, 3:54-9 (1987)

6. N. Palmgren, G. F. Jensen, K. Kaae, M. Windelin and H. C. Colov, "Low power laser therapy in rheumatoid arthritis", Lasers Med. Sci. 3:193-6 (1989)

7. M. Mayordomo, "Laser in painful processes of locomotor system: our experiences". In: G. Galleti (Ed.) "Laser" Monduzzi Editore, Bologna, p. 349 (1985)

8. K. C. Moore, N. Hira, P. S. Kumar, C. S. Jayakumar and T. Ohshiro, "A double blind crossover trial of low level laser therapy in the treatment of post herpetic neuralgia", Laser Therapy, (Pilot Issue): 7-9 (1988)

9. J. B. Walker, "Low level laser therapy for pain management: A review of the literature and underlying mechanisms", In: T. Ohshiro and R. G. Calderhead "Low Level Laser Therapy: A Practical Introduction", John Wiley & Sons, Chichester, pp. 43-56 (1988)

10. C. Shiroto, K. Ono and T. Ohshiro, "Retrospective study of diode laser therapy for pain attenuation in 3635 patients: detailed analysis by questionnaire", Laser Therapy, 1 :41-7 (1989)

11. K. C. Moore, "Low powered lasers for therapy", In: "Association of NHS Supplies Officers: Members Reference Book and Buyers Guide 1990", Sterling Publications Ltd., London, pp. 219-22 (1989)

PHOTODYNAMIC THERAPY: MECHANISMS AND DOSIMETRY

PHOTODYNAMIC THERAPY

J. S. Nelson

Beckman Laser Institute and Medical Clinic
University of California
Irvine, California, USA

The attack on cancer with drugs is based upon the thesis that it
should be possible to discriminate against cancer cells while having only
few or tolerable effects on normal cell populations. While many compounds
have been screened for such activity over the past forty years, unfortunately most solid cancers respond either not at all, or to a limited extent
only, to these selective agents.

Photodynamic therapy (PDT) by exposure of certain dyes to visible
light has been studied since the beginning of this century. The basic concept for the use of PDT in the treatment of malignant tumors is that certain molecules (natural or applied) can function as photosensitizers. The
presence of these photosensitizers in certain tumor cells makes the latter
vulnerable to light at wavelengths absorbed by the chromophore. The action
of photosensitizers is generally to absorb photons of the appropriate wavelength and intensity sufficient to elevate the sensitizer to an excited
state. The excited photosensitizer subsequently reacts (transfers its
energy) with a molecular substrate, such as oxygen, to produce highly reactive singlet oxygen which causes irreversible oxidation of some essential
cellular component. Uncertainty arises as to the exact targets of these
excited intermediates responsible for cell death although damage to the
cell membrane, mitochondria, lysosomes, microsomes and the nuclear material
have all been reported.

While numerous compounds have been tested as selective photosensitizers of malignant cells, considerable interest in the porphyrins was stimulated by early reports on inherent porphyrin fluorescence in large malignant tumors. While several porphyrins have been studied, hematoporphyrin
derivative (HpD) has received the most attention. Although the mechanism
of HpD's preferential localization and retention in malignant cells remains
uncertain, it is well established that the total time HpD is retained in
malignant tissue is much longer than in nonmalignant tissue from which it
is generally cleared in 24-72 hours. As a result, there is a "window" of
time wherein one can exploit the differences in HpD concentration to
achieve selective photodegradation of malignant tissue.

Clinically, PDT is carried out by a two-step procedure, HpD is first
administered intravenously as a 2-5 mg/kg bolus. After a delay of 24-72
hours (to allow for the accumulation of HpD in the tumor and for the clear-
ing from most normal tissues), the tumor is irradiated with visible red

light tuned to 630 nm. The radiation is generally obtained from an argon-pumped dye laser although more conventional light sources have been shown effective. The light may be delivered to the surface of the tumor, interstitially via optical fibers or endoscopically to deep tumors of the digestive, pulmonary, or urogenital tract. Topical application of HpD has also been proposed but only preliminary results have been reported. Shortly after PDT, the tumor becomes necrotic (usually within 24 hours) and when effectively treated, forms a nonpalpable scab, sloughed off within a few days. Histologically, the earliest changes occur in and around the tumor vasculature. Apparent internal hemorrhage with red blood cell extravasation is a common finding after PDT, not only in most experimental animal tumors but in tumors in patients as well. Furthermore, another study suggests that the effects of PDT are not the result of direct tumor cell kill but are secondary to destruction of the tumor microvasculature. Binding o photosensitizers to collagen and other fibers in the subendothelial zone o the tumor vessel wall, in combination with altered permeability and transport through the endothelial cell layer resulting from erythrocyte swellin and increased intraluminal pressure, may be the key features of the dye-sensitized photodynamic reaction leading to tumor destruction.

HpD-PDT has been shown effective in causing photodegradation of tumor tissue in experimental animal systems since 1972 and in clinical trials since 1976. A wide variety of tumors of varying histologic types have bee treated, including cancers of the skin, female genital tract, esophagus, lung, bladder, eye, breast, head and neck squamous cell carcinomas. The overall positive response rate as reported in the literature is greater than 70%. The high therapeutic ratio and relative lack of morbidity have made this a very attractive form of therapy. Treatment parameters have been refined such that therapy can be undertaken with a reasonable expecta tion of good results. While PDT can be used to eradicate relatively large tumors, it appears to be especially advantageous to patients with early disease or early recurrence. In some cases this therapy may be a viable alternative to debilitating surgery, and in others, the treatment of choice. In addition, previous surgery, radiation therapy, or chemotherapy do not preclude the use of PDT and many of the clinical studied reported t date have been on patients who have previously failed some, or all other, available therapies.

The fluorescent properties of HpD can also be used in the detection and localization of tumors not detected by more conventional techniques. Studies have shown that HpD fluorescence induced by the blue/violet 405 nm light of the krypton laser can be used successfully to detect occult lung tumors and delineate dysplasia and tumors in the bladder. In these clinical trials, the exciting light was delivered via a single quartz fiber in the biopsy channel of a bronchoscope or cystoscope fitted with appropriate filters and an image intensifier to observe the fluorescing (red) light which can be directly observed and recorded. Normal mucosa, and tumors in patients not receiving HpD, were seen not to be fluorescent.

Despite HpD's broad experimental application in clinical oncology, efforts have been hampered by the lack of a complete understanding of what active component in the HpD molecule is responsible for tumor uptake, retention, fluorescence and photosensitization. Even with apparently pure preparations of the individual HpD components, impurities have often complicated the interpretation of data. The active component has been described as a structural isomer of dihematoporphyrin ether or ester. Furthermore, the component responsible for photochemistry is not necessarily the same as the component responsible for fluorescence. Even though there is an increased tumor: neighboring tissue porphyrin content ratio followinc HpD administration, the amount retained by normal tissues such as skin, liver, spleen and kidney is clinically significant. The major drawback of this therapy is the potential for drug-induced sensitivity to sunlight. This effect is not trivial and may result in complications ranging from slight erythema and edema to extensive skin sloughing and necrosis. The foregoing problems as well as a relatively weak porphyrin absorption band and low tissue transparency at 630 nm, have resulted in considerable effort being devoted to developing new and more effective tumor localizing photosensitizers for PDT.

There is a recent report on the use of a chlorin compound, mono-L-aspartyl chlorin, as a photosensitizer for selective tumor necrosis. The chlorins are known to have strong absorption bands with high molar extinction coefficients at wavelengths longer that 650 nm thus providing an advantage over the lower tissue penetrance of 630 nm light used for HpD. Chlorin in combination with light at 664 nm was shown to be an effective tumor localizer and photosensitizer in Balb-C mice inoculated with EMT-6 tumor. Furthermore, it was suggested that this compound may not be retained in high skin concentrations thus minimizing the photosensitivity associated with HpD.

The phthalocyanines, used as industrial dyes and pigments, are porphyrin-like compounds capable of localizing and photosensitizing malignant tumors. These compounds are easily synthesized and purified and exhibit strong absorption in the 650-700 nm range. The metal atom complexed with the phthalocyanine ring is critical and studies have demonstrated that aluminum phtalocyanine (AlPc) is the most active photosensitizer. AlPc contains a mixture of isomers with varying degrees of sulfonation and the relationship between degree of sulfonation and tumoricidal activity is still under study. AlPc has been shown active "in vitro" and "in vivo" and, in at least one study, not to induce skin damage in the presence of ambient light.

While the studies on chlorins and phthalocyanines appear promising, they are still at an early stage. Neither of these compounds has been tested in humans and HpD remains the clinical standard for comparison. The greater tissue penetrance of the longer wavelength and the reduced residual skin photosensitivity give these compounds decided advantages over HpD. However, unanswered questions concerning these compounds include delineation of light and drug dosimetry parameters, mechanism of tumor uptake and retention, as well as possible uptake in other organ systems. It is hoped that future investigations will address these critical questions so that the role of other potential photosensitizing compounds in the management of cancer can be fully defined.

EXOGENOUS CHROMOPHORES FOR LASER PHOTOSURGERY (PDT)

L. Goldman

Dermatology Department
Naval Hospital
San Diego, California, USA

I. INTRODUCTION

After extensive research[1,2], over 10 years, and the treatment of thousands of patients, laser PDT surgery is here and for the future, The National Cancer Institute, USA, declares, with reason, that today, these are the priority cancers for PDT, primarily endobronchial cancer, early gastric cancer, and bladder cancer. This formal recommendation has not interfered with continued research and PDT for many other clinical types of cancer.

Also, as indicated previously, PDT has been used in many fields of biology and for many types of organisms[3+10]. Thus, there is a great future for the technology of exogenous chromophores and laser photosurgery. With renewed commercial interest and active financial support, unlike for DHE, there will be further extension of the use of well controlled needed exogenous chromophores, suitable for PDT, suitable for other forms of non-ionizing when needed, suitable economically, both for the chromophores and the specific laser system required. All this will make for greater use, in the whole oncology program.

The basic factors and the fields to which many questions continue to be addressed in the PDT program are:

1. What chromophores are still needed and exactly how are they to be used?
2. What special lasers are still needed and are they economical in use and service?
3. What special delivery systems are needed besides those developed by Doiron?
4. What is the real value of induced hyperthermia associated with PDT?
5. What is the value of the new microendoscopes in the PDT program?
6. Can you truly develop effective PDT deep in tissue?
7. Do you need these accessory instruments for PDT: a) instruments similar to those of Potter for detection of the chromophore; b) thermal sensors in the operative tissue field; c) a scanning instrument similar to the Laser Scanning Ophthalmoscope for scanning the operative field after PDT?
8. Do you need suitable vehicles for topical PDT when indicated?

There will be many additions to this list as PDT progresses. In this volume, Doiron and Potter, with their long involvement with PDT and as international authorities, present their current programs. This chapter is concerned with my research and my interests and as many will say, my dreams.

In our location, we are interested in some exogenous chromophores developed at the Navy's Ocean Systems Center by Pavlopoulos. These initial dyes were limited to 530-575 nm range and could be energized by a quartz lamp[6]. Tests were done on the standard murine breast cancer. Our present interests are with the flash pumped dye lasers, 577 and 585 nm, and Nd YAG CW and Q switched and the second harmonics, 532 nm, PDT energized with adequate quartz lamps, with appropriate filters, would decrease the expense of PDT in the cancer treatment program.

New exogenous chromophores have to be studied for standardization of the compound, unlike the early days of HpD. Also, perhaps, each commercial batch should have biological testing, for example, with one cell line often suggested, the murine fibrosarcoma cells (RiFi) with the identical PDT with HeNe. More details are needed for the specifications of the quartz lamps and their fiber optic or rod transmission.

II. LASER SYSTEMS REQUIRED

What laser system are required as related to specific exogenous chromophores? In the past, we preferred gold vapor (GVL) for the DHE therapy. As indicated, our interest is to try to use the flash pumped dye laser. With improvements in flash lamps, coaxial, and multiple prism technique, for narrow line width, for precision, adequate protection glasses also with magnification instrumentation on the lesion, the next concern relates to the dye chemistry, scientific or alchemic? More stable mixtures are now available with greater numbers of impacts. We are still waiting for the single small tablet to be dropped into the vat. Other lasers are the argon pumped, copper vapor (CVL), gold vapor (GVL), excimers, nitrogen, Nd YAG. The krypton is used mainly for diagnostics in PDT.

The main developments for the delivery systems in PDT have been done by Doiron. We are interested in the microendoscopes; these microendoscope can serve as light guides themselves or as carriers for fiber optics. With observation, they can also penetrate into tissue to open passages to cancer foci for PDT. The direction of the penetration in their search for the cancer focus is the great challenge to optics today. This will be reviewed later in the review of photon transmission through turbid media. The quartz lamp and its delivery systems are of interest now for PDT. These lamps also may relate to the microendoscopes.

With the known hyperthermia treatments for cancer, and the suspicion of thermal energy for some of the laser systems used for PDT, especially the Nd YAG, it is not surprising that hyperthermia was considered as adjunct therapy for PDT. Dougherty, with his work with veterinary oncologists, initiated this program. Hyperthermia for cancer may be used extensively as for cancer of the breast or locally for basal or squamous cancer of the skin for cancers in the lining of the various cavities of the body. For general hyperthermia, there are the extensive RF instrumentation and the older models of the special insulated IR body boxes as the Kettering Hypertherm. Also, hot baths and hot electric blankets have been used and microwaves. In experiments in rabbits, we used iron particles, sterile colloidal iron, and Interferon B with iron Dextron, all injected deep in tissue. These areas were exposed to microwaves, then to ruby laser impacts for double impacts for hyperthermia. A basal cell carcinoma in a patient was injected with sterile colloidal iron particles (Ferrofluids), then treated with ruby laser. The non-pigmented skin cancer cleared completely and showed no recurrence after years. Later, during our PDT clin-inical experiments, two patients with recurrent basal cancers of the face were treated with local injection of DHE in Azone[11,12] followed by GVL, after the PDT, local IR with quartz rods and Teflon tips was used over the laser treated area with thermal sensors fixed in the treated tissue. Uni-form tissue hyperthermia was found. No recurrences were noted, but no follow-up biopsies were taken again.

So, for local tissue induced hyperthermia, the following can be used: 1) hot water in test tubes or bottles; 2) IR rods ; 3) RF veterinary probe for local tissue hyperthermia; 4) microwaves ; and 5) ultrasonics. Excel-lent detailed studies for cancer hyperthermia with PDT including Doppler studies, since vascular dynamics in cancer tissue are important, are re-ported in References 8 and 9.

Dihematoporhyrin ether (or ester) is assumed to be a photocatalyst, energized by the laser and so emitting heat, light perhaps or transferring energy to form singlet oxygen which causes necrosis of the cancer cell. Through the vascular system of the cancer, DHE reaches its target in the cancer cell.

III. INVESTIGATIVE STUDIES WITH PDT

So, for energizing this DHE photocatalyst, the laser photons must reach this cell. How can the photons reach this target if the cancer is deep in tissue? We are very anxious to do this especially if the primary cancer is early or if the deep metastasis is also an early lesion. This our great challenge for PDT today! The early diagnostics are just as im-portant as the early treatment. In our current investigative studies, we are considering a number of ideas in repeated conferences with those whom we can expertise, the Naval Ocean Systems Center and the Optical Sciences Center of the University of Arizona in Tucson, Arizona. For baseline studies, we have studied laser transillumination of the skin, mucous membranes, body cavities with incoherent light, HeNe, krypton, and low output Nd YAG, including Q switched. The complicated alternatives, also as controls, include special probes with both visualization fiber optics and multiple laser photon transmission fiber optics. These provide both diagnostics and treatment. The diagnostics are for varying intensi-ties of induced fluorescence; the fiber-optic treatment photon are for the presumed target. Follow-up observations are necessary and also hopefull, effective confirmatory needle biopsies. Potter's chromophore detection instrumentation is important. For visualization just below the translucent surface, such as the lining of cavities and orifices, optical phase conju-gation technology is under study. Also, in these and in deeper areas, the so-called chrono-coherent imaging technology is considered[7]. The next technology is what is available from current and past studies, as unclassi-fied material, on the transmission of light through that which is called loosely, turbid media.

In our reviews of chromophores in tissues, we have asked if the intra-cellular chromophore could conjugate (through the nanotechnology[10]) with the cancer DNA. If so, would this combination be found also in the cancer DNA in the metastasis. It is assumed that if the circulation in all cancer foci is adequate, all cancer cells would have this DNA conjugation from the parenterally injected chromophore. Does this develop early in the early cancer and in the early metastasis? How early does the tumor angiogenic factor function to provide for the distribution of the photocatalyst? All we do know that it is time now to end speculations in the library chair and to start controlled bench work in the laboratory.

Are these additions and accessories, which we suggest now, truly
needed for a well controlled study? On a large TV screen, the laser scanning ophthalmoscope shows, vividly, the fluorescence in the eye of the
intravenous sodium fluorescein injection. We have shown this to a conference of laser neurosurgeons at the Roswell Park Memorial Institute to illu
trate how such scanning, after post-operative PDT neurosurgery, could show
residual fluorescence in the base of the tumor excision and in the adjacen
tissues. This post-operative laser scanning procedure can be adopted for
PDT surgery elsewhere.

Even, as shown before, with multiple fiber transmission, can we be
certain that there is adequate photon distribution throughout the tumor?
That is why, multiple thermal sensors should be distributed about the tumo
during laser surgery. Due precautions should be taken to avoid false
readings through direct impact on the sensor or failure to insert a sensor
Should liquid crystal sensors be used instead of metal sensors or fiberoptic sensors? Actually, how do we monitor the PDT reaction in tissue?

For local lesions in the skin and in the mucous membranes, Spinelli [12] and I have used topical PDT with Azone and alcohol to aid penetration
into a 2 mm, basal cell of the skin. We followed this topical medication
with the GVL laser. This was very effective especially for multiple basal
cell cancers of the skin. We agree that Azone is an excellent vehicle and
is used too little for local penetration and absorption. Azone formula is
1-dodesylaseryalchepton Azone. As is evident, topical application avoids
the phototoxic generalized reaction to sun exposure of DHE. Topical applications also provide test models for penetration and absorption and radioactive studies of labelled radioactive chromophore, spectroscopy and photoacoustical studies.

As we have indicated repeatedly, Potter's studies [13] for the detection
of the chromophore continue to be important. We continue to know the distribution patterns of the chromophore and its resultant induced fluorescence. We were not successful with attempts at biopsy sections with
fluorescence microscope.

IV. CONCLUSIONS

Therefore, detailed and well controlled studies of exogenous chromophores are necessary for diagnostic and treatment for the cancer program.
New chromophores and new associated laser systems will continue to be
needed and continue to be found. The chromophore must be needed, stable,
non-toxic except for the target, easily used in the manner needed, specifically related to the laser required, easily available commercially, and
economical so that this chromophore can be used widely. Exogenous chromophores have, as indicated now an important place in all phases of the
cancer program and will continue to in the future. It is possible that in
the future, other sources of so-called non-ionizing radiation may be used
with these chromophores. We are still at the beginning of this exogenous
chromophore technology. We must not forget that its applications extend
far beyond the cancer program.

V. ACKNOWLEDGEMENTS

The work reported herein was performed in part by the Navy Clinical
Investigation Program reports 84-16-1968-224 and 84-16-1968-225. The views
expressed in this article are those of the author and do not reflect the
official policy or position of the Department of the Navy, Department of
Defense, or the United States Government.

REFERENCES

1. L. Goldman, "Laser Cancer Research", Springer Verlag Inc., New York (1966)
2. T. J. Dougherty, J. E. Kaufman, A. Goldfarb, K. R. Weishauft, and D. Boyle, "Photoradiation Therapy for the Treatment of Malignant Tumors Cancer Research", 38:2628-2635 (1978)
3. G. Nath, personal communication
4. L. Goldman and R. Dreffer, "Microwave magnetic iron particles and lasers as a combined test model for investigation of hyperthermia treatment of cancer", Arch. Derm. Res., 257:227-232 (1976)
5. N. Levine, personal communication
6. L. Goldman, "Chromophores in Tissue for Laser Medicine and Surgery Lasers" in Medical Science, Baillière and Tindall, London (1990)
7. K. G. Spears, J. Serafin, N. H. Abramson, X. M. Zhu, H. Bjelkhagen, "Chrono-Coherent Imaging (CCI) for Medicine", IEEE Trans. Biomed. Eng. 36(12): 1210-21 (1989)
8. S. Anderson-Engels, "Photodynamic therapy and simultaneous near infra-red light inducing hyperthermia in human cancer tumors", ICALEO, San Diego, November (1987)
9. K. Svanberg, "Cancer Hyperthermia Cancer Res." 46:3803 1986
10. E. Drexler, "Nanobiotechnology", J. E. Dettling, O. E. Reports SPIE:51 1988
11. L. Goldman, R. O. Gregory, M. LaPlant "Preliminary Investigative Studies with PDT in Dermatologic and Plastic Surgery Lasers in Surgery and Medicine" 5:453-456 (1985)
12. P. Spinelli, Personal Communication
13. S. Potter,"Fluorescence Detection with background Subtraction for Localization of Tumors", ICALEO, San Diego (1987)

ENDOSCOPIC PHOTODYNAMIC THERAPY: CLINICAL ASPECTS

P. Spinelli, M. Dal Fante and A. Mancini

Endoscopic Department
National Cancer Institute
Milan, Italy

I. INTRODUCTION

Photodynamic therapy (PDT) is a selective, experimental treatment for solid tumors. PDT consists of the activation of a photosensitizing agent by light. The photodynamic reaction induced by light causes damage to the tissue containing the photosensitizer in the presence of oxygen.

The idea of treating tumors by photosensitizers is as old as the early 1900's; already in 1903, the topical application of eosin and exposition to sunlight was known to produce a response in skin tumors[1]. On the other hand, Policard, in 1924, reported reddish fluorescence in animal and human tumors observed under a Wood lamp. The presence of fluorescence was attributed to endogenous porphyrins accumulated after infection of the observed tissue by emolytic bacteria[2].

In 1942, Auler and Banzer[3] reported animal tumor fluorescence after systemic administration of Hematoporphyrine (HP) and in 1960 Lipson and coworkers prepared the Hematoporphyrin derivative (HPD), a mixture of porphyrins obtained by treating HP with acetic and sulphuric acids[4]. They demonstrated that HPD was selectively accumulated by malignant as well as by actively proliferating tissues, and demonstrated the first endoscopic diagnosis of malignant tissues by detection of fluorescence in the respiratory and in the upper digestive tract[5].

Since the development of the laser, diagnosis through fluorescence and particularly PDT have been intensively studied. Photodynamic reactions have different cellular targets: cross-linking of cellular membrane proteine[6], inactivation of mitochondrial membrane enzymes[7] and DNA damage[8] have been reported. However, in-vivo observations suggest that necrosis of malignant tumors may be secondary to a damage of the tumor vasculature[9].

The main parameters involved in PDT are: the photosensitizer, the light for activation and the selection of patients.

II. PHOTOSENSITIZERS

An ideal sensitizer should have low toxicity, specific absorption spectrum and tumor selectivity. Biologically photoactive agents can be

distinguished in (a) natural fluorochromes, such as porphyrins, (b) exogenous fluorochromes, such as acridine orange, fluorescein and rhodamine;
and, (c) endogenous fluorochromes, such as flavoproteins and keratine.
Most studies deal with the first group of natural fluorochromes and their
derivatives, because they are activated by a wavelength (600-690 nm) which
more deeply penetrates the biological tissues than the shorter wavelengths
necessary to activate other fluorochromes.

Next to the use of HPD, Di-Hematoporhyrin Ether or Ester (DHE) is the
most widely employed photosensitizer at present in clinical studies[10]; DHE
is considered the major active fraction of HPD. The drug is injected intravenously, and after an interval of 24 to 72 hours it is concentrated in
malignant tissues at a variable rate of 3-4 times more than in normal
tissue. HPD is administrated at dosages of 3 to 5 mg/kg body weight and
DHE at 1.5-3 mg/kg body weight.

These drugs present two kinds of limitations which restrict the procedure to an experimental stage: skin photosensitization, and low tissue penetration of the light at the wavelength used for the activation of the
sensitizer. The photosensitivity to sunlight, due to the drug retained by
the skin, may be present for 4 to 6 weeks after the injection. Precautions must be taken to avoid exposure to direct sunlight for 30 days, and
during this period patients are advised to stay indoors, cover exposed
parts and protect eyes from sun rays and strong fluorescent or incandescent
lighting.

Future efforts will be directed to improving the selectivity of photosensitizers, consequently allowing the use of smaller amounts of drugs, reducing cutaneous sensitization. An improved selectivity may be obtained by
means of inclusion of the drugs into liposomes[11], or their linkage with mo-
noclonal antibodies[12,13]. Furthermore, the possible use of new drugs, now
under investigation, having a high absorption coefficient in the near infrared, would improve light penetration into biological tissue, inducing
necrosis of larger volumes of tumor[13].

Among the new drugs, some are compounds resulting from modification of
porphyrins: modifying the structure of DHE by converting one or more of the
porphyrin rings to chlorin (DHEC), or linking HP to chlorin, or modifying
the ester and acid functions (Benzoporphyrine)[14]. Of great interest is
also the use of phtalocyanines, which are porphyrin-like compounds with a
main absorption band in the red; these have experimentally been demonstrated to be very efficient as photosensitizers. The action spectrum for
chloroaluminium phtalocyanine (ClAlPC) is a narrow band centered around 680
nm. ClAlPC appears to be about 50 times more efficient than HPD, and the
red-shifting of its action spectrum allows better light penetration into
irradiated tumors[13].

III. LIGHT FOR ACTIVATION

There are various possibilities for obtaining sufficient amount of
light to be useful for PDT. The activation of porphyrins is usually obtained with a 630 nm wavelength. This wavelength can be produced by fil-
tered lamps, for surface application[15], but when an intracavitary tumor
must be treated by endoscopic systems and light must be transmitted on fiberoptics, it needs special properties such as intensity, coherence and monochromacity[16], that are characteristic of lasers. Generally speaking, in
PDT applications lasers are the most suitable sources because the photobiological responses produced by laser-tissue interaction can be quantita-
tively and qualitatively superior to those caused by conventional light
sources[17].

The advantages of lasers can be summarized as follows: (a) intensity: the desired effects require large energy doses; by varying peak and average power separately, one can induce thermal or photodynamic effects. High peak powers can produce localized thermal damage, whereas high average powers are more likely to produce thermal damage over larger tissue volumes. Low energy density, with no or negligible thermal effect, is responsible for photodynamic action; (b) coherence: this is important for direction and focusing on small areas such as the proximal end of a flexible optic fiber; (c) monochromacity: this allows chromophore selection within tissues and selective photobiologic responses.

The lasers most frequently used for PDT are: 1) Ar-laser: (488-514 nm), limited penetration; 2) Dye lasers, namely Rhodamine-B laser (630 nm, tunable), the most extensively used for PDT; 3) Gold vapour laser (628 nm). A comparison of gold-vapour and dye-lasers for PDT has been done[18] and the gold-vapour laser appears to be simpler and easier to install and run, although it requires a larger diameter fiber for light delivery. This less flexible fiber, when inserted into the operative channel, can reduce the manoeuvrability of endoscopes. The wavelength of the dye laser is tunable, whereas that of gold vapour is fixed; it can be replaced by a copper vapour laser at 510 and 578 nm which is used to pump a dye laser.

New laser devices are also under study at the present time; there is special interest in tunable dye lasers, which allow the generation of different wavelengths of radiation. A future direction will be the development of diode lasers which are small, simple and easy to use[19, 20].

The major difficulty in light irradiation during endoscopic treatments is to convert the unidirectional laser beam into an isotropic illumination of the lesion. In endoscopic applications, irradiation of tumors can be done inserting the fiber into the tissue or keeping the fiber distant from the tissue itself. The evaluation of the energy delivered by the fiber is different in the two situations: if a sharp cut fiber is inserted into a tissue the energy is expressed simply in Joules; if a circularly radiating fiber is inserted into a tissue the energy is expressed in Joules/cm of inserted fiber; if the fiber is kept at a distance, the energy is expressed in Joules/cm^2.

Between the extremity of the fiber and the tissue, different light diffusing devices can be used, such as diffusing solutions, sapphire tips or microlenses[21, 22]. Diffusing solutions can be contained in diffusing balloons attached to the end of a fiber, or to the end of an endoscope. These devices have the aim of obtaining a homogeneous light distribution on the tumor surface.

IV. SELECTION OF PATIENTS

PDT has demonstrated usefulness in endoscopic treatment, especially in cases of superficial tumors with macroscopically undefined borders, or in cases of multicentric tumors. These conditions most often occur in the upper and lower digestive tract, in the bronchi and in the urinary bladder. Hayata, at Tokyo Medical College, began clinical endoscopic applications of PDT in 1980 and has, up to now, accumulated the widest experience in the world in this field[23].

In two international surveys conducted in 1984 and in 1986, we collected data from respectively 467 and 918 patients. Enquiries suggest that the number of centers working in the area of PDT is increasing. Geographically, these centers have expanded all over the world during the last few years. Up to 1984, 467 patients had been treated in 8 centers; between

1984 and 1986 the total number increased to 912 and the number of centers to 20.

Concerning laser sources, these surveys show that 4 groups are using the new gold vapour lasers and that activation by Nd:YAG laser photoradiation has been abandoned. Practically unchanged are: a) the photosensitizers used, b) the time interval between drug injection and irradiation, c) the modality of irradiation. Regarding the anatomical areas irradiated the number of bladder treatments is increasing. Regarding the stage, early tumors are receiving more attention than advanced ones; the power and the energy of treatments tend to decrease, probably because of a relative optimization of treatment parameters. The results of the 1986 enquiry show a complete response (CR) in 61% of early stage and in 7% of advanced tumors treated, partial response (PR) in 33% early and 80% advanced and no response (NR) in 6% early and 13% advanced[24].

Up to now, more than 3,000 patients have been treated, but mainly in uncontrolled trials. Results suggest that tumor histology plays a small role in determining response to PDT and that some lesions such as slowly growing, poorly vascularized and high pigmentated tumors do not respond well to this therapy[19]. Current indications for endoscopic PDT in the gastrointestinal tract include treatment of early stage cancers in high risk patients[25]. In the esophagus, lesions are usually treated with cylindrical diffusing fiber, but tumors located in the upper, as well as in the lower esophageal sphincters require particular devices. At these levels, the introduction of a Savary dilator or of an endoluminal light delivery system[26] allows the sphincter to be kept open; in this way it is possible to obtain a homogeneous distribution of the light.

Early gastric cancers are treated by PDT when they are type IIb or IIc, or type III according to the classification of the Japanese Society for Gastroenterologic Endoscopy[27]. Types I or IIa are better treated by Nd:YAG laser photocoagulation. In the stomach, the lesions are usually irradiated by bear fiber or sapphire micro-probe[21].

At present, the indications for PDT in the rectum are very limited. However, recently published results[28,29] have demonstrated that, after e.v. injection, adenomas present an uptake of HPD similar to that of carcinomas. This observation led us to consider the applicability of endoscopic PDT mainly in the case of flat colorectal adenomas. In fact, endoscopic resection or laser photocoagulation of these lesions is followed by a high rate of recurrence due to the presence of adenomatous tissue not macroscopically visible during the treatment[30].

Bronchoscopic PDT is performed as a conservative treatment of superficial cancers in high risk patients. Japanese authors[31] use PDT before surgery to reduce proximal extension of the tumor. In this way, some inoperable cases become resectable and more limited resection can be performed. Light is delivered to the surface of the tumor by cylindrical diffusing fiber that can be inserted into the lesion, or can be used for a surface illumination.

Cleanup bronchoscopy is necessary 72 hours after laser irradiation in order to remove necrotic tissue that could obstruct the airways.

In the urinary bladder, PDT is used in the treatment of carcinomas in situ; a whole bladder irradiation is performed because the tumor is usually multicentric. In order to obtain a homogeneous distribution of light dose, the use of a bulb tip fiber, or of a light scattering medium has been proposed[32,33]. In the former case, the bladder is filled with a normal saline solution and the bulb of the fiber is positioned at the center of the

viscerum. In the latter, bladder is filled by a lipidic physiological suspension (Intralipid 1:1000) and the position of the fiber tip is at a
quarter of the distance between the center and the neck of the bladder.

Side effects and complications after PDT are infrequent. Among the
first, pain and oedema at the irradiated site have been reported, mainly
in the esophagus and stomach where they can be responsible for a temporary
dysphagia or obstruction, respectively. NSAID are useful in these cases.

Major complications are hemorrhage and perforation due to extensive
wall necrosis.

Severe reduction of urinary capacity, due to bladder shrinkage, is the
most important complication after whole bladder irradiation; it seems to be
related to the light dose and to the pressure used to distend the bladder.
Light doses of 20 Joules/cm^2, pressure of less than 30 cm column of water
and the use of normal saline, or lipidic physiological suspension, can reduce the incidence of irreversible shrinkage and prolonged inactive bladder
symptoms [19].

V. CONCLUSION

Indications for PDT are changing compared with the first attempts at
its use. PDT seems to be more reliable in treating small cancer lesions,
superficially extended over large areas, and multicentric tumors and with
undefined borders. PDT can be used as a curative and as a palliative treat-
tment. It can treat cancer at various stages, from precancerous lesions,
to cancer both at early[34] and at advanced stages[35]. PDT seems to be the
ideal therapy for cancer, but many problems become evident upon detailed
study.

The selection of patients is very important for PDT, because for palliation of advanced cancers deeply infiltrating the walls of hollow viscera, thermal laser action is usually preferred to obtain a controlled and
quicker necrosis. In such a way, the incidence of the most serious complications such as a perforation and hemorrage can be limited.

Another consideration concerns the location of the cancer: the best
location for effective treatment is in large hollow viscera that can be
filled with a refractive medium, as for example the urinary bladder. A
quite homogeneous distribution of laser power can be obtained in this way.
On the contrary, in a narrow channel with a stenotic tortuous segment, as
in the esophagus or colon-rectum, light distribution is a problem.

The future tendency of this work is to develop treatments for early
cancer and precancerous lesions. At present, treatment protocols involving
combinations of traditional therapies are submitted for international
evaluation.

REFERENCES

1. H. Tappenier and A. Jesionek, "Therapeutische Versuche mit fluores-
   zierenden Stoffe", Münch. Med. Wochenschr. 1:2042 (1903)
2. A. Policard, "Etudes sur les aspects offerts par des tumeurs expéri-
   mentales examinées à la lumière de Woods", Cr. Soc. Biol. 91:1423
   (1924)
3. H. Auler and G. Banzer, "Untersuchungen über die Rolle der Porphine
   bei geschwulstkranken Menschen und Tieren", Z. Krebforsch. 53:65
   (1942)

4. R. L. Lipson, E. J. Baldes, "The photodynamic properties of a parti-
    cular hematoporphyrin derivative", Arch. Dermatol. 82:508 (1960)
5. R. L. Lipson, E. J. Balder and A. M. Olsen, "Hematoporphyrin deriva-
    tive: a new aid for endoscopic detection of malignant disease", J.
    Thorac. Cardiovasc. Surg. 42:623 (1961)
6. J. Moan, "Porphyrin-sensitized photodynamic inactivation of cells: a
    review", Lasers Med. Sci. 1:5 (1986)
7. R. Hilf, D. B. Smail, R. S. Murant, et al., "Hematoporphyrin deriva-
    tive-induced photosensitivity of mitochondrial succinate dehydro-
    genase and selected cytosolic enzymes of R3230 AC mammary adenocar-
    cinomas of rats", Cancer Res. 44:1483 (1984)
8. C. J. Gomer, "DNA damage and repair in CHO cells following hematopor-
    phyrin photoradiation", Cancer Lett. 11:161 (1980)
9. T. J. Dougherty, "Photodynamic therapy (PDT) of malignant tumors",
    CRC Crit. Rev. Oncol/Haemat. 2:83 (1985)
10. T. J. Dougherty, W. R. Potter and K. R. Weishaupt, "The structure of
    the active component of hematoporphyrin derivative", In "Porphyrin
    Localization and Treatment of Tumors (D. R. Doiron and C. J. Gomer,
    eds.), Alan R. Liss, Inc., New York, 301 (1984)
11. G. Jori, "Pharmacokinetics studies with hematoporphyrin in tumor-bear-
    ing mice", in "Photodynamic Therapy of Tumors and other Diseases",
    G. Jori, C. Perria eds., Progetto Publ., Padova, 159 (1985)
12. D. Mew, C. K. Wat, G. H. N. Towers and J. G. Levy, "Photoimmunother-
    apy: treatment of animal tumors with tumor specific monoclonal
    antibody-hematoporphyrin conjugates", J. Immunol. 130:1473 (1983)
13. R. Kol, E. Ben-Hur, E. Riklis, R. Marko and I. Rosenthal, "Photosen-
    sitized inhibition of mitogenic stimulation of human lymphocytes
    by aluminium phtalocyanine tetrasulphonate", Las. Med. Sci. 1:187
    (1986)
14. A. M. Richter, E. Sternberg, E. Waterfield, D. Dolphin and J. C. Levy,
    "Characterization of benzoporphyrin derivative, a new photosen-
    sitizer", Proc. SPIE 997:132 (1988)
15 T. J. Dougherty, J. E. Kaufman, A. Goldfarb, K. R. Weishaupt, D. G.
    Boyle and A. Mittleman, "Photoradiation therapy for the treatment
    of malignant tumors", Cancer Res. 38:2628 (1978)
16. P. Spinelli and M. Dal Fante, "Photodynamic therapy" in G. I. tract",
    Acta Endosc. 15:69 (1985)
17. J. A. Parrish, "Photomedicine potential for lasers. An overview from
    lasers in photomedicine and photobiology", R. Pratesi and C. A.
    Sacchi eds., Springer, Berlin, 2 (1980)
18. A. L. McKenzie and J. A. S. Carruth, "A comparison of gold-vapour and
    dye lasers for PDT", Laser Med. Sci. 1:117 (1986)
19. T. J. Dougherty, "Photodynamic therapy - New approaches", Sem. Surg.
    Oncol. 5:6 (1989)
20. R. Brancato, L. Giovannoni, R. Pratesi and U. Vanni, "New lasers for
    ophtalmology: retinal photocoagulation with pulsed diode lasers",
    Proc. SPIE 701:365 (1986)
21. P. Spinelli and M. Dal Fante, "Contact laser endoscopic surgery with
    sapphire microprobes", Proc. SPIE 701:331 (1986)
22. V. Russo, "Optical fiber delivery systems for laser medical appli-
    cations" in "Photodynamic Therapy of Tumors and other Diseases",
    G. Jori, C. Perria eds., Progetto Publ. Padova, 371 (1985)
23. Y. Hayata, H. Kato, C. Konaka, J. Ono and N. Takizawa, "Hematopor-
    phyrin-derivative and laser photoradiation in the treatment of lung
    cancer", Chest 81:269 (1981)
24. P. Spinelli and M. Dal Fante, "PDT - State of the art", in "Laser Opto
    electronics in Medicine", W. Waidelich and P. Kiefhaber eds.,
    Springer Verlag, 609 (1987)
25. P. Spinelli, S. Andreola, R. Marchesini, E. Melloni, V. Mirabile, P.
    Pizzetti and F. Zunino, "Endoscopic HpD-laser photoradiation ther-
    apy (PRT) of cancer", in "Porphyrins in Tumor Phototherapy", A. An-
    dreoni, R. Cubeddu eds., Plenum Publishing Corporation, 423 (1984)

26. J. T. Allardice, A. C. Rowland, N. S. Williams and C. P. Swain, "A new light delivery system for the treatment of obstructing gastrointestinal cancers by photodynamic therapy", Gastrointest. Endosc. 35:548 (1989)

27. Japanese Study for Gastroenterological Endoscopy. Cited by T. Murakami, "Pathomorphological diagnosis, definition and gross classification of early gastric cancer", T. Murakami ed., Tokyo, University of Tokyo Press (1971)

28. M. Dal Fante, G. Bottiroli and P. Spinelli, "Behaviour of haematoporphyrin derivative in adenomas and adenocarcinomas of the colon: a microfluorometric study", Lasers Med. Sci. 3:165 (1988)

29. R. M. Cothren, R. Richards-Kortum, M. V. Sivak, Jr., et al., "Gastrointestinal tissue diagnosis by laser-induced fluorescence spectroscopy at endoscopy", Gastrointest. Endosc., 128 (1990)

30. E. M. H. Mathus-Vliegen and G. N. J. Tytgat, "Nd:YAG laser photocoagulation in colorectal adenoma. Evaluation of its safety, usefulness, and efficacy", Gastroenterol. 90:1865 (1986)

31. H. Kato, C. Konaka, J. Ono, et al. "Preoperative laser photodynamic therapy in combination with operation in lung cancer", J. Thorac. Cardiovasc. Surg. 90:420 (1985)

32. R. C. Benson, Jr., "Laser photodynamic therapy for bladder cancer", Mayo Clin. Proc. 61:859 (1986)

33. D. Jocham, G. Staehler, C. Chaussy, et al., "Integral dye-laser irradiation of photosensitized bladder tumors with the aid of a light-scattering medium", Prog. Clin. Biol. Res. 170:249 (1984)

34. H. Tajiri, N. Daizukono, S. N. Joffe and Y. Oguro, "Photoradiation therapy in early gastrointestinal cancer", Gastrointest. Endosc. 33:88 (1987)

35. J. S. McCaughan, T. E. Williams and B. H. Bethel, "Palliation of esophageal malignancy with photodynamic therapy", Am. Thor. Surg. 40:113 (1985)

TARGETS AND MECHANISMS OF ACTION ASSOCIATED WITH LASER

MEDIATED PHOTOSENSITIZATION

A. Ferrario, N. Rucker, S. Wong, M. Luna and C. J. Gomer

Childrens Hospital of Los Angeles
University of Southern California School of Medicine
Los Angeles, California, USA

I. INTRODUCTION

Photodynamic therapy (PDT) is the treatment of malignant lesions with visible light following the systemic administration of a tumorlocalizing photosensitizer. Hematoporphyrin derivative (HpD) and a purified component called Photofrin II are currently used in clinical PDT and this therapy continues to show promise in the treatment of solid tumors. However, it is clear that PDT is still at an early stage in its development. In this chapter, we will examine molecular, cellular, and in-vivo mechanisms related to PDT.

II. SUBCELLULAR TARGETS AND MOLECULAR RESPONSE ASSOCIATED WITH PDT

Numerous studies have demonstrated that PDT is extremely effective in generating cytotoxic damage to subcellular organelles and biomolecules[1]. Photodegradation of lipids, proteins and nucleic acids can routinely be observed following porphyrin photosensitization. Membrane damage can lead to inhibition of transport of amino acids and nucleosides, increased permeability and rupture of lysosomes, as well as marked contraction and rupture of mitochondria[2]. Enzymes bound to hydrophobic regions of mitochondria (membrane bound) are extremely sensitive to PDT[3]. At the level of nucleus PDT can produce single strand breaks in DNA, sister chromatid exchange and chromosome aberrations[4,5]. Even though a large spectrum of specific types of subcellular damage have been documented following PDT, the actual target site(s) for cytotoxicity has not been identified[6].

II.1. Differential Cell Photosensitivity Following Porphyrin
      Photodynamic Therapy

Interestingly, the mechanism of action for PDT in experimental tumors is thought to involve both direct tumor cell kill and direct tumor vessel damage[7-9]. Vascular damage is also implicated in PDT-induced damage to normal tissues such as brain, intestine and skin[9-11]. Several studies have indicated that malignant and normal cells accumulate similar levels of porphyrin during in-vitro incubation[12-14]. Differences in cell photosensitivity are usually not observed for normal and transformed cell lines[13]. Recently, differential cell photosensitivity following PDT was evaluated by

direct comparison of level of photosensitization in cells with varying
levels of DNA repair properties, as well as in cells which make up various
components of the vasculature[15]. Interestingly, while the human DNA repair
deficient fibroblasts ataxia telangiectasia (AT) and xeroderma pigmentosum
(XP) expressed extreme hypersensitivity to ionizing radiation and to ultraviolet radiation respectively compared to normal fibroblasts, survival
curves for PDT resulted in similar levels of photosensitivity. These findings support the premise that non-nuclear damage, such as that induced in
mitochondria and/or the plasma membrane, is of primary importance in terms
of PDT-induced cytotoxicity[6]. Porphyrin-induced photosensitization does
not induce mutagenic[5] or transformation[16] activity in mammalian cells and
these observations in cellular photosensitivity have been observed when
components of the vasculature were examined[15]. Specifically, bovine cells
of endothelial, smooth muscle and fibroblasts origin were compared for porphyrin retention and photosensitivity. Bovine endothelial cells were considerably more sensitive than smooth muscle or fibroblast cells treated
under identical conditions when assayed for viability using clonogenicity.
The increased photosensitivity observed in endothelial cells can not be ac-
counted for on the basis of cellular porphyrin content at the time of the
treatment since all bovine cells accumulated similar levels of porphyrin.
The results indicate that endothelial cell photosensitivity may play a role
in the vasculature damage observed following porphyrin photodynamic
therapy.

II.2. Stress Protein Production Following Photodynamic Therapy

Eucaryotic cells respond to a transient stress, such as heat shock, by
inducing the synthesis of a specific set of highly conserved proteins known
as the heat shock proteins (HSP)[17]. These proteins are generally thought
to play a role in protecting cells from subsequent stresses and/ or in en-
hancing the recovery of injured cells. Some of these proteins are produced
constitutively, while others are only synthesized under the influence of a
variety of cellular stresses. It is become increasingly clear that expression of heat shock genes is not limited to cells that are undergoing acute
stress[18]. The eucaryotic genome also encodes proteins which are closely
related in sequence to HSP-70, but which appear to be regulated distinctly.
The synthesis of this second stress-responsive group of proteins, the
glucose regulated proteins (GRP), is induced under a variety of conditions
including glucose deprivation, anoxia, treatment of cells with glycosylation inhibitors or the calcium ionophore A23187[19].

Singlet oxygen generated via a Type II photochemical mechanism is
thought to initiate most damage following Photofrin II mediated PDT[20, 21].
The treatment can induce stress proteins of both the heat shock (HSP) and
glucose regulated (GRP) families. A comparison of the kinetics and characterization of photosensitizer-induced stress proteins with proteins induced
by other oxidative stress, may suggest several complementary modes of action for PDT and provides a novel method for identifying oxidative species
involved in PDT. An extensive time-dependent increase in GRP-78 gene expression was observed in RIF-1 cells first incubated with Photofrin II for
16 hours and then exposed to a dose of visible light which resulted in
20-30% survival level[22]. The significance of this finding is currently unclear, although it is known that GRP-78 expression is strongly induced by
agents which affect posttranslational processing events in the endoplasmic
reticulum[19]. The product of GRP 78 gene is a nonglycosylated protein with
an apparent molecular weight of 78,000 daltons and is found in the lumen of
the ER of most cell types. GRP-78 has been detected in association with
proteins when their folding or assembly is incomplete or improper[23]. The
GRP-protein complex may prevent the secretion of abnormal proteins presumably maintaining them in soluble form until the stress condition is re-

moved. Experiments are currently ongoing to determine whether the major stress proteins induced by PDT can play a role in detoxification or protection from oxidative damage. Preliminary data utilizing a standard clonogenicity assay, showed that the activation of GRP genes by the calcium ionophore A23187 protected the Chinese hamster lung V-79 fibroblasts from Photofrin II mediated photosensitization. Induction of the GRP genes by the calcium ionophore A23187 treatment may be ascribed tl the movement of intracellular Ca^{++} ions out of intracellular stores. In particular depletion of intracellular Ca^{++} store from the endoplasmic reticulum, which is the site where N-glycosylation of nascent polypeptides takes place, might interfere with the activity of the dolichol kinase enzyme localized to the ER which catalyzes (with relatively high calcium requirement) the rate-limiting step in protein glycosylation[24]. Moreover, the temperature sensitive mutant hamster cell line K12 was less sensitive to PDT when the cells were shifted for the 16 hours of Photofrin II incubation from 35 °C (permissive temperature) to 39.5 °C, temperature that activates the expression of GRP-78 gene in K12 cells[25]. Interestingly, the induction of GRP-78 in K12 cells detected at 39.5 °C is also observed under other conditions such as at 35 °C in medium depleted of glucose[25]. A large number of stress conditions which induce the GRP's interfere with protein glycosylation. This observations suggest that a common transacting regulatory factor(s) can be generated by different primary inducers and interact with common or overlapping domains in the promoter region to initiate GRP transcription. It is interesting that transcriptional and translational activation of stress proteins varies as a function of invitro PDT treatment conditions[22]. Short porphyrin incubation time periods (less than one hour) followed by light exposure lead to damage associated primarily with the plasma membrane while photosensitizer incubation for extended periods (up to 24 hours) followed by light exposure induces an increased amount of damage to cytoplasmic organelles and enzymes. Increased synthesis of GRP-78 was observed for cells receiving an extended porphyrin incubation, while little or no increase in GRP-78 was induced in cells incubated with porphyrin for only 1 hour prior to light treatment even though both the short and the long incubation conditions were followed by light doses which induced comparable levels of cell killing. While the significance of these findings is currently unclear, it is reasonable to suggest that different subcellular targets are being effected by the two porphyrin incubation protocols. It is possible that damage to only select biomolecules will induce the synthesis of GRP-78. In addition, photoactivation of mono-l-aspartyl chlorin e6 a new photosensitizer that appears to localize preferentially in lysosomes upon long incubation conditions[26], did not induce any of the major stress proteins normally overexpressed following porphyrin photo-sensitization. The kinetic analysis of stress protein synthesis of Photofrin II and NPe6 mediated PDT in RIF-1 cells was performed at comparable levels of cell killing. It is therefore reasonable to assume that different subcellular targets may be affected by the two treatments. Mitochondria are the major subcellular site in the localization of porphyrins and are thought to be major targets in the cytotoxic effects of porphyrin mediated PDT as identified by fluorescence microscopy[27], while lysosomes may be the initial target in chlorin induced photosensitization. Interestingly, short incubation time periods of the two photosensitizers (condition associated with plasma membrane bound drug) followed by isoeffect doses of light generated a similar pattern of stress protein.

Questions also remain as to whether the actual in-vitro targets for PDT can account for the mode of action of PDT in-vivo. Methods to enhance the effectiveness of PDT in-vivo, in fact, may be effected by the grp-PDT induced state. Studies have shown that a poor response to chemotherapy with a significant resistance to topoisomerase II-targeted drugs (i.e. VP-16, adriamycin) may be obtained by exposing cells to anoxia and other stresses which specifically lead to the induction of GRP's[28]. Hyperthermia can potentiate the toxicity induced by PDT in a manner qualitatively similar to that observed when ionizing radiation or chemotherapy is combined with heat[29]. Interestingly, there are a number of similarities related to possible subcellular targets and in-vivo mechanisms of action of hyperthermia and PDT[30]. A comparison of thermal sensitivity and porphyrin photosensitivity was performed to determine whether resistance to hyperthermia was related to resistance to PDT[31]. A Chinese hamster cell line (3012) which overexpresses a 70-KDa heat shock protein (HSP-70), has been compared to its parent cell line (HA-1) in terms of sensitivity to PDT and heat. In addition, the mouse radiation induced fibrosarcoma cell line (RIF-1) was compared to the three heat tolerant cell strains (TR-4; TR-5; TR-10) in which the thermal resistance may be associated with membrane fluidity alterations. While quantitative cell survival curves documented a significant differential resistance to hyperthermia (45 °C for various exposure periods) for the 3012 vs. the parent HA-1 and for the thermal resistant RIF cell strains vs. the parent RIF-1 cell line, no cross resistance to porphyrin mediated PDT was observed. This phenomenon can not be ascribed to a different accumulation of amount of porphyrin in control and thermal resistant cells since comparable levels of Photofrin II were measured by absorption spectroscopy. These results suggest that the mechanisms of cytotoxic action for PDT (using Photofrin II incubations which would lead to either membrane or subcellular damage to organelles) and hyperthermia are different. Therefore, while HSP-70 may modulate thermal sensitivity it is unlikely that HSP-70 is associated with modification of cellular photosensitivity.

Oxidative stress caused by porphyrin mediated PDT induces a transient increase in mRNA levels of heme oxygenase. Heme oxygenase is a microsomal enzyme of 32-34 kDa inducible by a variety of cellular perturbants: UV radiation, the oxidizing agent hydrogen peroxide, menadione, heavy metal salts, the sulphydryl reagent sodium arsenate and heme analogues[32]. This enzyme can be rapidly mobilized in response to certain stress conditions even in tissues not normally sites of heme catabolism and catalyzes the oxidative degradation of heme to biliverdin[32]. Among all the heme oxygenase inducers, only hydrogen peroxide and menadione directly generate active oxygen intermediates in cells[33]. However, since many of the potential cellular forms of the product of heme catabolism react efficiently with peroxyl radicals, it is possible that the activity of this enzyme may play a direct role in cellular defense against oxidant damage[32]. The induction of heme oxygenase may therefore be the result of a protective pathway specifically activated under conditions of oxidant stress. In addition, agents that induce heme oxygenase (i.e. sodium arsenate) interact with and probably reduce the level of available glutathione[32]. Glutathione red/ox cycle is known to play an important role in quenching reactive oxygen intermediates[33]. Thus the signal for the induction of heme oxygenase may be related to modified or reduced levels of the cellular nonprotein thiol, GSH. Kinetics studies of heme oxygenase expression in V-79 cells showed thet long incubation (16 hours) with Photofrin II alone leads to a substantial increase in the base level of the enzyme, possibly due to a substrate-effect phenomenon. This finding rose the question whether the heme oxygenase induction observed upon photoactivation of Photofrin II should be ascribed to the same mechanism turned on by the chemical structure of the porphyrin molecule. Interestingly, photosensitization of Rose Bengal (a xanthene dye with a chemical structure different than porphyrins) which photochemically produces singlet oxygen[34], caused a transient increase in mRNA and heme oxygenase protein levels in V-79 cells, as well. This similarity suggests that common mechanisms of action for both Rose Bengal and Photofrin II mediated PDT (singlet oxygen-target sites), may initiate similar pathways of induction.

Oxidative stress caused by porphyrin mediated PDT leads also to an initial induction of c-fos and a subsequent activation of metallothionein genes in RIF-1 celles. It is not completely clear, at the moment, how cytotoxic damage to subcellular organelles and biomolecules generated by oxidative stress could be transmitted to the nucleus leading to the expression of these genes. There are strong indications suggesting that a possible "cascade" of gene activity may be induced by Photofrin II mediated PDT. Rat heme oxygenase gene, recently isolated and sequenced, contains in the 5'-flanking region several potential binding sites for different transcription factors[35]. It contains two copies of core sequences of the metal regulatory elements found in metallothionein gene, binding sites of a heat shock transcription factor and a sequence very similar to the binding of TPA-modulated trans-acting factor Fos/Jun complex originally termed AP-1[36]. The complexity of the heme oxygenase promoter region, as well as interactions between regulator factor(s) and their target DNA sequences may therefore explain the sequential pattern of gene activity (c-fos ⟶ metallothionein-heme oxygenase) induced by PDT.

III. IN VIVO MECHANISM OF ACTION AND TISSUE RESPONSE ASSOCIATED WITH PDT

While the photochemical properties of photodynamic therapy indicate that singlet oxygen is a primary agent in the induction of tumoricidal effects, the actual targets of PDT at the tissue level have not been identified. A number of investigators have proposed that both normal and tumor vasculature can be severely damaged following photodynamic therapy[21].

III.1. Physiological and Immunological Response to Porphyrin Mediated PDT

Examination of tumor metastasis induced by PDT was evaluated because this treatment is being suggested for use in early-stage clinical malignancies and because of the continued observation of direct vascular injury following PDT. Studies were performed to determine whether localized PDT treatment of subcutaneously growing Lewis lung carcinoma affected the formation of distant metastasis[37]. Mice exposed to localized PDT had equal or decreased number of metastatic lung colonies when compared to controls or to mice treated with localized surgical tumor excision. These results indicate that local PDT did not enhance the metastatic spread of Lewis lung carcinoma following treatment.

Interestingly, localized porphyrin PDT procedures have been shown to induce an immunosuppressive action towards T-cells, documented in hypersensitivity reactions[38], and NK cells[31]. We examined the natural killer (NK) cell activity in non-tumor bearing mice treated with localized photodynamic therapy. While Photofrin II alone at concentrations of 10 and 25 mg/kg induced a significant increase in spleen size, there was no change from control values for NK activities. However, a rapid decrease in spleen size and NK cell activity was observed following PDT. Within 24 hours, the NK cell activity was reduced to levels near zero. Both of these changes were transient in nature, with NK cell activity and spleen size returning to control values within three days. The observed decrease in NK cell activity following PDT appears to be a systemic effect, since only the hind limb of the experimental animals received the light treatment[31]. These immunological effects may be mediated in part by increased prostaglandin levels induced by localized PDT. In addition, PDT treatment of the hind leg of rats, whether free of, or implanted with chondrosarcome was found to induce the release of thromboxane into the serum[39]. Moreover, a recent investigation examining the systemic toxicity induced by localized PDT (in which only the hind leg of the experimental mice was irradiated) suggests that endogenous mediators of microcirculatory disruption such as prostaglandins, thromboxane and histamine are activated by the treatment[40]. Doses of PDT which induced lethality (10 mg/kg Photofrin II, 200-500 J/cm^2) were in the

range of doses required to obtain murine tumor cures. The percentage of
lethality was proportional to the total light dose but inversely proportional to the dose rate of delivered light. Similar levels of PDT-induced
lethality were observed for pigmented and albino mice as well as for mice
transplanted with subcutaneous melanomas. Agents that can reduce the action of endogenous mediators of microcirculatory disruption such as indo-
methacin and aspirin (antiinflammatory drugs which inhibit cyclooxygenase
and therefore decrease prostaglandin and thromboxane synthesis), warfarin
(anticoagulant) and antihistamine were able to decrease PDT-induced lethality. Hystological profiles obtained 24 hours following localized PDT demonstrated vascular congestion in liver, kidney, lung and spleen. Significant decreases in removable blood volume, core temperature and spleen
weight were also observed within 24 hours of localized PDT treatment.
These results indicate that PDT-induced lethality is consistent with a
traumatic shock syndrome. Interestingly the lethality observed in mice
following PDT is not unique to Photofrin II. We have observed comparable
effects with chlorin photosensitizers and other investigators have observed
lethality following localized PDT by using phtalocyanine[41]. The acute
lethality observed in mice does not appear to be clinically relevant; the
doses of PDT currently used in larger animals and humans do not elicit any
observable systemic toxicity[42], and therefore it is possible that systemic
factors associated with localized PDT in mice will be greatly reduced or
eliminated following treatment in larger animals and humans. The phenomenon, however, should be taken into account during the in-vivo screening of
new tumor photosensitizers using mouse models.

REFERENCES

1. G. Jori and J. D. Spikes, "Photobiochemistry of porphyrins", in:
    "Topic in Photomedicine", K. C. Smith, ed., Plenum Press, New
    York (1984)
2. J. Moan, "Porphyrin-sensitized photodynamic inactivation of cells: a
    review", Laser Med. Sci., 1:5 (1986)
3. R. Hilf, D. B. Smail, R. S. Murant, P. B. Leakey and S. L. Gibson,
    "Hematoporphyrin derivative-induced photosensitivity of mito-
    chondrial succinate dehydrogenase and selected cytosolic enzyme of
    R3230 AC mammary adenocarcinomas of rats", Cancer Res., 44:1483
    (1984)
4. C. J. Gomer, "DNA damage and repair in CHO cells following hematopor-
    phyrin photoradiation", Cancer Lett. 11:161 (1980)
5. C. J. Gomer, N. Rucker, A. Banerjee and W. F. Benedict, "Comparison
    of mutagenicity and induction of sister chromatid exchanges in
    Chinese hamster cells exposed to hematoporphyrin derivative,
    photoradiation, ionizing radiation, or U.V. radiation", Cancer
    Res. 43:2622 (1983)
6. D. Kessel, "Sites of photosensitization by derivatives of hemato-
    porphyrin", Photochem. Photobiol. 44:489 (1986)
7. S. H. Selman, M. Kreimer-Birnbaum, J. E. Klaunig, P. J. Goldblatt,
    R. W. Keck and S. L. Britton, "Blood flow in transplantable
    bladder tumors treated with hematoporphyrin derivative and light",
    Cancer Res. 441924 (1984)
8. B. W. Henderson, S. M. Waldow, T. S. Mang, W. R. Potter, P. B. Malone
    and T. J. Dougherty, "Tumor destruction and kinetics of tumor cell
    death in two experimental mouse tumors following photodynamic
    therapy", Cancer Res. 45-572 (1985)
9. W. M. Star, H. P. A. Marijnissen, A. E. van der Berg-Blok, J. A. C.
    Versteeg, K. A. P. Franken and H. S. Reinhold, "Destruction of rat
    mammary tumor and normal tissue microcirculation by hematoporphyrin
    derivative photoradiation observed in-vivo in sandwich observation
    chamber", Cancer Res. 46:2532 (1986)

10. M. C. Berenbaum, G. W. Hall and A. D. Hayes, "Cerebral photosensitization by hematoporphyrin derivative. Evidence for an endothelial site of action", Br. J. Cancer 53:81 (1986)

11. S. H. Selman, M. Kreimer-Birnbaum, P. J. Goldblatt, T. S. Anderson, R. W. Keck and S. L. Britton, "Jejunal blood flow after exposure to light in rats injected with hematoporphyrin derivative", Cancer Res. 45:6425 (1985)

12. C. Chang and T. J. Dougherty, "Photoradiation therapy: kinetics and thermodynamics of porphyrin uptake and loss in normal and malignant cells in culture (abstract)", Radiat. Res. 74:498 (1978)

13. J. Moan, H. B. Steen, K. Keren and T. Christensen, "Uptake of hematoporphyrin derivative and sensitized photoinactivation of C3H cells with different oncogenic potential", Cancer Lett. 14:291 (1981)

14. B. W. Henderson, D. A. Bellnier, B. Zirig and T. J. Dougherty, "Aspects of the cellular uptake and retention of hematoporphyrin derivative and their correlation with the biological response to PRT in-vitro", in: "Porphyrin Photosensitization", D. Kessel and T. J. Dougherty, eds., Plenum Press, New York (1983)

15. C. J. Gomer, N. Rucker and A. L. Murphree, "Differential cell photosensitivity following Porphyrin photodynamic therapy", Cancer Res. 48:4539 (1988)

16. C. J. Gomer, N. Rucker and A. L. Murphree, "Transformation and mutagenic potential of porphyrin photodynamic therapy in mammalian cells", Int. J. Radiat. Biol. 53:651 (1988)

17. M. J. Schlesinger, "Heat shock proteins: the search for functions", J. Cell. Biol. 103:321 (1986)

18. G. N. Teodorakis, D. J. Zand, P. T. Jotzbauer, G. T. Williams and R. I. Morimoto, "Hemin-induced transcriptional activation of the HSP70 gene during erythroid maturation in K562 cells is due to a heat shock factor-mediated stress response", Mol. Cell. Biol. 9:3166 (1989)

19. S. S. Watowich and R. I. Morimoto, "Complex regulation of heat-shock and glucose-responsive genes in numan cells", Mol. Cell.Biol. 8:393 (1988)

20. K. Weishaput, C. J. Gomer and T. J. Dougherty, "Identification of singlet oxygen as the cytotoxic agent in photo-inactivation of a marine tumor", Cancer Res. 36:2326 (1976)

21. J. P. Keene, D. Kessel, E. J. Land, R. W. Redmond and T. G. Truscott, "Direct detection of singlet oxygen sensitized by hematoporphyrin and related compounds", Photochem. Photobiol. 43:117 (1986)

22. C. J. Gomer, N. Rucker, A. Ferrario and S. Wong, "Properties and applications of photodynamic therapy", Radiat. Res. 120:1 (1989)

23. A. J. Dorner, M. G. Krane and R. J. Kaufman, "Reduction of endogenous GRP78 levels improves secretion of a heterologous protein in CHO cells", Mol. Cell. Biol. 8:4063 (1988)

24. I. A. S. Drummond, A. S. Lee, E. Resendez and R. A. Steinhardt, "Depletion of intracellular calcium stores by calcium ionophore A23187 induces the genes for glucose-regulated proteins in hamster fibroblasts", J. Biol. Chem. 262:12801 (1987)

25. A. S. Lee, "The accumulation of three specific proteins related to glucose-regulated proteins in a temperature-sensitive mutant cell line K12", J. Cell. Physiol. 106:119 (1981)

26. W. G. Roberts, F.-Y. Shiau, J. S. Nelson, K. M. Smith and M. W. Berns, "In vitro characterization of monoaspartyl chlorin e6 and diaspartyl chlorin e6 for photodynamic therapy", J. Natl. Cancer Inst. 80:330 (1988)

27. M. W. Berns, A. Dahlman, F. M. Johnson, R. Burns, D. Sperling, M. Guiltinan, A. Siemens, R. Walter, W. Wright, M. Hammer-Wilson and A. Wile, "In-vitro cellular effects of hematoporphyrin derivative", Cancer res. 42:2325 (1982)

28. J.-W. Shen, J. R. Subjeck, R. B. Lock and W. E. Ross, "Depletion of topoisomerase II in isolated nuclei during a glucose-regulated stress response", Mol. Cell. Biol. 9:3284 (1989)

29. S. M. Waldow and T. J. Dougherty, "Interaction of hyperthermia and photodynamic therapy", Radiat. Res. 97:380 (1984)

30. C. J. Gomer, N. Rucker and S. Wong, "Porphyrin photosensitivity in cell lines expressing a heat resistant phenotype", accepted for publication: Cancer Res. 1990

31. C. J. Gomer, A. Ferrario, N. Hayasji, N. Rucker, B. C. Szirth and A. L. Murphree, "Molecular, cellular, and tissue responses following photodynamic therapy", Lasers Surg. Med. 8:450 (1988)

32. S. M. Keyse and R. M. Tyrrel, "Heme oxygenase is the major 32-kDa stress protein induced in human skin fibroblasts by UVA radiation, hydrogen peroxide, and sodium arsenate", Proc. Natl. Acad. Sci. USA 86:99 (1989)

33. S. M. Keyse and R. M. Tyrrel, "Both near ultraviolet radiation and the oxidizing agent hydrogen peroxide induce a 32-kDa stress protein in normal human skin fibroblasts", J. Biol. Chem. 262-142821 (1987)

34. D. K. Luttrull, O. Valdes-Aguilera, S. M. Linden, J. Paczkowski and D. J. Neckers, "Rose Bengal aggregation in rationally synthetized dimeric systems", Photochem. Photobiol. 47:55 (1988)

35. J. Alam, S. Shibahara and A. Smith, "Transcriptional activation of heme oxygenase gene by heme and cadmium in mouse hepatoma cells", J. Biol. Chem. 264:6371 (1989)

36. H. Kageyama, T. Hiwasa, K. Tokunaga and S. Sakiyama, "Isolation and characterization of a complementary DNA clone for a M_r 32,000 protein which is induced with tumor promoters in Balb/c3T3 cells", Cancer Res. 48:4795 (1988)

37. C. J. Gomer, A. Ferrario and A. L. Murphree, "The effect of localized photodynamic therapy on the induction of tumor metastasis", Br. J. Cancer 56:27 (1987)

38. C. A. Elmets and K. D. Bowen, "Immunological suppression in mice treated with hematoporphyrin derivative photoradiation", Cancer Res. 46:168 (1986)

39. V. H. Fingar, T. J. Wieman and K. W. Doak, "Role of thromboxane and prostacyclin release on photodynamic therapy-induced tumor destruction", Cancer res. 50:2599 (1990)

40. A. Ferrario and C. J. Gomer, "Systemic toxicity in mice induced by localized porphyrin photodynamic therapy", Cancer Res. 50:539 (1990)

41. J. van Lier, personal communication

42. T. J. Dougherty, Photodynamic therapy (PDT) of malignant tumor", CRC Crit. Rev. 2:83 (1984).

PHOTODYNAMIC THERAPY OF TUMORS

A REVIEW OF PDT DOSIMETRY

W. R. Potter

Department or Radiation Biology
Roswell Park Cancer Center
Buffalo, New York, USA

I.   INTRODUCTION

The field of photodynamic therapy (PDT) has been the subject of a
number of recent comprehensive review articles. The aim of this chapter is
to deal in some depth with the dosimetry of PDT.

PDT is an experimental cancer treatment. It consists of an intravenous injection of the photo-sensitizing drug Photofrin II. The drug is
retained by the tumor at a higher level than the surrounding normal tissue.
24 to 72 hours later the tumor and surrounding normal tissue are exposed to
630 nm light. This results in the selective destruction of malignant
tissue. Significant levels of drug also are retained by cells of the reti-
culoendothelial system such as the Kupffer's cells of the liver and the
macrophages of the skin and spleen[1]. At the currently prescribed drug
doses (1 or 2 mg/kg) this results in post treatment cutaneous photosensitivity of at least 4 to 6 weeks. Consequently the patient suffers the incon-
venience of being forced to remain indoors during daylight hours for about
one month. Drug de-escalation protocols are being carried out at Roswell
Park as a part of phase two dosimetry studies[2]. Although these studies are
motivated by an attempt to improve the therapeutic ratio it is hoped that a
peripheral benefit of lower drug dose will be a reduction in the severity
and longevity of skin photosensitivity[3].

Since 1977 about 4,000 patients have been treated by photodynamic
therapy. Most of this work has been on late stage patients and was palliative in nature. We are just beginning to see the results of the treatment
of early stage lung cancer from Japan. A group at the Tokyo Medical College[4] and one at the Mayo Clinic[5] have data on a small but growing number
of cases with a five years follow up.

Phase III trials (randomization of patients into a standard therapy or
PDT) are underway. When the results of these studies are reviewed by the
FDA the drug will no longer be restricted to experimental protocols and
will be approved for routine use in oncology in the United States.

An application for regulatory approval has been filed with the Canadian Health Protection Board based on the accumulated clinical data in the
treatment of lung and bladder cancer. A similar filing has been made in
Sweden.

The effect of PDT (when the dosimetry is correct) is to cause the tumor tissue necrosis and spare the surrounding normal tissue although both receive the same light dose. Until recently the light dose which could be safely delivered was limited by normal tissue drug levels. This in turn limited the depth to which a satisfactory therapy could be delivered. The wish to improve the clinical results lead to a study of PDT dosimetry.

II. REVIEW OF THE DOSIMETRY LITERATURE

The development of PDT dosimetry has proceeded from purely empirical rules toward a theoretical framework relating the clinical parameters of injected drug dose and light dose to a predictable quantitative result. In the process insight into the physics of light propagation in tissues was necessary together with the relationship of tissue levels of sensitizer to injected dose.

The requirement for oxygen to be present for PDT to occur was demonstrated in mice[6]. The detection of the 1.27 micron fluorescence from the radiative decay of singlet oxygen generated by PDT in a mouse has been reported[7]. The measurement of singlet oxygen fluorescence is extremely difficult and has not as yet been reported by others[8].

Finally the biological response must be related to the effects produced by the combination of sensitizer and light. There is as yet no theoretical way to predict the response of tissue. Thus the dosimetry is only useful if we quantify theoretically the photodynamic dose used to achieve a measured biological end point (e.g. depth of tissue necrosis). Then we use this empirical response parameter (the minimum dose required to necrose tissue) in the theory to choose another set of drug and light doses which will reliably necrose tumor to any desired depth (assuming knowledge of the attenuation of the tisasue and within the practical limits imposed by the penetration of light). This is useful because as we shall see different drug and light combinations which result in the delivery of the same dose to the desired depth are not equivalent at any other depth. This is true because of the effects of photobleaching. Photobleaching, the photochemical destruction of sensitizer, results in the limitation of the PDT dose in the tissue to a finite value which depends only on the amount of sensitizer present at the start of treatment and the bleaching rate (a single constant for all tissue).

III. LIGHT IN TISSUE

L. O. Svaasand was one of the first to recognize that because propagation of red light in tissues is dominated by scattering it is best described as a process of diffusion[9]. The transport theory of Chandrasekhar has been extremely useful in the description of the diffusion of photons through tissue[10]. The early European workers in this field tended to employ the Kubelka-Munk theory of light transport which had several disadvantages when dealing with this problem. It required the determination of light fluxes traveling in several directions in tissue which were not accessible to measurement. It has been recently demonstrated that the Kubelka-Munk parameters are connected to those of diffusion theory[11]. The great advantage of diffusion theory is that it describes light transport using only two parameters, a diffusion coefficient and an absorption coefficient. These parameters are measurable by noninvasive techniques and can be determined in the clinical situation (see for example the References 12 and 13). Furthermore once the light is completely diffused (i.e. isotropic) as the result of multiple scattering is described by a single parameter, the total attenuation coefficient, which can be determined by the noninvasive technique described in Reference 14.

The transition from collimated to fully diffuse light creates a problem in determining the proper boundary value to use for the space irradiance at the point just below the tissue surface where diffusion theory becomes an accurate description of the transport process. Flock has approached this problem using Monte Carlo calculations which generate random numbers describing the scattering of each photon (both path length and direction). When carried out on a statistically significant number of photons an exact description of the space irradiance is produced[15]. Unfortunately calculating more than a few dozen such scatterings for a large number of photons taxes the capacity of all but the largest computers and is very time consuming. Profio and Doiron have employed the discrete elements method of computer modeling to solve this problem for some clinically interesting geometries[16]. A more detailed review of these developments can be found in Reference 17.

IV. SENSITIZER LEVELS

The determination of tissue sensitizer levels has proceeded using tritiated and carbon 14 labeled sensitizer (HpD)[18]. This together with recent work[19] using mice has led to the conclusion that the tissue drug levels (although quite different from each other) are all proportional to injected dose over a wide range (0.1 mg/kg to 50 mg/kg) for Photofrin II. In mice a linear relationship between the tissue fluorescence and the tissue concentration of Photofrin II as determined by chemical extraction has been found[14]. Again the relationship must be determined for each tissue[14].

Reflection spectroscopy has been used[12] to determine the concentration of Photofrin II in tissue models such as Neutralipid solutions or ground meat to which known concentrations of sensitizer had been added. Theory and experiment were in good agreement.

V. PDT DOSIMETRY: THE EFFECTS OF SENSITIZER AND LIGHT

V.1. The Current Clinical Situation

The optical properties of tissue are not entirely uniform. The distribution of sensitizer in tissue is also not completely uniform. The observed clinical success in producing differential necrosis of tumor is dependent upon a good therapeutic ratio in the tumor uptake of the sensitizer (Photofrin II) and the ability to deliver a light dose which raises the total photodynamic dose above the threshold for tissue necrosis. Thus perfect knowledge of the distribution of light and drug in tissue is not required for situations where sufficient experience exists to guide the choice of dosimetric parameters. Indeed the current clinical situation is such that no measurements to determine tissue optical properties or drug levels are ever used to guide the planning of treatment. This is unfortunate because in the future when novel treatment sites or new drugs are involved the development of clinical dosimetry will be limited to a purely empirical approach unless appropriate (non invasive) instrumentation is developed to produce information relevant to treatment planning. It has been suggested[20] that fluorescence measurements could be used for determination of sensitizer levels in PDT dosimetry.

V.2. Theory of PDT Dose

As the drug light product represents the probability of useful event (i.e. the generation of an excited sensitizer molecule that generates

singlet oxygen through collision with an oxygen molecule), the first definition of PDT dose was the drug light product[21]. The concept of absorbed
dose in explicit fashion with the inclusion of the sensitizer molar absorption coefficient directly into the theory of dosimetry has been introduced
in 1986[22]. Later on the effects of photobleaching of the sensitizer have
been incorporated into the dosimetry theory[23,24].

Photofrin II acts by absorbing light. This energy is then transferred
by nonradiative processes to the molecular oxygen dissolved in the target
tissue. The product of this transfer is singlet oxygen, a highly reactive
and toxic form of the oxygen molecule.

Until about 1985 the general view of PDT dosimetry held that the drug
(Photofrin II or its earlier less purified form Hematoporphyrin derivative
or HpD) was photochemically stable. Thus the singlet oxygen produced
(given a sufficient supply of oxygen from the circulating blood) was proportional to the number of joules absorbed per unit volume of the tissue by
the drug. This was in turn proportional to the number of Joules delivered
and the concentration of the Photofrin II. Thus one expected the same biological end point as long as long as the drug light product was constant.

At the same time it was apparent from a variety of drug uptake studies
that the amount of drug present in any tissue was proportional to the
amount of drug injected[14]. As long as the same tissues were studied one
could substitute the injected dose for the actual tissue level and observe
the same correspondence of result and drug light product. Indeed there is
a range of drug and light values for which reciprocal changes in drug and
light produce the same effect[25].

About 1986 it became apparent (from fluorescence measurement and
chemical extraction) that Photofrin II is photobleached in-vivo during
PDT[26,27]. That is the molecule is altered by photochemistry so that it no
longer absorbs light and thus can no longer generate singlet oxygen. Photobleaching occurs slowly and becomes significant only if large light doses
are employed.

The theory of dosimetry was significantly altered by the inclusion of
the effect of photobleaching. One of the triumphs of the revised theory
was the prediction (based on measurements of the photobleaching rate) that
reciprocity must begin to fail at a particular low drug and high light
level. When the reciprocity experiments were extended to this region the
predicted failure occurred[28].

A second more significant prediction of the theory is that there exists a finite upper limit on the photodynamic dose that can be delivered by
an infinite amount of light. This limit is directly proportional to the
initial drug level and inversely proportional to the bleaching rate (the
bleaching rate is constant in-vivo)[23]. The motive for a dosimetry which
exploits photobleaching is to protect the normal tissue by bleaching essentially all of the drug initially present before the threshold for tissue
necrosis is exceeded. This requires a sufficiently low injected dose. It
must also be possible to achieve a therapeutic effect at these drug levels.
This requires a higher drug level in the tumor than in the surrounding normal tissue. The clinically effective level of Photofrin II has been reduced from 2 mg/kg to 0.57 mg/kg in patients being treated for breast cancer
metastatic to the chest wall. At injected doses of 1 mg/kg or less it appears that skin levels of the drug are low enough to prevent skin necrosis
even at light doses greater than 200 Joules/cm^2.

The result of such an approach to dosimetry should be the ability to
use light doses determined by the depth of the desired effect and not lim-

ited by concern for normal tissue in the field. This can be illustrated for Photofrin II by comparing the safe skin dose of 36 J/cm^2 at 2 mg/kg with the safe dose of greater than 200 J/cm^2 at 1 mg/kg. Traumatized skin and areas of compromised lymphatic drainage are less tolerant at all drug levels.

It remains to be seen whether all of these hoped for results can be realized with the current drug, Photofrin II.

REFERENCES

1. P. J. Bugelski, C. W. Porter and T. L. Dougherty, Cancer Research (1981) 41:4605-4612
2. M. Schue, U. O. Nseyo, W.R. Potter, T. L. Dao and T. J. Dougherty, "Photodynamic therapy for the palliation of locally recurrent breast carcinoma", Journal of Clinical Oncology vol. 5, 11:1766-1770 (1987)
3. D. G. Boyle and W. R. Potter, "Photobleaching of Photofrin 2 as a means of eliminating skin photosensitivity", Photochemistry and Photobiology vol. 46, 6:997-1001 (1987)
4. C. Konaka, H. Kato, Y. Hayata, "Lung cancer treated by photodynamic therapy alone: survival for more than three years", Lasers in Medical Science 2:10-17 (1987)
5. D. A. Cortese and J. H. Kinsey, "Hematoporphyrin derivative photo-therapy in the treatment of broncogenic carcinoma", Chest 86:8-13 (1984)
6. C. J. Gomer and T. J. Dougherty, "Determination of ^{3}H and ^{14}C-hemato-porphyrin derivative in normal and malignant tissue", Cancer Res. 39:146-151 (1979)
7. J. G. Parker, "Optical monitoring of singlet oxygen during photo-dynamic treatment of tumors", IEEE Circuits and Devices magazine Jan. 1987, 10-21 (1987)
8. M. A. J. Rogers, "On the problems involved in detecting luminescence from singlet oxygen in biological specimens", J. Photochem. and Photobiol. 1:371-373 (1988)
9. L. O. Svaasand and R. Ellingsen, "Optical properties of the human brain", Photochem. and Photobiol. 38:293 (1985)
10. L. I. Grossweiner, "Optical dosimetry in photodynamic therapy of neo-plastic disease", in "Photodynamic Therapy of Neoplastic Disease" CRC Press, Inc. Boca Raton Fl (1989)
11. M. J. C. Van Gemert and W. M. Star, "Relations between the Kubelka-Munk and the transport equation models for anisotropic scattering", Lasers in the Life Sciences, 1: 287-1987 (1987)
12. B. C. Wilson, M. S. Paterson and D. M. Burns, "Effects of photo-sen-sitizer concentration in tissue on the penetration depth of photo-activating light", Lasers in Medical Science 1:235-243 (1986)
13. B. C. Wilson, M. S. Patterson and S. T. Flock, "Indirect vs direct techniques for the measurement of the optical properties of tissues", Photochem. and Photobiol. 46:601-608 (1987)
14. W. R. Potter and T. S. Mang, "Photofrin II levels by in vivo fluores-cence photometry", in "Porphyrin Localization and Treatment of Tumors" pp. 177-186, D. R. Doiron and C. J. Gomer eds, Published by Alan Liss Inc. (New York, 1983)
15. S. T. Flock, B. C. Wilson and M. S. Patterson, "Hybrid Monte Carlo-diffusion modeling of light distribution in tissue", SPIE O-E LASE'88, Los Angeles, CA, 1, 10-17 (1988)
16. A. E. Profio and D. R. Doiron, "Transport of light in tissue in photo-dynamic therapy", Photochem. and Photobiol., 46:591 (1987)
17. L. I. Grossweiner, "Optical dosimetry in photodynamic therapy of neo-plastic disease", in "Photodynamic Therapy of Neoplastic Disease" CRC Press Inc., Boca Raton Fl (1989)

18. C. J. Gomer and N. J. Razum, "Acute skin response in albino mice following porphyrin photosensitization under oxic and anoxic conditions", Photochem. and Photobiol. 40:435-439 (1984)

19. D. A. Bellnier, Y. Ho, R. K. Pandey, J. R. Missert and T. J. Dougherty, "Distribution and elimination of the tumor localizing component of hematoporphyrin derivative in mice", Photochem. and Photobiol., 51:97-101 (1990)

20. A. E. Profio and J. Sarniak, "Fluorescence of HpD for tumor detection and dosimetry in photodynamic therapy" in "Porphyrin Localization and Treatment of Tumors", pp. 177-186, D. R. Doiron and C. J. Gomer eds., Published by Alan Liss Inc., (New York, 1983)

21. A. E. Profio and D. R. Doiron (1981), "Dosimetry considerations in phototherapy", Med. Phys. 10:10 (1981)

22. L. I. Grossweiner, "Optical dosimetry in photodynamic therapy", Lasers Surg. Med. 6:462-466 (1986)

23. W. R. Potter, T. S. Mang, and T. J. Dougherty, "The theory of photo-dynamic therapy dosimetry: Consequences of photodestruction of sensitizer", Photochem. and Photobiol., 46:97-101 (1987)

24. P. A. Cowled and I. J. Forbes, "Photocytotoxicity in vivo of haemato-porphyrin derivative components", Cancer Lett. 28:111-118 (1985)

25. V. H. Fingar, W. R. Potter and B. W. Henderson, "Drug and light dose dependence of photodynamic therapy: a study of tumor cell clono-genicity and histological changes", Photochem. and Photobiol., 45, 5:643-650 (1987)

26. J. Moan, "Effect of bleaching of porphyrin sensitizers during photo-dynamic therapy", Cancer Letters, 33:45-53 (1986)

27. T. S. Mang, T. J. Dougherty, W. R. Potter, D. G. Boyle, S. Somer, and J. Moan, "Photobleaching of porphyrins used in photodynamic therapy and implications for therapy", Photochem. and Photobiol., 45:501-506 (1987)

28. M. S. Patterson, B. C. Wilson, J. W. Feather, D. M. Burns and W. Puska, "The measurement of dihematoporphyrin ether concentration in tissue by reflection spectrophotometry" Photochem. and Photobiol., 46, 3:337-343 (1987)

PHOTODYNAMIC THERAPY: EXPERIMENTAL RESULTS

PHOTODYNAMIC THERAPY VARIATION IN BIOLOGICAL ACTIVITIES

OF HPD AND BIOLOGICAL EFFECTIVENESS WITH RIF-1

F. A. H. Al-Watban

King Faisal Specialist Hospital and Research Centre
Biomedical Physics Dept.
Riyadh, Saudi Arabia

I. PHOTOMEDICINE AND LASERS

Photobiology applied to medicine is one of the oldest interests of
men. Since ancient times, human beings have had a belief in the health
giving properties of sunlight and its biological effects on all life on
earth. Accordingly, heliotherapy (or sun exposure) was prescribed before
any modern drugs for the treatment and prevention of cutaneous tubercu-
losis, rickets and maintaining sterility (i.e. killing germs); further evolution is shown in Figure 1.

The growing interest in advancing photomedicine has been encouraged by
the observed photochemical alterations in biomolecules, which affect cell
viability and is due to exposure to visible and ultra-violet (U.V.) radiation. Many artificial light sources have been manufactures to serve this
purpose, but during the last decade, a growing interest in photomedicine
has emerged from the creative use of lasers. U.V. and visible light as a
laser beam can now be used to treat chronic common skin diseases, prevent
certain forms of blindness, perform bloodless surgery and diagnose and
treat a variety of human diseases including cancer. Each year we learn
more about this mode of treatment.

Historically speaking, the principles necessary for the laser were
established as early as the 19th century. Einstein in 1917 proposed the
concept of stimulated emission, and thus laid the foundation for Schawlow
and Townes in 1958, to describe the principles of the laser. Later, Maiman
in 1960 registered the first observation of stimulated emission in a visible portion of the electromagnetic spectrum by exciting a ruby rod. Since
that time laser beams have been generated from different mediums such as
gas, liquid and solid, and the numbers of lasers are counted in hundreds,
and cover the radiation from visible to near ultraviolet. They are characterized by different wavelenghts of radiation and modes of emission. From
these scientific prototype lasers emerged lasers suitable for military,
industrial and medical applications.

What is the laser? Laser is an acronym which stands for Light Amplification by Stimulated Emission of Radiation. The Laser is amplified to
extremely high intensity by an atomic process called stimulated emission
and the word radiation in this context means energy transfer, and does not
mean radioactive material or ionizing radiation, since energy moves from

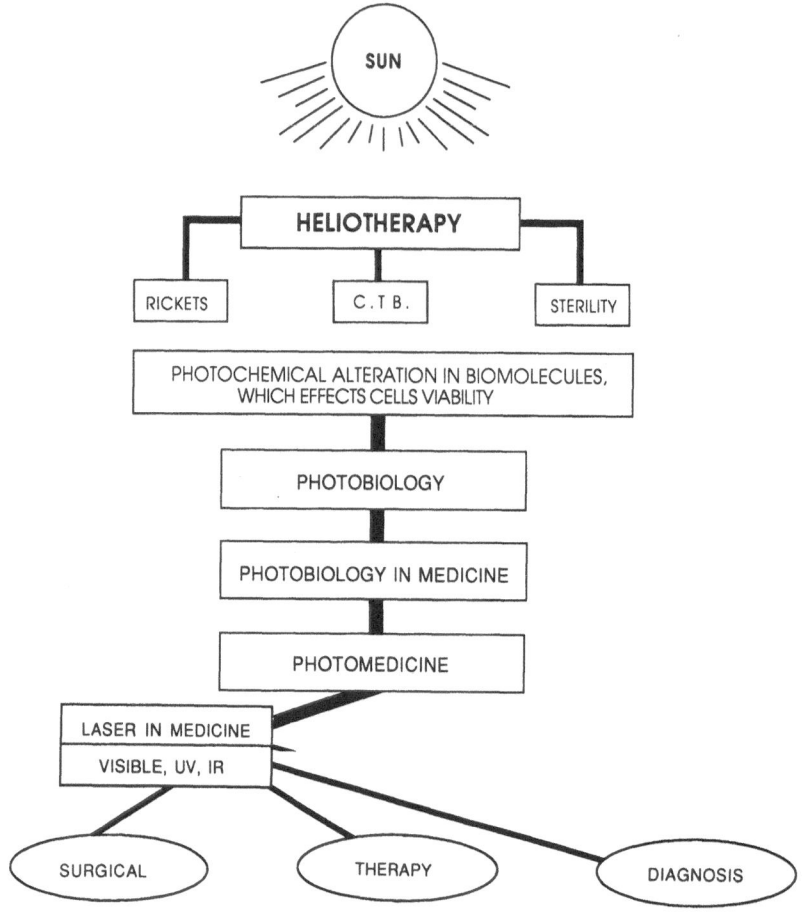

Fig. 1. Heliotherapy and photomedicine

one location to another by conduction, convection and radiation. Lasers in the optical and infrared regions produce non-ionizing radiation; whereas lasers in the ultraviolet and X-ray region produce ionizing radiation.

All lasers have the characteristic of emitting beams of light which are extremely intense, small, and nearly non-divergent, consisting of mono-chromatic radiation with all waves in phase (.e. Coherent) unlike conven-tional light which is made up of light of various frequencies and is highly divergent. Accordingly, laser light is capable of mobilizing immense heat and power when focused at close range, and therefore is extremely useful as a tool in surgical procedures.

Laser applications in medicine principally use three different kinds of laser: the Argon laser in the visible region, the YAG laser in the near infrared region, and the CO_2 laser in the infrared region. Every day more lasers are being manufactured to supply the overwhelming demand in photome-dicine.

Advancement of photomedicine has been encouraged by the observed photochemical alterations in biomolecules due to exposure to light. Modern photomedicine emerged from the creative use of lasers, and the interaction of laser light with living tissues. There can be basic photochemical, pressure, electromagnetic, and heat effects which can interact individ-ually or in combination depending on the laser type. With the photochem-ical effects initiated, photodynamic therapy (PDT) [which refers to the combined use of a photosensitizer and laser] selectively destroys the malignant cells. To address the basic principles of PDT, a Hematoporphyrin derivative (as the photosensitizer) and an Argon pumped dye laser are described, together with discussion on neoplastic tissue volume limitations and response using a resistant tumor cell type (murine fibrosarcoma). The clonogenic survival rates and tumor growths following PDT were studied as a function of tumor volume. After PDT, all tumors (> 90 mm^3) responded with immediate swelling and skin discoloration, followed by necrosis, tumor delay, and regrowth. The survival rates (< 0.03%) corresponding to a volume lower than 50 mm^3 gave complete tumor eradication. PDT with oncology has the potential to advance photomedicine beyond the year 2000.

II. LASER TISSUE INTERACTION

The interaction of laser light with living tissues can basically exhibit four different kinds of effect: heat, photochemical, pressure and electromagnetic. In every application one of these effects is more dominant than the others. The most important feature for determining biological effects is the capability of the beam to be absorbed, selectively, according to the color (i.e. frequency or wavelength); see Figure 2.

Since the laser beam is of an electromagnetic nature, it is capable of developing free radicals. The heat effect is used to cut, coagulate, carbonize, and vaporize tissues and blood in surgical applications referred to as the laser scalpel. Lasers immediately convert most kinds of medical endoscopes to therapeutic as well as diagnostic devices.

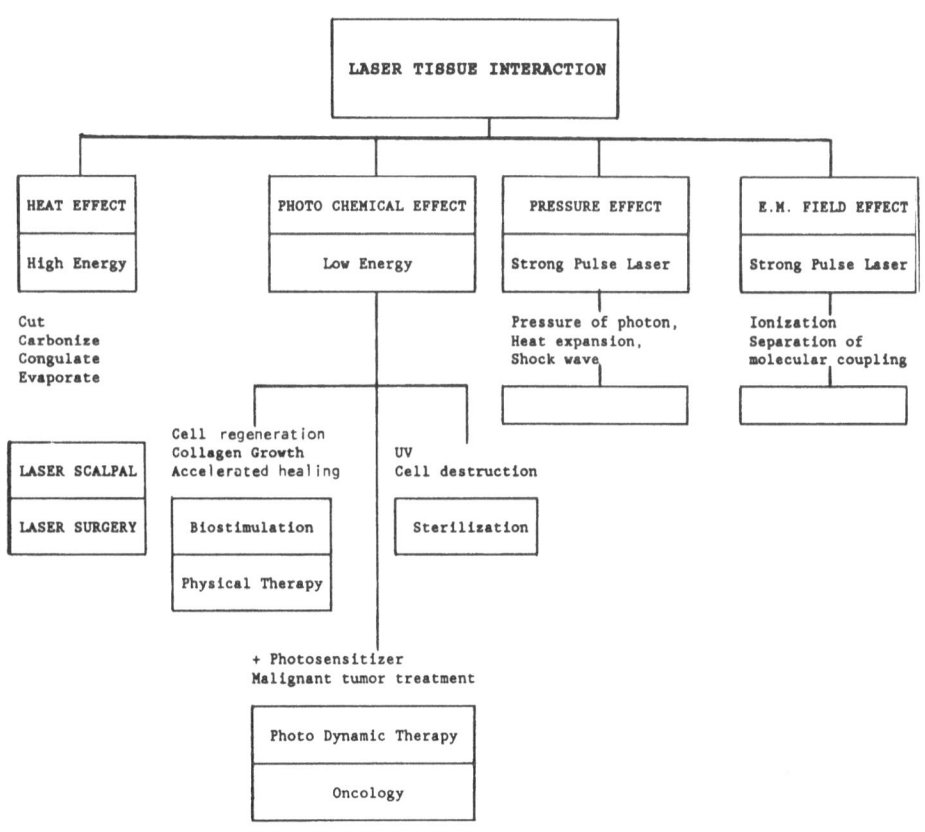

Fig. 2. Laser tissue interaction

Here we are primarily interested in the photochemical effects of the laser, which arise from the selective absorption of laser radiation by frequency (or color). These effects can be used for cell destruction, (e.g. using the Argon laser blue light to destroy cells containing red pigmentation in treatment of nevi, red birth mark and portwine stains). They also exhibit cell regenerative capabilities which accelerate the healing process, possibly through stimulating the hydroxyproline, which is one of the amino acid compounds of collagen, and they can also be used for sterilization.

To initiate the photochemical effects of the laser a low energy laser is required. Generally speaking, this can activate and stimulate biological molecules by using the laser alone (a new discipline called biostimulation uses such a technique). Alternatively, using a low energy laser to activate a drug inside the cells to produce selective effects was originally called phototherapy and/or photochemotherapy, and has recently become known as photodynamic therapy.

III. PHOTODYNAMIC THERAPY

The term photodynamic therapy (PDT) refers to the combined use of drugs (i.e. a photosensitizer) plus non-ionizing electromagnetic radiation (laser) to treat and diagnose malignant cells. In the doses usually used, the photosensitizer alone or the laser alone has no effect. Hematoporphyrin derivative (HPD) is one example of a photosensitizer that selectively remains in higher concentration in malignant tissue, and has therapeutically been used in combination with the laser, to treat cancer. The HPD is also fluorescent and can be used to detect tumor masses. The depth of therapeutic action in tissues is influenced by the properties of the stimulating photons (laser). Thus, careful selection of the photosensitizer, wavelength, timing, and power of exposure can produce specific or selective effects on the viability and function of target cells that are accessible to laser beam directly or by using fiberoptics through an endoscope[1÷3].

PDT is a new treatment modality in which specific chemical reactions can be produced in a controlled manner at certain tissue depths in selected body sites, using laser photons as the activation energy for the photosensitizer. Figure 3 shows the basic procedure for PDT.

PDT employs molecules that act as photosensitizers (e.g. HPD photoprin I or DHE photoprin II) and have two types of electronically excited states. On illumination, the ground state sensitizer S is converted to a singlet excited state ^1S, which typically has a very short lifetime. In most efficient PD sensitizers the excited singlet state undergoes conversion to another energy rich form, the triplet state ^3S. This state has a much longer lifetime, and therefore has a much greater probability of undergoing chemical reaction, either Type I by electron transfer or Type II by energy transfer. Typical results are on oxidized form of the substrate (cellular constituent) and ground state sensitizer as the final products[4÷6].

$$S + \text{light energy (h}\nu\text{)} \longrightarrow {}^1S$$
$${}^1S \longrightarrow {}^3S$$
$${}^3S + {}^3O_2 + S$$
$${}^1O_2 + \text{substrate*} \longrightarrow \text{oxidation}$$

_____

(*) cellular constituent

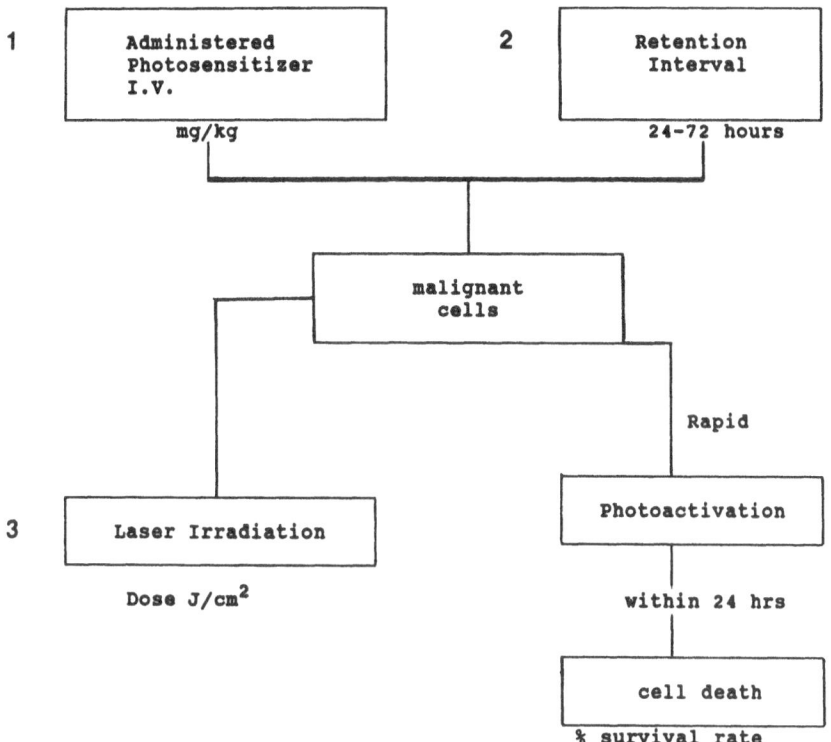

Fig. 3. Photodynamic therapy. In this treatment a photosensitizer
(e.g. HPD), is first administered intravenously (I.V.).
Then, after a retention interval of typically 24-72 hr
to allow accumulation of HPD in the tumor tissue, the tumor
volume is irradiated with light (e.g. dye laser at 630 nm
when HPD is used). The HPD is thereby photoactivated, generating reactive products which cause cell death. The tumor
response is usually rapid.

The most vulnerable target for PDT photodamage is the cell membrane.
DNA and Mitochondrial effects have also been observed[7]. The direct cause
of tumor cell death could be cell membrane oxidization and/or disruption of
the tumor vasculature.

PDT has been applied to a wide variety of malignant tumors, generally
after failure of other therapies. To date more than 2000 patients have
been treated, including those with lung, bladder, esophagus, gynaecological, eye and skin cancer[8].

The therapeutic objective in PDT is to deliver an adequate light dose
(laser energy density, J/cm^2) to the photosensitizer (e.g. HPD) which has
already been injected and left for a duration (retention interval) to
localize the malignant cells, and to eradicate the tumor tissue (Figure 4).

The dosimetric relationship[9÷11] between the laser energy density (light
dose, J/cm^2), the HPD dose (mg/kg), tumor weight, and HPD retention intervals (hr) have been carried out with empirical optical dosimetry, based on
experience with tumors of comparable type and size, and the laser light
dose calculated from the total power density (dose rate) mW/cm^2 and the
exposure times (sec). In summary, to be more specific in the understanding of PDT we should address the following topics:

        i)   The tumor model

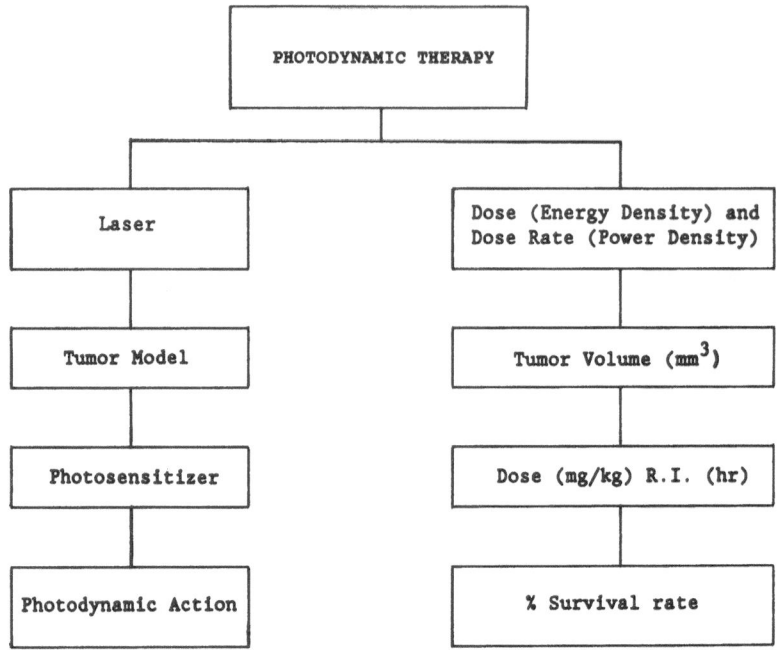

Fig. 4. Photodynamic therapy: Dosimetric parameters.

<pre>
        ii)   The laser system
       iii)   The photosensitizers
        iv)   The PDT tumor response.
</pre>

IV. TUMOR MODEL - MURINE FIBROSARCOMA RIF-1

The RIF-1 tumor was provided by Peter Twentyman at Stanford University and was prepared as described by him[12]. This was made available to us in March 1984, and was chosen for our studies because of several favorable characteristics: (a) reproductive integrity can be assayed in vitro after either in vitro or in vivo exposures; (b) has a low frequency of spontaneous metastases, thus allowing a long follow-up period after the treatment; (c) for all practical purposes, is currently being used in several laboratories worldwide, allowing for comparison of results. The RIF-1 was grown in the alternating in vitro and in vivo culture protocol, as outlined by Twentyman. Inoculation of experimental tumors was done from in vitro passages, by subcutaneous injection of viable tumor cells in the thigh region of C3H mice[13].

The RIF cultures used in clonogenic assays were incubated at 37°C in 5% CO_2, 95% air undisturbed for 7 days, at which time they were fixed, stained and macroscopic colonies of more than 50 cells counted. The relative survival rates were calculated as the number of colonies per gm tissue tumor (exp.) multiplied by 100 and divided by the number of colonies per gm tissue tumor (Control).

The volume of flank tumor was estimated by taking two caliper measurements, both at right angles, using the approximation volume = $1/2$ L.W^2, where L is the major axis and W is the minor axis. Comparison with saline water displacement (and tumor mass) data indicates that this approximation probably gives the true volume, from 50 to 500 mm^3. Outside this range, the volume tends to be under-estimated and over-estimated for larger and smaller sizes respectively, (taking into account that all tumors were shaved and hair removal cream applied). Uncertainty in viable cell counts and/or inaccuracy in inoculation of RIF cells intramuscularly at the base of the gastrocnemius muscle of the leg (because of the small flank of the mice) resulted in tumor volume variations; therefore, prior to PDT for every group, tumor variations larger or smaller than the normal established growth curve[14] were isolated from treatment (in addition, the general criteria of spheroid type tumors were a basis for selection[12], see Figure 5.

V.   LASER SYSTEM

An argon pumped tunable dye laser (Coherent Radiation) using DCM dye was used as a source of red light with wavelength 630 nm.

A modified spectrophotometer (Gilford Instruments) was used to check the wavelength. The addition of a fiber interface and a shutter box with an automatic timer system allowed for control of the exposure time. Laser power was measured from the fiber tip with a Coherent 210 power meter and integrating sphere calibrated for 630 nm (Laserguide). The energy per unit area (light dose) was calculated from the power per unit area (dose rate).

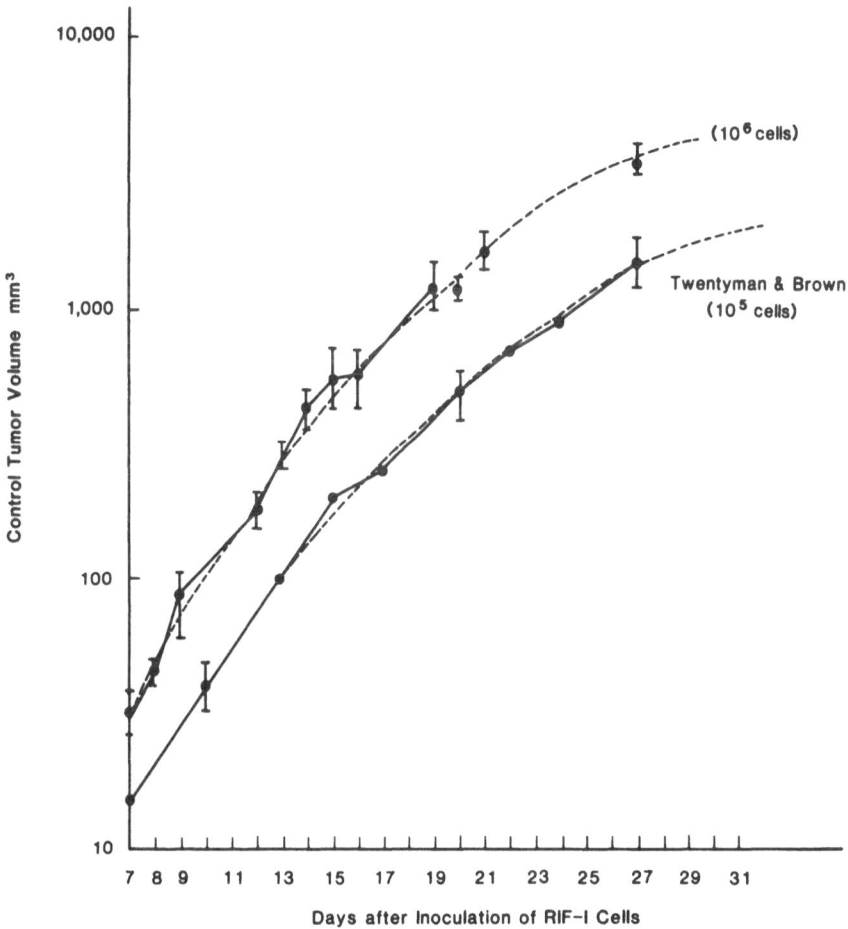

Fig. 5.   Tumor volume versus days after inoculation of RIF-1 cells shows the inoculation of 10^6 and 10^5 cells.

Tumors were given localized external laser light at 300 mW giving a 132 mW/cm^2 light dose at 14 min. This was directed along a 600 μm quartz fiber (Fibres Optiques Industries, France) through an unreflecting stainless steel cylinder, whose diameter was fixed according to tumor size, with an allowance for a margin of normal tissue. The optical fiber was fixed in the cylinder perpendicular to the treated area. The light spot was measured at a metallic O-ring guide attached to the end of a stainless steel cylinder, which was adjusted in accordance with the visible mass of the tumor bulk to ensure a fixed power density applied to every tumor size.

For interstitial irradiation as in Figure 6 (as a second treatment), 200 mW laser radiation was measured from the fiber tip, achieving a 100 J light dose in 8.34 min (optimum dose based on previous unpublished results), and directed along a 400 μm quartz fiber (General Fiber Optics Inc., N.J.) through a No.19G butterfly needle (which penetrated into the center of the tumor mass). For superficial and interstitial irradiations, the temperature variations were measured by a needle-like platinum thermocouple (Doric-450, Doric Scientific, CA) implanted into the visual surface and visual center of the tumor bulk, respectively.

VI. THE PHOTOSENSITIZERS (HPD)

The concept of PDT for malignant tumors is based upon the action of the photosensitizer. This is produced by the absorption of a photon with a wavelenght sufficient to promote singlet oxygen capable of killing malignant cells. Interaction with the cell membranes and/or vascular damage thus causes tumor cell death. In spite of various studies on the complex structure of HPD, as yet, there has been no conclusion about the components directly responsible for its biological activity. Furthermore, very few comparative studies on different preparations of HPD have been done.

When work on PDT commenced in Saudi Arabia, the requirements of our Research Centre and the FDA made it preferable for us to prepare, rather than import, HPD. It was therefore important that we have an acceptable

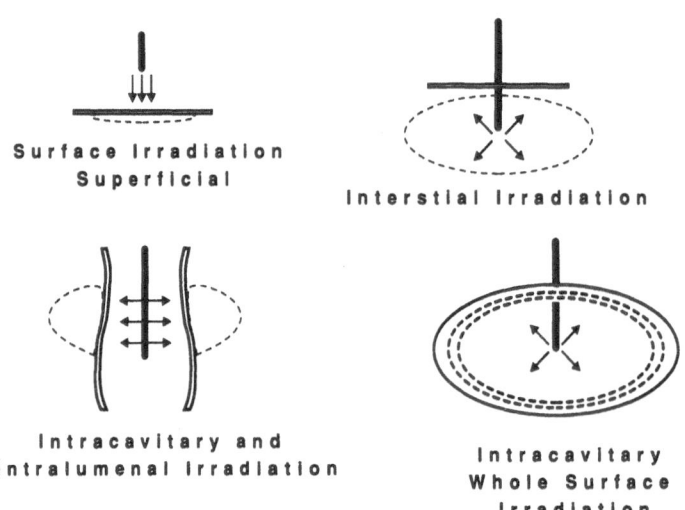

Fig. 6. Examples of the irradiation geometries used in PDT. The dashed area represents the tumor, and the thick lines represent fiber optics.

assay, enabling comparison between our material and the product used elsewhere.

Accordingly, all batches of HPD were prepared by the acetylation of hematoporphyrin dihydrochlorides (Alpha Products or Calbiochem Corp.), with a mixture of sulfuric and acetic acid as by Dougherty's modification of Lipson's method. This solution (HPD) was filter sterilized and stored in darkness at -20°C until used [15].

An attempt was made to compare tumor response from several HPD samples and to analyze their variations in biological effectiveness in vitro and in vivo using RIF-1 in C3H mice.

HPD biological variations were 2 to 10% in vitro (Figure 7) and 4 to 14% in vivo giving 10 to 16.5 days in growth delays with RIF tumor achieving 20 to 62% tumor response, in comparison with 20 to 80% tumor response for SMT-F [16].

VI.1. Absorption Spectra

All samples of HPD were diluted to 25 μg/ml with isotonic saline and scanned from 250 to 700 nm with a Varian Model Cary 210 spectrophotometer. The absorption spectra of HPD batches were identical with a major absorption band ~374 nm and four minor bands of decreasing magnitude at ~ 500 nm, ~ 533 nm, ~566 nm and 626 ~ 629 nm (Figure 8).

The HPD clears most normal tissues rapidly but is retained by essen-

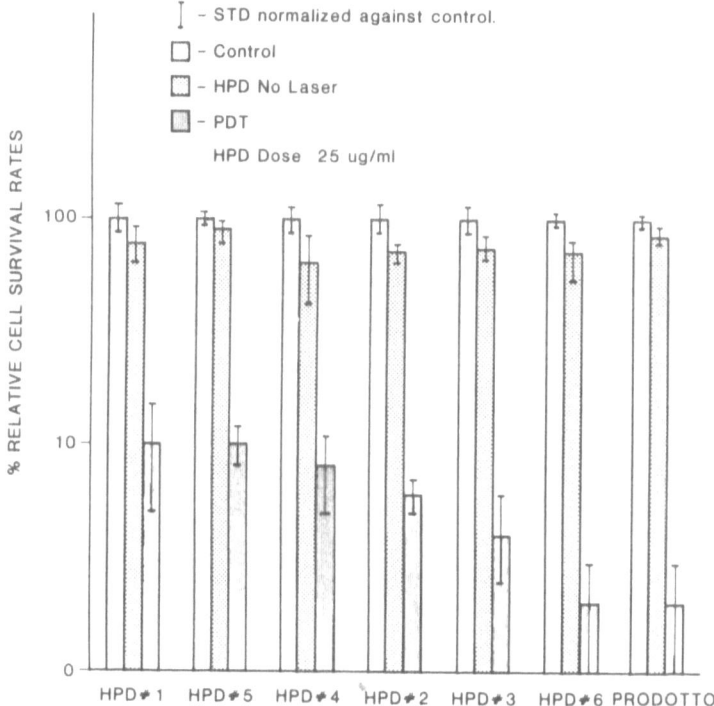

Fig. 7. In-vitro clonogenic assay data of RIF cells for several HPD samples prepared by the same method put with different HP dihydrochloride brand, also with the imported ready injectable solution HPD Prodotto from Genoa, Italy. The relative survival rates varied between 2 to 10%.

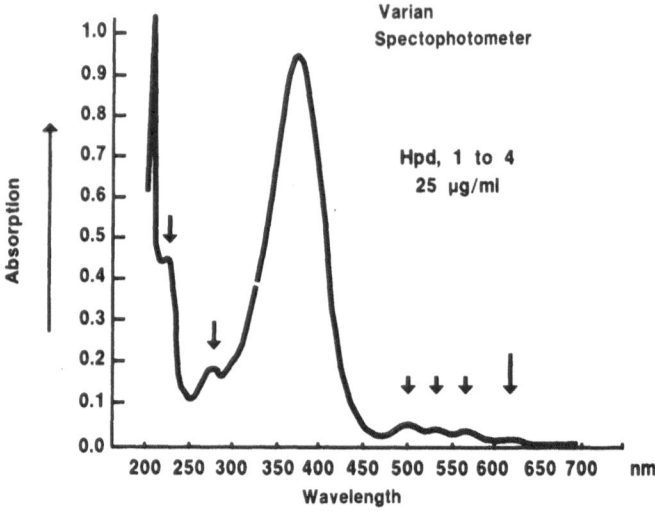

Fig. 8.   Absorption spectra of some HPD samples.

tially all tumors. The HPD must also be activated by light which has good tissue penetration. HPD is activated by < 600 nm red light for induction of light-catalyzed photodamage.

VI.2. Thin Layer Chromatography (TLC)

Using the ethylacetate: water (1:1) solvent system and stationary phase aluminium-backed silica gel sheets, each batch showed 3 components with the same RF values of 0.3, 8.2 and 1.0.

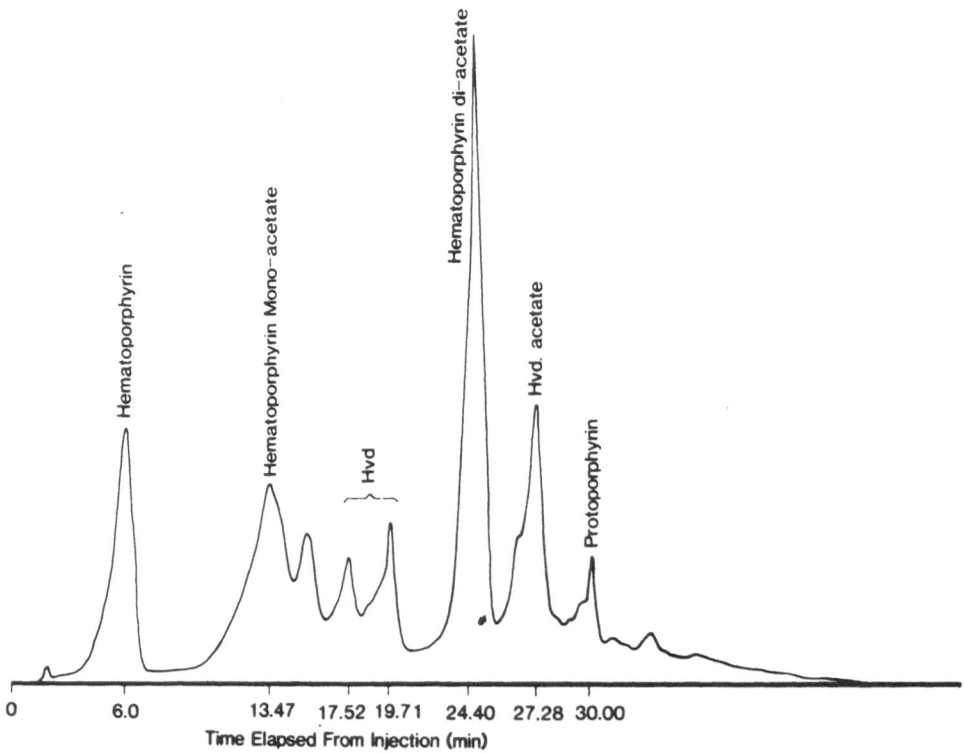

Fig. 9. HPLC chromatograph of acetylated HP.

VI.3. High Performance Liquid Chromatography (HPLC)

Initially, the feasibility of separating the porphyrin mixture was examined using conventional TL chromatography. These separation methods were clearly not satisfactory, so attention was turned toward HPLC methodology. Due to the lack of instrumentation and control, four samples of HP acetate and the injectable solutions of HPD represents all prepared batches sent to Roswell Park Memorial Institute (RPMI) for analysis. The column chromatography was carried out as described in the Ref's 17 and 18. The HPLC analysis showed that acetylated hematoporphyrin (HPA) was very similar to the control (Figure 9).

It should be noted that some variations in retention time and resolution occur in the chromatogram with different columns and/or as the columns age.

VI.4. HPD Dosage

The dose of HPD administered (20 mg/kg body weight) has been empirically derived, the upper limit being set by skin photosensitivity and necrotic tissue prior to laser applications.

The 50% lethal dose (LD_{50}) at 24 hr in mice, in the absence of laser radiation, is 150 mg/kg. Although HPD has proved to be an adequate photosensitizer there are limitations which might be overcome by alternative drugs. 630 nm is the longest useful wavelength for HPD. Longer wavelengths with other photosensitizers (e.g. pathalocyanines, chlorins and purpurins) could give more effective light penetration in tissue, and allow treatment of larger tumor volumes.

VII. PHOTODYNAMIC THERAPY TUMOR RESPONSE

Inoculation of experimental tumors was done from previous in vitro passages in two to five cycles by subcutaneous injections of 1×10^5 or 5×10^5 viable tumor cells in the thigh region of C3H mice. Treatment time commenced after 7 to 23 days (when tumor sizes had appreciable differences) after RIF injection.

For superficial irradiation, the noncontrol tumor-bearing mice were given an intraperitoneal injection of HPD at 20 mg/kg body weight (optimum for a large tumor). After a retention interval of 40 hr, the animals given HPD were restrained without anesthesia during treatment in specially designed holders with sheets of black paper covering the surrounding area.

Following each treatment, mice from both treated and control groups (which had comparable tumor volumes) were sacrificed immediately and the tumors subjected to in vitro clonogenicity assays. The remaining mice from each group were returned to the cages and kept under normal conditions for daily measurement of tumor size and assessment of skin response.

For interstitial irradiation (as a second treatment), 0.1 ml (corresponding to 20 mg/kg) HPD was injected into the tumor, with a retention interval of 20 hr.

VII.1. Superficial PDT

A total of 33 tumor-bearing mice were used for this study. When tumor size was approximately between 50 and 700 mm^3, 24 mice were injected with HPD; nine (without HPD and laser) served as controls. The HPD-injected mice were divided into six groups (each individual group having comparable tumor volumes) and treatment begun after the retention interval. Average tumor size (T), as measured at the time of treatment, were, for small tumors, $T_1 = 90$ mm^3, $T_2 = 157$ mm^3, and $T_3 = 214$ mm^3; for medium tumors, $T_4 = 335$ mm^3; and, for large tumors, $T_5 = 887$ mm^3 and $T_6 = 1.014$ mm^3. The controls (nine mice) were formed into three groups. Average tumor volumes (measured at the time the treated group received laser irradiation) were, for small tumors, $C_1 = 170$ mm^3; for medium tumors, $C_2 = 317$ mm^3; and, for large tumors, $C_3 = 1.014$ mm^3.

Tumor growth. Results of the superficial irradiation (Figure 10, Table 1) showed an immediate response of swelling in the leg, followed by skin discoloration, which later developed into skin necrosis (varying in size and depth) with occasional scab and burn formation. Tumor delay between 1 and 18 days for medium and small tumors, respectively, was noted (no appreciable delay occurred for the large tumors), followed by regrowth. Regrowth rates were slower in the treated groups compared with the control groups (Figure 11) and were as seen in Table 1. A complete eradication of

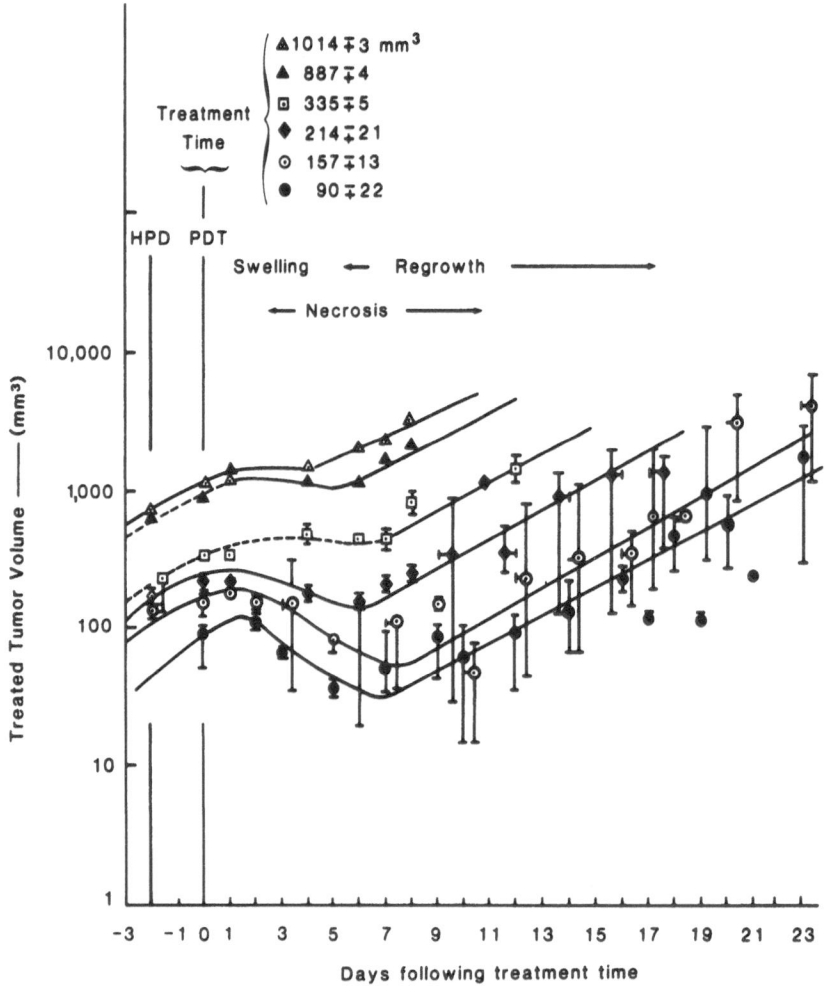

Fig. 10. The superficial treated tumor volume versus days following PD treatment time (marked 0) for different tumor volume. HPD injected prior to treatment time (marked –2). The curve also shows the swelling, necrosis and regrowth periods.

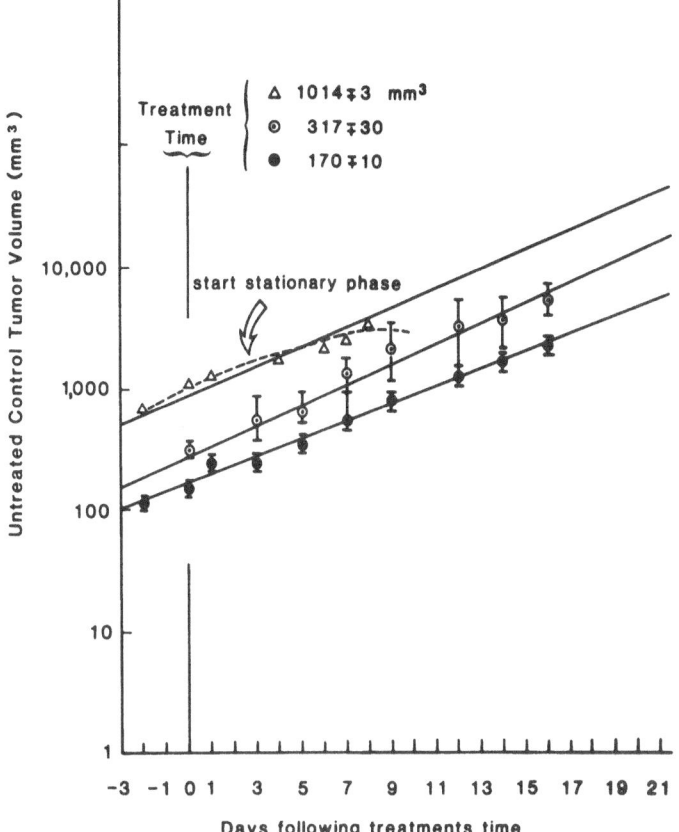

Fig. 11. The control tumor (without HPD and Laser) volume
versus days following treatment time for different
tumor volume. --- shows the start of the stationary phase in large size tumors.

tumor following PDT was noticed in tumor volumes ranging between 15 and
52 mm^3 (approximately 4-6 days after injection 10^5 of RIF cells). In general, the <u>first doubling time</u> and the <u>average doubling time</u> (averaged over
the entire period) decreases upon increasing initial tumor size at day 0,
in comparison to the control average (Table 1, Figure 10).

VII.2. <u>Interstitial PDT</u>.

Two groups of mice were followed up by a second interstitial irradiation starting from day 8 following the first PDT treatment. The average
tumor size (I) in these groups were as follows: for small tumors,
I_1 = 187 mm^3; and for medium tumors, I_2 = 335 mm^3, where I was measured at
the time of the first treatment. The controls for these treatments were
T_3, T_4 (superficial PDT) and C_1, C_2 (without HPD and laser) for small and
medium tumors, respectively.

<u>Tumor growth</u>. Table I and Figure 12 show the second treatment follow-
ing initial superficial irradiation. Seven days after the second treatment
(which was 16 days after the first treatment), the regrowth rates for small
(I_1) and medium (I_2) tumors were approximately two times slower than the
first rate of regrowth (the ratios were 2.21 and 2.25, respectively). Leg
swelling was also noticed after the second treatment, which lasted for
several days.

Table 1. RIF-1 Tumor Responses for C3H mice after PDT.

Tumor Size	No. of Animals	Delay Period	1st Doubling Time	Average Doubling Time	Growth or Regrowth Rates
mm^3		Days	Days	Days	mm^3/day
Superficial PDT					
T1 90 ± 22	6	11.75 ± 6.5	14.4	6.83	13 ± 21
T2 157 ± 13	8	11.75 ± 4.0	14.4	6.43	20 ± 10
T3 214 ± 21	4	08.8 ± 2.0	10.5	4.75	73 ± 15
T4 335 ± 5	2	03.4 ± 3.4	8.5	4.75	100 ± 7
T5 887 ± 4	2	0	7.3	5.25	214 ± 5
T6 1014 ± 3	2	0	5.5	5.20	357 ± 4
1st Treatment / 2nd Treatment Superficial PDT / Interstitial PDT					
I1 187 ± 25	2	7.5/0	9.8/4.0	2.5/4	133±21/60+19
I2 335 ± 5	2	6.5/0	8.25/4.5	2.0/4	300+10/133+25
Control without HPD and Laser					
C1 170 ± 10	3	0	4.0	3.12	134 ± 11
C2 317 ± 30	3	0	4.0	3.26	227 ± 9
C3 1014± 3	3	0	4.25	4.0	485 + 7

Note: All delay periods and 1st tumor volume doubling time measured with reference to PDT time (Day 0). The growth or regrowth rates for small, medium and large tumor size measured at days 15, 11, and 7, respectively. The regrowth
rates for superficial followed by interstitial treatments
both measured at day 15, where the average doubling time
was measured with reference to Day 9.

Tumor Cell Doubling time. For comparison purposes, the ratios for the
first doubling time for superficial PDT (day 0) to the first doubling time
for interstitial PDT (day 9) were 2.45 and 1.83 for I_1 and I_2, respectively
(Table 1), and the ratios of the average doubling times (both measured with
reference to day 9) were 0.62 and 0.5 for I_1 and I_2, respectively, taking
into account that the dashed curve in Figure 12 represents the regrowth for
I_1 and I_2 not followed by a second interstitial treatment (based on the
data in Figure 10. In general, the regrowth rates and average doubling
time ratios show enhancement of tumor response if followed by a second interstitial treatment.

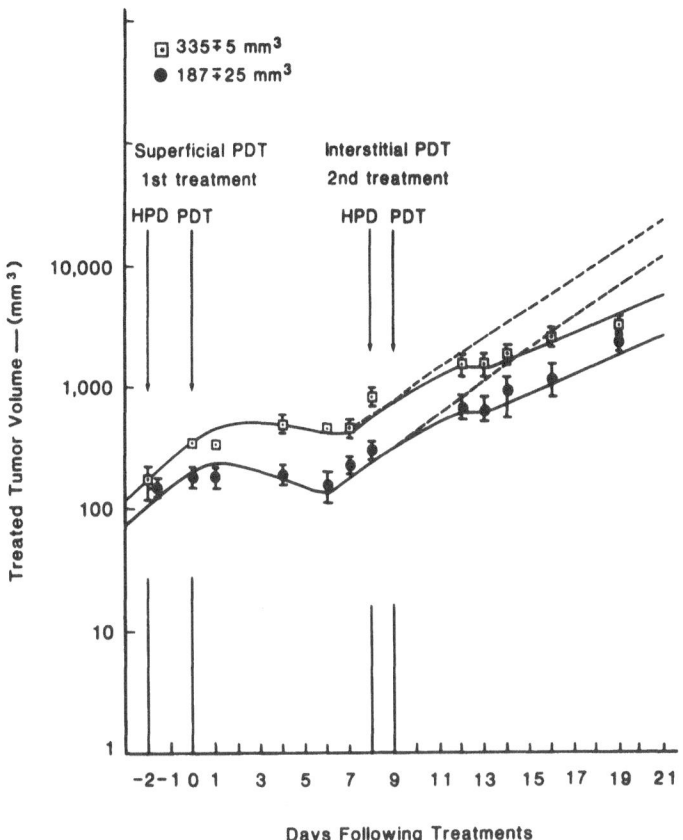

Fig. 12. The superficial tumor volume followed by the
interstitial treatment after 9 days as a function of days following both treatments.

VII.3. Clonogenic Assay

Forty-eight tumor-bearing mice were used for this study; 38 were
treated with HPD and laser (superficial irradiation only), and 10 untreated
were controls (without HPD and laser). Treatment commenced when tumor size
was between 15 and 1.746 mm^3. Tumor tissue was excised immediately and the
tumors subjected to in vitro clonogenicity assay. Tumor of the control
group (10 mice) were comparable in size to those of the treated group at
the time the mice were sacrificed.

Tumors from two mice were pooled, weighed, and minced with scissors.
The in vitro colony-forming assay was used to determine the fraction of
clonogenic cells in tumors. Colonies were counted under a binocular
dissecting microscope; those containing more than 50 cells were scored as
clonogenic. No correction was made for cell multiplicity.

The results of the clonogenic assay were plotted in percentages of
clonogenic survival rates (No. of colonies per gram of tumor after PDT with
respect to the number of colonies per gram of control tumor, multiplied by
100) as a function of tumor volume on a semilog scale (Figure 13). For the
control tumor tissue (Table 2), the average number of viable cells per gram
was $(4.34 \pm 4) \times 10^7$, and the average no. of clonogenic cells per gram was
$(1.198 \pm 1.06) \times 10^7$, where percentage of plating efficiency was 26.6 ± 2.5.
The mean value for the percentage plating efficiency following PDT was
16.44 ± 11.78 for initial tumor values from 100 to 1,746 mm^3.

$$Y = \exp\left[C_1 + (C_2 V - C_3)^{-1}\right]^{-1}$$

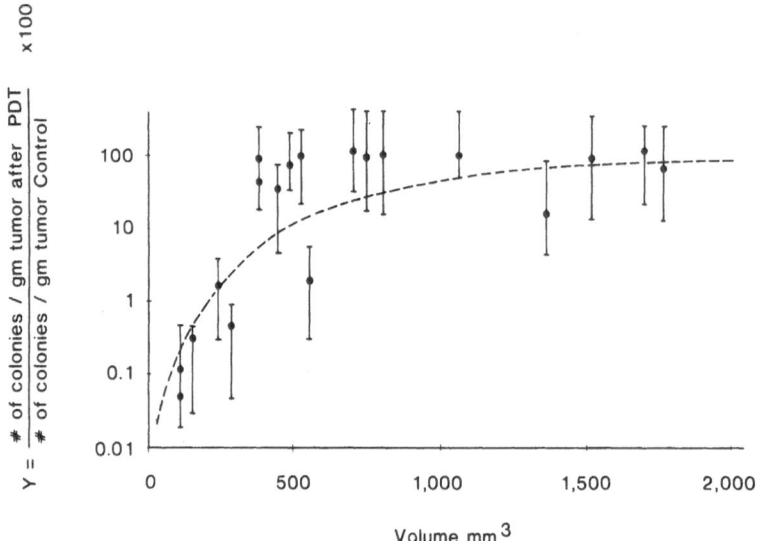

Fig. 13. The percentage clonogenic survival rates as a
function of tumor volume. --- represents the
curve which follows the formula indicated above.

The clonogenic survival rate for the group showed a saturation level (stationary phase) approaching 99.8% survival rate with a tumor size larger than ~1,900 mm^3. When the volume was smaller than 150 mm^3, the curve became a straight line with survival rate up to 0.3%. For a tumor, between 150 and 900 mm^3, the survival rate was 0.3% to 30%.

Complete tumor eradication was noticed in tumor volumes between 15 and 52 mm^3, corresponding to less than 0.03% survival rates. The 90 to 225 mm^3 tumor size resulted in a survival rate of 0.1% to 1.0%. For volumes larger than 500 mm^3, the treatment resulted in greater than 10% survival rates. The assumption of continuity refers to the belief that there is only one smooth curve that hould be fitted to the obtained experimental points (dashed curve in Figure 13). The equation was fitted to the data using a nonlinear least squares algorithm[19]. The resultant equation was as follows:

$$\% \text{ survival rate } Y = \exp\left[C_1 + (C_2 V - C_3)^{-1}\right]^{-1} \tag{1}$$

where C_1, C_2, and C_3 had values of 0.169, 0.012 mm^{-3}, and 2.788, respectively. This equation is limited to a volume up to 1.920 mm^3.

VIII. CONCLUSION

This study highlighted the advantage of using interstitial illumination following superficial irradiations (Figure 12, Table 1) to decrease the regrowth rate further. It also suggested that the interstitial treatment would be more effective, particularly when there is skin necrosis. This is valuable for the palliative treatment of a tumor having partial

Table 2. Tumor cell survival parameters in C3H mice without
HPD or light (i.e. control)

Tumor Size * mm^3	No. of Animals	No. of Exper.	Mass/Vol gm/mm^3	No. of viable cells per gm tumor tissue	No. of clono- genic cells per gm tumor tissue	Percent of Plating Efficiency
015±09	1	6	1.2x10^{-3}	0.476x10^7	-	-
256±14	1	6	1.21x10^{-3}	3.22	0.21x10^7	29±2.2
335±16	5	10	1.23x10^{-3}	0.75	0.934	28±2.3
733±7.5	1	6	1.26x10^{-3}	5.05	1.18	23.4±2.1
1136±04	1	6	1.22x10^{-3}	11.4	2.98	26.1±3.3
1416±17	1	6	1.23x10^{-3}	5.8	0.689	26.54±2.6
Average			1.22x10^{-3} ± 0.2	4.34 x 10^7 ± 4	1.198x10^7 ± 1.06	26.6 ± 2.5

Note: The tumor tissue excised and disaggregated, and clonogenicity of
tumor cells was assessed by in vitro colony formation assay.
(*) Uncertainty in tumor size (±) refers to the tumor volume
measured before and after excision.

response to this modality. This technique has the disadvantage of possible
blood clogging at the fibertip, causing a reduction in light energy. However, we added 0.9% NaCl via a syringe with "T" extension set to a no.19G
butterfly needle, which equalized the needle's capillary action of blood
suction, thus reducing blood clogging to a minimum.

For each tumor three parameters were recorded (Tables 1-3): the delay
period, the regrowth rate, and the clonogenic survival rate. These were
convenient indices of tumor response following PDT for different irradiation techniques and initial tumor volume. The growth delay was more
suitable than the first or average doubling time for tumors that shrink
significantly after PDT.

The regrowth pattern in Figure 10 shows that, for every tumor size
from 90 to 1,000 mm^3, leg swelling appears first (swelling was observed in
all animals with inconsistency in time span and size; Figure 10 only shows
the approximate time span), followed by skin discoloration and necrosis
(varying in depth and area) before regrowth, which was dependent on initial
tumor volume. The delay period was longer for smaller tumors (90 mm^3) and
was completely eradicated (no growth of bulk tumor) when the volumes were
between 15 and 52 mm^3 (corresponding to less than 0.03% survival rates;
Figure 13. The delay period prior to regrowth appears to decrease depending on the initial tumor size (Figures 10, 12, and Table 1). When tumor
volumes ranged between 1,000 and 90 mm^3, there was a delay from 1 to 18
days, respectively, and correspondingly there were 40% to 0.09% survival
rates. The necrotic areas appearing on the normal skin surrounding the
tumor may be due to the high concentration of HPD in the skin and also to

the stretching of the skin (smaller thickness) caused by the bulk of tumor growth, making the necrosis deeper and larger around a large tumor. Also, the random nature of HPD distribution (in the tumor and skin) contributes to the difficulty of measuring the relationship between dose and depth of necrosis.

The pronounced skin pigmentation of the C3H mice, which transmits only (37 ± 6)% of incident light, and the use of HPD rather than photofrin II (a more purified form of HPD) required the upward adjustment of HPD; based on these, to achieve the optimum values for large tumor volume (~ 750 mm^3), use of 20 mg/kg with a 40 hr retention interval was required. This dose gave an increasing area of necrosis unrelated to depth for small and medium tumor volumes. Comparing values in previous studies, a lower dose/retention interval could have been used to achieve optimum values had photofrin II been available.

For tumor volumes larger than ≈ 900 mm^3, the high clonogenic survival rates (above 30%; Figure 13) indicated poor light utilization of 630 nm mainly because of the limited tissue penetration and partially because of tumor hypoxia. This corresponds to no delay (or limited response; Figures 10, 12). Absence of abundant central necrosis - even for large tumors in RIF-1 cells - will not entirely exclude the partial results of tumor hypoxia for a volume $> 1,416$ mm^3 (Table 2), and the existence of focal necrosis effects in these cells cannot be ignored. A final conclusion requires further study. On the other hand, the swelling and necrosis noticed give a low therapeutic ratio (large necrosis in comparison to treated tumor survival rate). Below ≈ 150 mm^3, the survival rates (less than ≈ 0.3%) exponentially increased with the volumes, and the tumor growth delays reached

Table 3. The tumor response as a function of tumor half depth and light energy received.

Tumor size mm^3	Tumor center depth mm	Light energy received at the center J/cm^2	Percent survival rates	Tumor response
15 very	1.55	41.3	0.0095	complete
50 small	2.32	26.58	0.032	eradication
90	2.82	19.98	0.095	tumor
157 small	3.39	14.39	0.36	delay
214	3.77	11.62	0.85	
335 medium	4.37	8.22	2.98	possible delay
887 large	6.05	3.16	30.70	No
1014	6.32	2.7	39.37	delay
1921	7.83	1.14	100.0	

Note: Calculation of tumor center depth is based on the spheroid's growth of RIF-1 cells. The data for light energy received calculated from the information that the base of 3 mm RIF tumors received (18 ± 4)% of incident light. Superficial PDT, 100 J/cm^2, 20 mg/kg and 40 hr retention intervals for HPD.

up to 18 days (for less than 100 mm^3 tumor sizes). Small tumors (less than $\approx$ 50 mm^3) showed complete tumor eradication (followed up to 40 days), and the survival rates were less than 0.03%, which probably means that the dosimetric parameters used were optimum to those sizes in the RIF-1 tumor.

Henderson showed that the dark skin of C3H mice transmitted (37 ± 6)% of incident light and that the base of the 3 mm RIF tumor received (18 ± 4)% of incident light. Accordingly, the core of $\approx$108 mm^3 and the base of the $\approx$ 14 mm^3 tumor sizes received only (18 ± 4) J/cm^2. Also, the core of $\approx$ 900 mm^3 and $\approx$ 50 mm^3 tumor volumes received 3 and 26.58 J/cm^2, respectively (thicknesses for these tumors were $\approx$ 12.16 and $\approx$ 4.64 mm, respectively) calculated in accordance with the spheroid nature of the RIF tumor (Table 3).

The regrowth rate of the tumor bulk (ranging from 90 to 1.000 mm^3) appears to be dependent on initial tumor size (at the time of treatment) and increased relatively with it (Figure 10). The survival rates higher than 0.09% may be due to the combination of inhomogenous light (within the limit of 630 nm energy deposition) and drug distribution, as well as the weak immune nature of C3H mice to murine fibrosarcoma RIF-1 in vivo. This condition makes it difficult to choose the optimum laser-HPD dose for complete tumor mass removal and possibly explains the wide variation in survival rates for this region (Figure 13).

There is no clear evidence of thermal contribution caused by small intratumor temperature variations $[(4 \pm 1)$ and $(7.0 \pm 1)°C]$ and treatment time (14 and 8.34 min) for superficial and interstitial PDT, respectively. This does not exclude the possibility that a 7°C temperature rise could have a significant additive or synergistic effect on PDT. We plan to study this further.

The values $[C_1, C_2,$ and $C_3]$ of the empirical relation in Figure 13 could be related to the laser-HPD dosimetric parameters and tissue and tumor properties[21]. Further work is required to identify these quantities. The data for percentage of survival rates as a function of the estimated light energy received at the center of the tumor bulk (Table 3) shows that complete eradication of the tumor corresponding to depositing light energy at the core of the tumor was in excess of $\sim$ 26% ($\leq$ 4.7 mm tumor depth) of the incident light where the regrowth delay required more than $\sim$ 8% ($\leq$ 8.8 mm tumor depth) of the incident light. The actual inhomogenous scattering of light penetration in tissues complicates this picture. We are planning to conduct more experiments to study the survival rate relationship with initial tumor size following PDT, using malignancies which have a higher response to this modality.

REFERENCES

1. T. J. Dougherty, C. J. Gomer and K. R. Weishaupt, "Energetics and efficiency of photoinactivation of murine tumor cells containing hematoporphyrin", Cancer Res, 36:2330-2333 (1976)
2. T. J. Dougherty, D.G. Boyle and K. R. Weishaupt, "Photoradiation therapy of human tumors", in The Science of Photomedicine, J. D. Regan, J. A. Parrish (eds), New York: Plenum Press, 625-638 (1982)
3. M. W. Berns, G. S. Berns, J. Coffey and A. G. Wile, "Exposure tables for hematoporhyrin derivative photoradiation therapy", Lasers Surg. Med., 4:107-131 (1985)
4. J. Moan, "Porphyrin photosensitization and phototherapy", Photochem. Photobiol., 43:681-690 (1986)
5. J. Moan, "Porphyrin-sensitized photodynamic inactivation of cells: A review", Lasers Med. Sci., 1:5-12 (1986)
6. B. C. Wilson and M. S. Patterson, "The physics of photodynamic therapy", Phys. Med. Biol., 31, 4, 327-360 (1986)

7. S. Chang-Ho, E. Duzman, J. Mellot, L. Liaw and M. W. Berns, "Spectro-scopic, morphologic and cytotoxic studies on major fractions of hematoporhyrin derivative and photofrin II", <u>Lasers Surg. Med.</u>, 7:1179 (1987)

8. T. J. Dougherty, "PDT: Present and future", <u>Laser Med. Sci. Abstr. issue</u>, Proceedings of the International Conference on PDT and Medical Laser Applications, London, p.8 (1988)

9. D. R. Doiron, L. O. Svaasand and A. Profio, "Light dosimetry in tissue: Application to photoradiation therapy", <u>in</u>: D. Kessel, D. J. Dougherty (eds): "Porphyrin Photosensitization", New York: Plenum Press, 63-76. (1983)

10. L. I. Grossweiner, "Optical dosimetry in photodynamic therapy", <u>Lasers Surg. Med.</u>, 6:462-466 (1986)

11. L. O. Svaasand, "Thermal and optical dosimetry for photoradiation therapy of malignant tumors", <u>in</u>: A. Andreoni, R. Cubeddu (eds): "Porphyrin in Tumor Phototherapy", New York: Plenum Press, 261-279 (1984)

12. P. R. Twentyman, J. M. Brown, J. W. Gray, A. J. Franko, M. W. Scales and R. F. Kallman, "A new mouse tumor model system (RIF-1) for comparison of end-point studies", <u>JNCI</u>, 64:595-604 (1980)

13. F. A. H. Al-Watban, G. King, R. Brannon and T. Altamero, "Photodynamic therapy: Optimal treatment parameters in murine fibrosarcoma", <u>Lasers Med. Sci.</u>, 2:55-62 (1987)

14. F. A. H. Al-Watban, "Photodynamic therapy: Tumor volume limitation and tumor response for murine fibrosarcoma", <u>Lasers in Surg. Med.</u>, 10:2 (1990)

15. T. J. Dougherty, G. B. Grindey, R. Fiel, K. R. Weishaupt and D. G. Boyle, "Photoradiation therapy II. Cure of animal tumors with hematoporphyrin and light, <u>JNCI</u>, 55:115-121 (1975)

16. F. A. H. Al-Watban and W. R. Harrison, "Biological activity of prepara-tion of HPD", <u>ICALEO Proc. Laser Res. Med.</u>, 60:75-81 (1987)

17. D. Kessel, "Tumor localization and photosensitization by derivatives of hematoporhyrin: A review", <u>IEEE J. Quantum Electron.</u>, WE23:1718-1720 (1987)

18. T. J. Dougherty, "Photosensitizers: Therapy and detection of malignant tumors. Yearly review", <u>Photochem. Photobiol.</u>, 45:879-889 (1987)

19. D. W. Marquardt, "An algorithm for least squares estimation of non-linear parameters", <u>J. Soc. Ind. Appl. Math.</u>, 2:431-441 (1963)

20. B. W. Henderson, S. M. Waldow, T. S. Mong, W. R. Potter, P. B. Malone and T. J. Dougherty, "Tumor destruction and kinetics of tumor cell death in two experimental mouse tumors following photodynamic therapy", <u>Cancer Res.</u>, 45:572-576 (1985)

21. F. A. H. Al-Watban, "Dosimetric parameter relationship of PDT in RIF-1: Clonogenic survival rate as a function of tumor volume", <u>Laser Med. Sci. Abstr. issue</u>, Proceedings of the International Conference on PDT and Medical Laser Applications, London, p.166. (1988).

A COMPARISON OF LASER AND ARC-LAMP SPECTROSCOPIC SYSTEMS FOR IN-VIVO PHAR-

MACOKINETIC MEASUREMENTS OF PHOTOSENSITIZERS USED IN PHOTODYNAMIC THERAPY

D. A. Russell, P. Nadeau and R. H. Pottier (*)
G. Jori and E. Reddi (**)

(*) The Royal Military College, Kingston, Ontario, Canada
(**) University of Padova, Italy

I. INTRODUCTION

The treatment of neoplastic tissue using photodynamic therapy is currently undergoing a worldwide concerted research effort, enabling human treatment of such tissue using the drug hematoporphyrin derivative (HPD), see for example[1]. Current research is centered on the development of new photosensitizers, new instrumentation for delivering the required radiant energy, an understanding of processes involved in the selective biodistribution of photosensitizers as well as the mechanism by which the induced cytotoxic reaction is produced.

In order to assess the suitability of potential photosensitizers with regards to their ability to selectively biodistribute in tumor tissue, in-vitro pharmacokinetic studies are normally adopted. This traditionally involves the injection of numerous mice with the drug of interest. At fixed times after drug administration the mice are sacrificed, several tissues are removed, the photosensitizer drug is biochemically extracted and its concentration determined by fluorescence spectroscopy[2]. An in-vivo pharmacokinetic method, based on fluorescence spectroscopy, has recently been developed and compares favourably with the traditional in-vitro method[3,4]. The in-vivo technique makes use of a dye laser pumped by a nitrogen laser in order to select the excitation wavelength of the photo-sensitizer. Both excitation and emitted light were channelled via two optical fibers.

In this chapter two excitation sources are examined for their suitability for such in-vivo pharmacokinetic studies. For this type of study it is preferable that little or indeed no photodynamic effect is observed, so that any excitation light is of sufficient energy to probe the photosensitizer in the internal organs of the mouse without causing any tissue damage. To this end a dye laser pumped by a nitrogen laser has been compared with an arc-lamp as possible excitation sources. These two types of sources are compared based on their individual characteristics and the suitability for use in the in-vivo pharmacokinetic study of the model photosensitizer hematoporphyrin IX (HP).

Two spectroscopic systems, one using a dye laser pumped by a nitrogen laser, the other using a high pressure xenon arc-lamp as an excitation source, are compared for in-vivo pharmacokinetic measurements of the photosensitizer hematoporphyrin IX. Pharmacokinetic data were obtained from intravenously injected hairless mice over a period of up to 18 hours. The two spectroscopy systems are compared based on stability and reproducibility of fluorescence signal as well as the pharmacokinetic data obtained. The arc-lamp system has a greater stability and reproducibility of fluorescence signal, whereas the laser system is capable of higher sensitivity. However, the in-vivo pharmacokinetic data suggest that either system may be successfully used for the routine monitoring of drugs in-vivo.

II. MATERIALS AND EXPERIMENTAL PROCEDURE

II.1. Drug Preparation

Hematoporphyrin IX dihydrochloride (HP) obtained from Porphyrin Products, Logan, Utah, USA, was used as supplied. The HP was dissolved in a minimum amount of 0.1 M sodium hydroxide solution. Phosphate buffered saline (PBS) solution of pH 7.3 was added so that the final concentration of HP obtained was 0.5 mg/ml.

II.2. Mice

Skh:HR-1 female mice, aged between 6-8 weeks, were used for these experiments. These mice are genetically hairless and relatively docile.

II.3. Instrumentation

The overall layout of the instrumentation is shown in Figure 1. The figure shows the nitrogen laser / dye laser configuration. The same instrumentation was employed when the arc-lamp was used as the excitation source. In both instances the excitation light was focused by means of a 10 cm focal length quartz lens onto a single 1 mm diameter polymer clad silica optical fiber (Fiberguide, NJ, USA). This optical fiber was part of an in-house built probe shown in Figure 2. A total of 7 similar optical fibers are arranged such that 6 detection fibers surrounded the single excitation fiber in a concentric manner. The optical fibers are mounted in a brass holder using epoxy-resin. The probe head which has a 13 mm diameter quartz window on the front, was fitted to be able to slide along the optical fiber holder. This may be positioned such that the maximum intensity

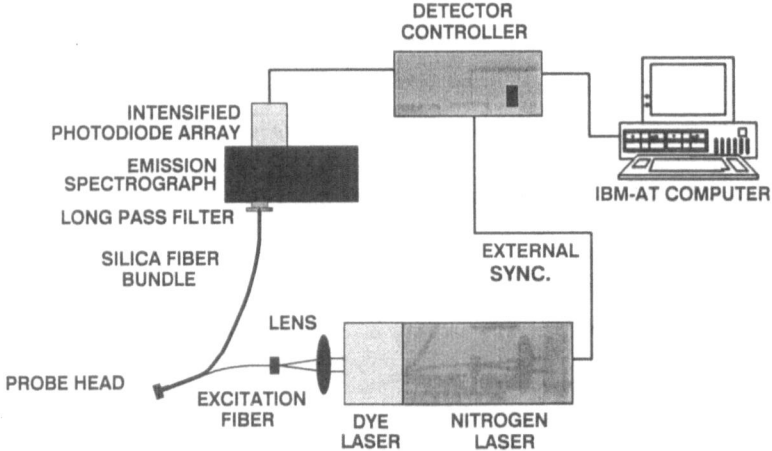

Fig. 1. Schematic diagram of the nitrogen laser/dye laser spectro-scopic system used for the collection of pharmacokinetic data. For the arc-lamp based system, the lasers were re-placed by a 75 W Xe arc-lamp.

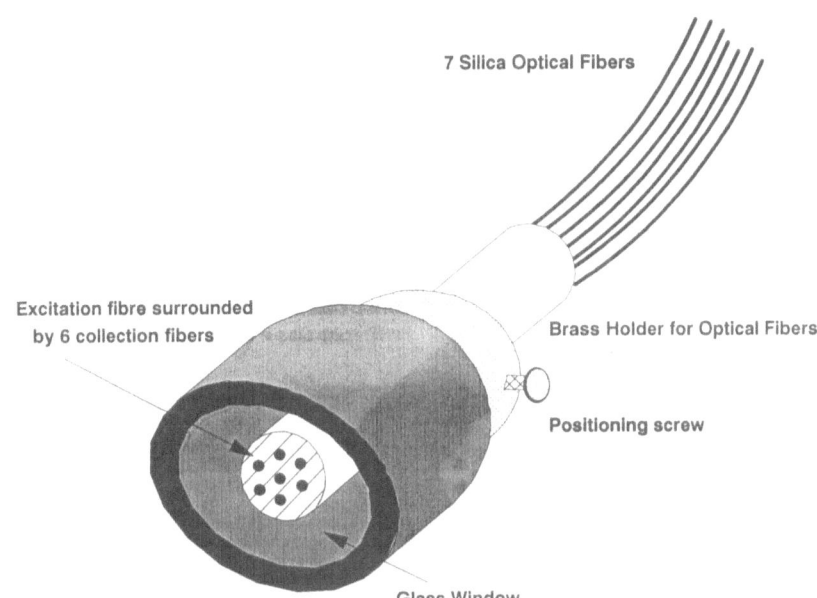

7 Silica Optical Fibers

Excitation fibre surrounded
by 6 collection fibers

Brass Holder for Optical Fibers

Positioning screw

Glass Window

Fig. 2. Schematic diagram of the optical fiber probe used
in conjunction with both spectroscopic system.

from a fluorescence reference standard contained in a spectroscopic cuvette
can be obtained. The design was chosen so that a lager surface area of the
measurement site on the mouse would be probed by the excitation light.
This approach was used because of the inhomogeneity of the mouse skin. For
homogeneous samples, such as often found in in-vitro studies, a more focused light source i.e. laser, would result in a higher fluorescent signal.
However, due to the varying pigments and blemishes of the skin, an integrated signal over a wider area would lead to more reproducible pharmacokinetic results. The fluorescence collected by the 6 detection optical fibers, which are mounted in a vertical array, was focused directly onto a
100 μm entrance slit of a Jarrell-Ash, 0.27 m monochromator (Monospec -
27,100 grooves/mm grating blazed at 600 nm). In order to reduce the intensity of any excitation light reflected from the surface of the mouse,
a 525 nm long pass cut off filter (Opticon, LWP-525) was placed in front
of the entrance slit of the monochromator. A complete fluorescence spectrum, from approximately 560-800 nm, was obtained using an intensified photodiode array (Princeton Instruments Inc. Model No. IRY-512) as a detector.
The recording of each spectrum was controlled via the photodiode array
detector controller (ST-120) by an IBM-AT computer. Wavelength calibration
of the photodiodes was achieved by the use of a neon pen-lamp. Two excitation sources were used. A nitrogen laser [Photon Technology International (PTI), PL2300] which was used to pump a dye laser (PTI, PL201). The
dye laser was tuned for emission at 400 nm. The dye used was the known
2-(1,1',Biphenil)-4-yl-6-phenyl benzoxazole (emission maximum = 400 nm). A
high pressure, high brilliance 75 W xenon arc-lamp (Model A1010, PTI) was
also used as an excitation source. The output of this arc source was held
constant by the use of a photodiode feedback system (Model No.LPS002, PTI).
The light from the arc-lamp was focused onto the excitation optical fiber
by use of a 10 cm focal length quartz lens and the ex- citation wavelength
of 400 nm was selected by using an interference filter of 10 nm bandpass.

II.4. Drug Injection

Two mice were used for each pharmacokinetic study. One served as a
control while the other became the experimental mouse. These were both

weighed and then anaesthetised using Ketamine (Ketalar; Park Davis, Ont. Canada). The experimental mouse was intravenously injected in the femoral vein with the HP solution so that a concentration of 5 mg/kg was introduced. The control mouse was likewise injected with PBS solution of the same total volume as that of the experimental mouse.

II.5. In-vivo Pharmacokinetic Measurements

Prior to injection of the mice with the HP or saline solutions, the background fluorescence was taken from the selected anatomical sites on the mouse. The measurement sites chosen were the stomach and the skin-fold. Before each group of measurements, the fluorescence intensity from a standard solution of HP contained in a quartz cuvette was measured. This ensured that throughout the pharmacokinetic study, a constant signal from each of the optical systems was obtained. The mice were hand held against the optical probe, which was clamped in position, and three fluorescence measurements were taken for both anatomical sites. The average of these three measurements was used as the fluorescence intensity value as a function of elapsed time post injection. To ensure that no erroneous fluorescence intensities were obtained from contamination of the quartz window by the mouse, the window was periodically cleaned using an ethanol dampened tissue. For each optical system, pharmacokinetic data were collected for up to 18 hours. For the first two hours, data points were obtained on each mouse every 5 minutes. This was reduced to every 20 minutes for the remainder of the study.

III. RESULTS AND DISCUSSION

For both spectroscopic systems the stability and the reproducibility of intensity of the fluorescence signal was measured by taking 100 spectra from a standard solution. Once the nitrogen laser / dye laser excitation

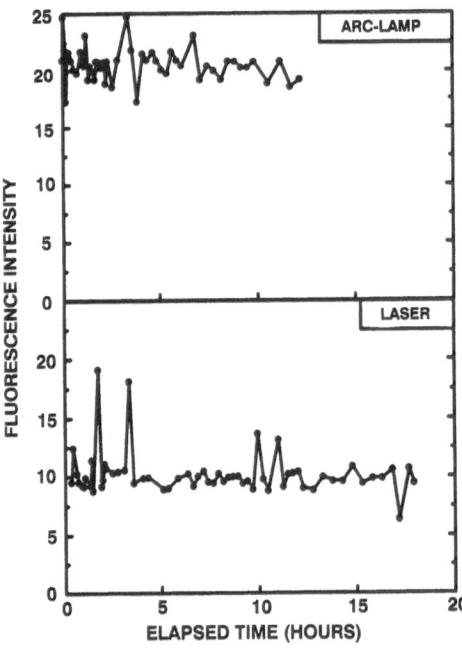

Fig. 3. Variation of the fluorescence intensity of an hematoporphyrin standard solution over the period of a pharmacokinetic study.

source had been optimized for use, the reproducibility of the pulses obtained had a relative standard deviation of 2-4%. The comparable stability
of the arc-lamp was of the order of 0.8%, suggesting that the relative
standard deviation ratio of laser:lamp was at best 2.5 and at worst 5.0.
Another measure of the reproducibility and stability of the spectroscopic
systems is shown in Figure 3. These data were obtained by taking the fluorescence intensity obtained from an HP standard, measured prior to each
batch of in-vivo measurements for the period of a pharmacokinetic study.
It can be seen that the overall reproducibility and stability of fluorescence intensity of the arc-lamp spectroscopic system is superior to that of
the laser system. The higher stability obtained from the arc-lamp system
is due to the optical feedback device which critically controls the output
of the lamp.

Figures 4 and 5 show typical in-vivo pharmacokinetic data that can be
obtained from the spectroscopic system for the stomack and skinfold measurements sites respectively. In-vivo fluorescence measurements from the
stomach area of each mouse can be considered to be an integrated measure of
the concentration of the HP in internal organs such as the liver, as well
as the skin. The fluorescence obtained from the skinfold however, would be
due solely to the skin. The laser excited fluorescence maximum from the
stomach area occurred at 1.4 hrs post injection. The total residency time
of the HP in this area was approximately 10 hrs. The equivalent HP fluorescence intensity maximum and residency times measured using the arc-lamp
source were at 1.0 hrs and 11 hrs respectively. For the skinfold measurement site the maximum and residency times were; (a) for the laser source,
0.75 hr and 7.5 hrs and (b) for the arclamp source, 0.85 hrs and 11 hrs
respectively.

Overall the pharmacokinetic data shown in Figures 4 and 5 suggest that

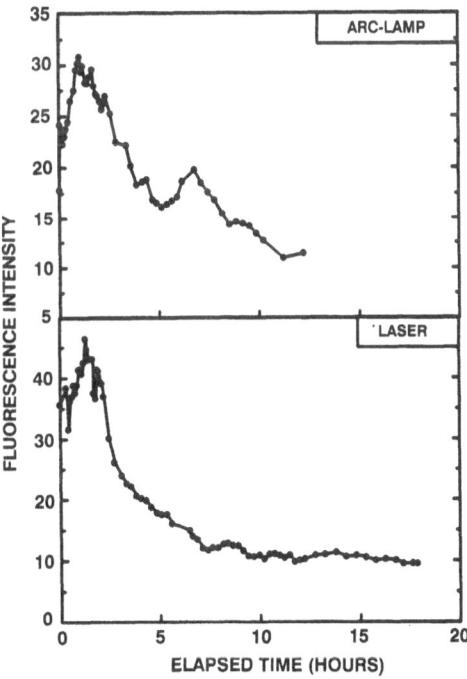

Fig. 4. In-vivo pharmacokinetic data taken from the stomach
         area of the mice using both laser and arc-lamp spectro-
         scopic system.

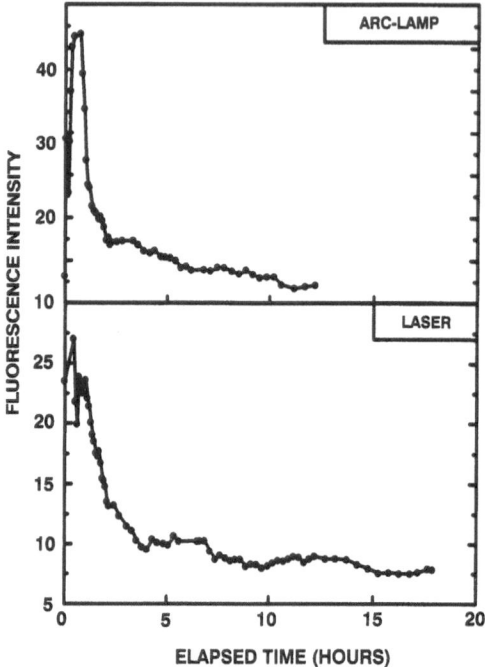

Fig. 5. In-vivo pharmacokinetic data taken from the skinfold
area of the mice using both laser and arc-lamp spectroscopic systems.

similar data can be obtained using either the arc-lamp or the laser based
spectroscopic systems. Within the experimental error, the times of fluorescence maxima and residency of the drug from the different measurement
sites is equivalent for both systems. An important point to note is that
in these measurements HP is excited at 400 nm, which is well separated fr
the fluorescence wavelength region. A problem may arise when such spectr
scopic systems are used to monitor a photosensitizer with a short Stoke's
shift, for example zinc(II)-phtalocyanine. For this example, excitation
usually induced at 600 nm and the fluorescence emission is monitored at 6
nm. For such a case, a spectroscopic system which has a narrow band widt
of excitation, i.e. a laser source, would have obvious advantages. Howev
for the in-vivo HP pharmacokinetic data measurements, although the arc-la
system has a greater stability and better reproducibility of fluorescence
signal than the laser based system, the results obtained suggest that
either spectroscopic system may be employed for the collection of such
data.

IV. ACKNOWLEDGEMENTS

The authors would like to thank the Canadian Department of National
Defence for financial support of this work.

REFERENCES

1. "Photosensitizing Compounds: Their Chemistry, Biology and Clinical
    use", Ciba Foundation Symposium 146, Wiley, Chichester (1989)
2. E. Reddi, G. Lo Castro, R. Biolo and G. Jori, "Pharmacokinetic studi
    with zinc (II)-phtalocyanine in tumor bearing mice", Br. J. Cance
    56:597 (1987)

3. R. M. Pottier, Y. F. A. Chan, J.-P. LaPlante, T. G. Truscott, J. C. Kennedy and L. A. Beiner, "Non-invasive technique for obtaining fluorescence excitation and emission spectra in-vivo", Photochem. Photobiol., 44:679 (1986)
4. R. M. Pottier and J. C. Kennedy, "Photodiode array fluorescence technique for measuring drug clearance rates in-vivo, J. Photochem. Photobiol., B3:135 (1989).

EXPERIMENTS ON LASER MEDIATED PHOTOSENSITIZATION:

PRECLINICAL EVALUATION OF MONO-1-ASPARTYL CHLORIN e6 (NPe6)

A. Ferrario and C. J. Gomer

Childrens Hospital of Los Angeles
University of Southern California School of Medicine
Los Angeles, California, USA

A renewed interest in the synthesis and evaluation of "second gene-ration" photosensitizers for use in PDT has developed due to the several photophysical restrictions of Photofrin II. Photofrin II is a poorly de-fined mixture of monomeric and aggregated porphyrins and has extremely weak absorption at wavelengths above 600 nm (where light transmission through tissue is greatest). In addition the drug is retained in numerous normal tissues (such as skin) for extended periods of time. Photosensitizers with preferential tumor tissue retention, rapid clearance from normal tissues, increased extinction coefficients and absorption peaks shifted to higher wavelengths, should allow for greater utilization of delivered light which has improved tissues transmission properties.

A recent investigation was designed to document pharmacologic and photosensitizing properties of mono-1-aspartyl chlorin e6 (NPe6) in a mouse-tumor model[1]. NPe6 is an effective tumor localizer that possesses properties such as chemical purity and a major absorption band at 664 nm which are potentially exploitable for photodynamic therapy. The C3H BA mammary carcinoma transplanted subcutaneously in the flank of C3H/HeJ mice was used as the tumor/animal model. Tumor response (defined as the percen-tage of mice without tumor recurrence at least 40 days following treatment) was evaluated as a function of NPe6 dose, total light dose and time inter-val between drug administration and light treatment. In the initial pro-tocol, in which light treatments were delivered 24 hours following drug administration (up to 50 mg/kg), a 100% tumor recurrence rate was observed. Tumor cures were obtained using lower NPe6 doses (2.5-7.5 mg/kg), when light treatments were delivered 4 to 7 hours after drug administration. These NPe6-PDT parameters were as effective as standard Photofrin II medi-ated PDT in which the photosensitizer the photosensitizer is administered 24-26 hours prior to light treatment. A quantitative scoring system for acute skin photosensitization reactions was utilized to document the degree of damage induced by either NPe6 or Photofrin II mediated PDT. Interest-ingly, the level or PDT induced normal skin damage in albino mice injected with either NPe6 or Photofrin II and then exposed to localized light treat-ments at time intervals of 4-6 hours, was significantly greater for Photo-frin II than for NPe6 at comparable drug and light doses. The difference in magnitude if skin damage increased when a 24-26 hours time interval was used between drug administration and light treatment.

Pharmacologic properties of NPe6 were also investigated by analysis of the quantitative tissue localization and distribution of ^{14}C-NPe6, radio-isotopically labeled on the ring structure of the chlorin molecule. Eighteen tissues and fluid samples were analyzed: plasma, whole blood, tumor, brain, heart, liver, kidney, adrenal glands, spleen, skin, intestine, muscle, lung, esophagus, stomach, bladder, feces and urine. Seventeen time points were assayed in experiments related to drug uptake as a function of time (up to 96 hours) after a 5 mg/kg I.V. injection of ^{14}C-NPe6. NPe6 pharmacokinetics showed that tumor tissue concentrations of the photosensitizer were higher than in all tissues except liver, kidney, spleen, adrenal glands at intervals greater than 4 hours following drug administration, while the lowest levels of ^{14}C-NPe6 concentrations were found in brain, muscle and esophagus. Excretion of the photosensitizer was primarily through the bile-gut pathway. Computer stripping analysis of NPe6 pharmacokinetics indicates that a two compartment, bioexponential model fits the current data. The half-lives for all tissues were greater than those obtained for plasma. Interestingly, the plasma half-lives of Npe6 (0.5 hr for the first compartment and 11.6 hr for the second one) were shorter than those observed for porphyrin compounds [2]. Analysis of ^{14}C-NPe6 to that of ^{14}C-NPe6 labeled in the aspartyl residue in a mouse/tumor model, did not show any significant difference in tissue localization and distribution of ^{14}C-activity. These findings indicate that NPe6 did not undergo metabolic transformation and they were confirmed by HPLC analysis of tissue extracts [3]. Interestingly data from NPe6 pharmacologic studies combined with the informations on PDT time interval requirements, suggest that plasma concentrations of NPe6 may be a more important predictive factor than tumor tissue levels of the photosensitizer for the production of PDT mediated tumor cures. Therefore, vascular damage may be a primary target of the treatment in-vivo NPe6 mediated PDT. The results of this investigation indicate that NPe6 is an effective tumor photosensitizer with in-vivo clearance properties that eliminate the side effect of prolonged normal skin photosensitization.

REFERENCES

1. C. J. Gomer and A. Ferrario, "Tissue distribution and photosensitizing properties of mono-l-aspartyl chlorin e6 (NPe6) in a mouse tumor model", accepted for publication Cancer Res. (1990)
2. D. A. Bellnier, Y.-K. Ho, R. K. Pandey, J. R. Missert and T. J. Dougherty, "Distribution and elimination of Photofrin II in mice", Photochem. Photobiol. 50:221 (1989)
3. D. Kessel, personal communication.

PHOTODYNAMIC THERAPY OF TUMORS:

DEFINITION OF PDT DOSE AND RESPONSE

W. R. Potter

Department of Radiation Biology
Roswell Park Cancer Center
Buffalo, New York, USA

I. INTRODUCTION

The PDT dose at any point in tissue can be generally defined as the integral of the Photofrin II concentration over the time integrated space irradiance at that point. We know the dependence of Photofrin II concentration (i.e. photobleaching) as a function of the incident Joules/cm^2 for a thin superficial layer of tissue. If the space irradiance is also available as a function of the incident light for this geometry we can express the photobleaching of the Photofrin II as a function of the time integrated space irradiance in this layer (we assume that the space irradiance is constant in this thin layer). At this point knowledge of the space irradiance as a function of the incident flux for other simple geometries allows us to determine the PDT dose within the tissue for the clinically interesting cases of an implanted cylindrical source, a water filled bladder with a spherically isotropic source at the center, an implanted spherical source and the case of a uniform surface illumination.

Using the end point of depth of tissue necrosis and available clinical data for the case of surface illumination of skin[1], we can calculate the minimum photodynamic dose required for tissue necrosis using the generalized definition of dose. This value can then be used to predict the necrotic depth obtained in the other simple geometries. The available clinical data is in good agreement with these predictions.

Determination of the dependence of Photofrin II concentration on the time integrated space irradiance in tissue allows the definition of Photodynamic dose to be formulated as an integral of sensitizer concentration over the time integrated space irradiance at each point in tissue. This integral, together with a measurement of the depth of necrosis produced by the clinically measurable sensitizer and light doses, and the tissue optical properties allows the calculation of the expected necrotic depth for any similar tissue. These predictive calculations are carried out for various dosimetric conditions and a variety of clinically interesting sources e.g. long cylinders or spherically symmetric sources imbedded in the tissue, uniform surface illumination over a wide field or a spherical source at the center of a water filled bladder. The results are presented in graphical form and are based on published clinical data.

II. GENERALIZED PHOTODYNAMIC DOSE

The dose integral, D, has been defined by:

$$D = \int_0^J C(J)dJ \qquad (1)$$

C(J) is the concentration of Photofrin II in a thin slab of tissue[2]. From measurements of extractable Photofrin II as a function of J, the incident J/cm^2, we know that:

$$C(J) = C_o \exp(-\beta J) \qquad (2)$$

where C_o is the initial Photofrin II concentration and $\beta = 0.36$ J/cm^2 is the photobleaching rate constant: the photobleaching rate was determined in[2] human tumors by measurement of the decrease in tissue fluorescence during treatment. The loss of fluorescence correlates well with the loss of extractable drug in mice[3].

Fortunately the determination of space irradiance within tissue as a function of the incident flux is a subject which has received both theoretical and experimental study[4,5]. We will use the results of Reference 4 for all but the case of the water filled bladder where the results of Reference 5 will be employed.

Photodynamic therapy is mainly concerned with the use of 630 nm wavelength light where scattering is the dominant interaction with tissue. Space irradiance (the total light flux through an infinitesimal surface at a point in tissue) is best described by the results of diffusion theory which can be applied to light which is scattered multiply before absorption. As this scattering initially increases the space irradiance at the tissue surface the determination of space irradiance within the tissue is not a trivial problem because the clinically measurable quantity is the incident flux. The computer modeling techniques used to study the scattering of neutrons in the design of nuclear reactors have been applied in Reference 4 to produce equations for the space irradiance for some clinically interesting source geometries.

The result for a planar surface illuminated over a large area with a uniform dose J_o (J/cm^2) is

$$\Phi t = 4J_o \frac{\exp(-\alpha z)}{1 + \xi \alpha} \qquad (3)$$

where ξ is the diffusion coefficient, t is time in sec, ϕ is space irradiance in W/cm^2 and α is the tissue attenuation coefficient. Equation (3) is correct for a diffuse source. In PDT we are measuring the flux from a collimated source. However in all of the cases treated here the light passes through the skin which acts to provide significant diffusion. Thus we will use these results with the understanding that we assume the light becomes diffuse with negligible loss in the first $200\,\mu$ of the skin surface. Thus this description of the space irradiance applies from just below the skin surface (the plane of z = 0). Substituting $\alpha = 3.3$ cm^{-1}, and $\xi = 0.07$ cm in equation (3) gives

$$\phi t = 2.7 \ J_o \ \exp(-\alpha z) \qquad (4)$$

Writing equation (2) in terms of space irradiance for z = 0, gives

$$\beta J_o = \beta * \phi t \qquad (5)$$

where, using equation (3),

$$\beta* = \beta/2.7 = .0127 \tag{6}$$

is the photobleaching rate constant when the light dose expressed in terms of the time integrated space irradiance, ϕt. The dose integral is now generalized as

$$D' = C_0 \int_0^{\phi t} \exp(-\beta*\phi t)d(\phi t) \tag{7}$$

or

$$D' = (C_0/\beta*)[1-\exp(-\beta*\phi t)] \tag{8}$$

As we know only the injected dose and not the tissue level we will use the injected dose, C_{inj} instead of the tissue level, C_0. The result will be denoted $D*$ where

$$D* = (C_{inj}/\beta*)[1-\exp(-\beta*\phi t)] \tag{9}$$

Again we have from Reference 4 expressions for ϕt.

For the case of a cylindrical source of radius a, implanted in tissue delivering a total dose of J_L (J/cm) where r is the distance from the cylinder (note that the cylinder is long compared to r)

$$\phi t = 2J_L K_0(\alpha r)/\pi a[K_0(\alpha a) + 2\xi\alpha K_1(\alpha a)] \tag{10}$$

A spherically symmetric point source of radius a, implanted in tissue results in a time integrated space irradiance of

$$\phi t = J_0\{\exp[-\alpha(r-a)]\}/4\pi\xi r(1+\alpha a) \tag{11}$$

where J_0 is total energy delivered.

If the minimum dose required for tissue necrosis, $D*_{min}$ is known for one geometry then it is possible to solve equation (9) for the space irradiance, ϕt_{min}, at the maximum depth of necrosis z_{max}

$$\phi t_{min} = -(1/\beta*)\ln(1-D*_{min}\beta*/C_{inj}) \tag{12}$$

Note that for finite ϕt_{min}

$$1 > D*_{min}\beta*/C_{inj} \tag{13}$$

Using equations (10) or (11) for ϕt in equation (12) it is possible to find the expected depth of necrosis for the case of an implanted cylinder fiber or an implanted spherical diffuser. The case of planar illumination where the field diameter is large compared to z will be used as an example.

Substituting the right hand side of equation (3) for ϕt_{min} in equation (12) gives

$$4J_0\{[\exp(-\alpha z_{max})]/(1+\xi\alpha)\} = -(1/\beta*)\ln(1-D*_{min}\beta*/C_{inj}) \tag{14}$$

This equation can be solved for J_0 in terms the tissue optical parameters, the injected dose, the desired depth of necrosis z_{max} and the minimum dose needed for necrosis $D*_{min}$. Alternatively z_{max} can be found as a function of these dosimetric variables and tissue parameters

$$J_0 = -[(1+\xi\alpha)/4\beta^*] \exp(\alpha z_{max}) \ln(1-D^*_{min}\beta^*/C_{inj}) \qquad (15)$$

or

$$z_{max} = \ln[-4\beta^* J_0/(1+\xi\alpha)\ln(1-D^*_{min}\beta^*/C_{inj})] \qquad (16)$$

The required value of D^*_{min} is available from Reference 1. Similar clinical response data is available for other light source geometries and tumor types and locations. In analogous fashion the expected depth of necrosis can be calculated for the other source geometries as a function of the tissue optical parameters, the injected drug dose and the light dose and D^*_{min}, the response of the tissue to PDT.

III. RESULTS OF APPLICATION OF DOSIMETRY THEORY TO THE AVAILABLE CLINICAL DATA

The results of careful measurements performed at autopsy on the depth of necrosis produced by PDT in a breast cancer patient who was treated at a number of sites on the chest wall by a uniform planar light field have been published in 1983[1]. Death occurred within a week of treatment and the results allow the determination of the minimum PDT dose necessary to achieve tumor necrosis, D^*_{min}. Similar data where obtained by measuring the depth of necrosis resulting when colon cancer was treated using a cleaved fiber implanted in the tumor. In these patients ultrasound was used to determine the depth of necrosis[6]. Finally we have treated an amelanotic melanoma which was scheduled for surgical removal by the implantation of a 2.5 cm long cylindrical diffusing fiber. The depth of necrosis produced by this treatment was measured. The minimum dose required to produce necrosis was calculated from the data in each of these clinical results.

The tissue optical properties were assumed identical and the Photofrin II uptake by the various tumors was assumed to be proportional in the same manner for each case. This is a significant and questionable assumption because we know that different tissues take up different amounts of Photofrin II although the level of any given tissue is proportional to inject dose.

The clinical data and the calculated values of D^*_{min} are given in Table 1, produced from the values of z_{max}, published in Reference 1 and assuming that $\alpha = 3.3$ cm^{-1}. We will use the value $D^*_{min} = 33$, the average of these data.

C_{inj} and D^*_{min} may be used to plot z_{max} (the maximum depth of necrosis multiplied by the attenuation coefficient, α, of the tissue) as a function of the delivered light dose. The results for surface illumination appear in Figure 1.

Each clinically interesting light source geometry results in a family

Table 1. Clinical data and D^*_{min} values.

Number of lesions	light dose (J/cm^2)	z_{max}	D^*_{min}
6	21.6	0.61	37.8
11	43.2	1.0	27

206

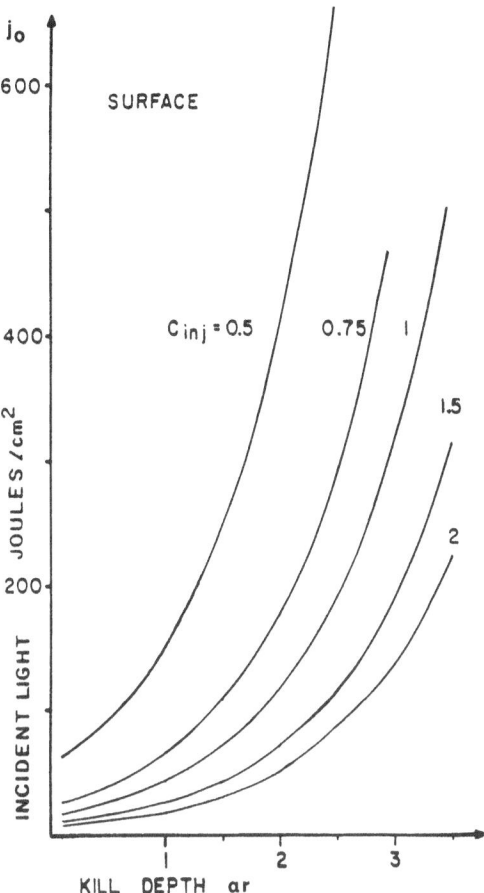

Fig. 1. Incident light dose, J_O, required for depth of necrosis,
(αr), with various injected doses C_{inj}, of Photofrin II.
The light field diameter must be large compared to r,
otherwise the depth of necrosis will be less than predicted by these curves.

of such curves (one curve for each injected dose). Data on the depth of
necrosis produced by a point source implanted in a series of eight colorectal cancer patients have been published[6]. A 0.4 mm diameter fiber was
implanted 1.2 mm into the tumors and the value of z_{max} was measured by en-
dosonography at one week post treatment. The light dose was 50 J in each
case. Table 2 summarizes the results of the dosimetric calculations, where
we assume that the implanted fiber approximates a spherical point source
and the $\alpha = 3.3$ cm^{-1}. The injected drug dose was 2.5 mg/kg of HpD which is
approximately equivalent to an injection of 1.25 mg/kg of Photofrin II.
Using this value of D^*_{min} the calculated depth of necrosis under a variety
of dosimetric conditions is plotted for an implanted point source in Figure 2.

Table 2. Dosimetric calculations.

Number of lesions	Light dose (J)	αr_{max}	D^*_{min}
8	50	1.78±0.28	39.4

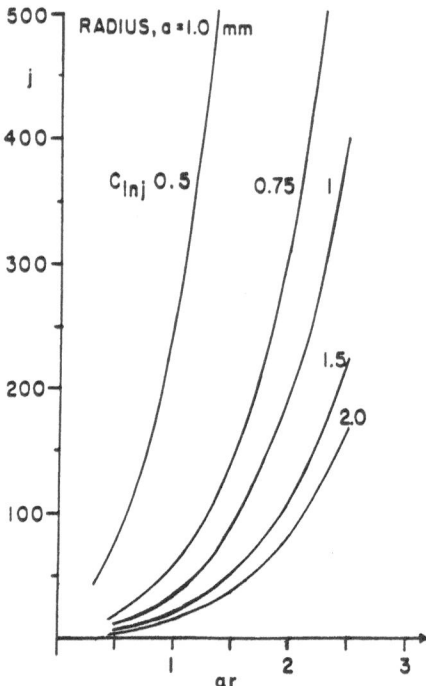

Fig. 2. Total light dose, J (in Joules), from an implanted
spherical source, required for depth of necrosis (αr),
with various injected doses, C_{inj}, of Photofrin II.

Finally a single amelanotic melanoma was treated at Roswell Park by
implanting a 2.5 cm long cylindrical diffuser. The light dose was 200 J/cm^2
and the injected dose was 2.5 mg/kg of HpD. The tumor was removed surgically three days after treatment. R_{max} was 0.9 cm. The predicted depth of
necrosis for an implanted cylinder under a wide variety of dosimetric
conditions appears in Figure 3.

All of the calculated dosimetric parameters appear in Table 3. The
results of these calculations have been rounded to two significant figures
to be consistent with the data which has a standard deviation of about 15%.

The case of an isotropic point source at the center of a water filled
spherical bladder is analyzed by using the graphs of $\phi(\alpha r)$ published in
Reference 5. $D^*_{min} = 39$ (the average value of the Table 3 results) will be
used. As before the depth of necrosis expected for various drug and light
doses is plotted in Figure 4. In this case we use $\alpha = 4.58$ cm^{-1} the
value measured in dog bladder[5].

Table 3. Calculated dosimetric parameters.

Geometry	Lesion	Number	t	D^*_{min}	Origin
planar	Breast Ca	16	42	33	Tokyo
sphere	Colorectal	8	40	39	London
cylinder	Amelanotic melanoma	1	39	46	Buffalo

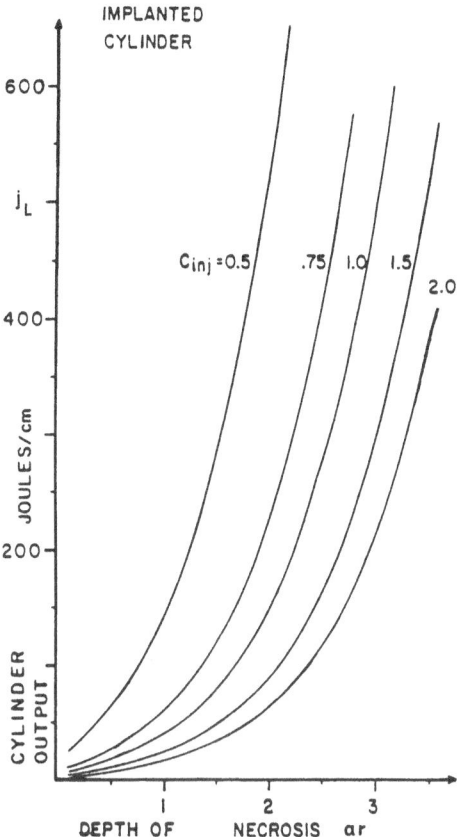

Fig. 3. Light dose required using an implanted cylinder (radius a = 1.0 mm) to necrose to a distance r measured perpen dicular to the axis of the cylinder. The total length of the cylinder must be large compared to r, otherwise the results will be less than this graph indicates.

IV. DISCUSSION

Under the assumptions of identical optical properties and Photofrin II uptake, clinical data on the depth of tissues necrosis gives remarkably similar values for the minimum photodynamic dose, D^*_{min}, required to pro duce tumor necrosis. The average value of D^*_{min} from Table 3 is 39. The values of D^*_{min} in Table 3 all lies within 15% of this average. The per cent standard deviation of the z_{max} values used to calculate the D^*_{min} were about ± 15%.

The value of 0.07 cm for the diffusion coefficient, ξ, was based on Reference 4. The diffusion coefficient varies by only a small amount for most tissues.

The value of the total attenuation coefficient, α, of 1/3 mm was based upon some in vivo clinical measurements in breast cancer recurrent to the chest wall. The technique was the non invasive method[3]. The use of this value for the colon cancer and the amelanotic melanoma was arbitrary. Thus we can not conclude with certainty that all of the tumors took up the same amount of Photofrin II for each injected mg/kg. Although the injected drug doses were different and the illumination geometries were different it is still possible that the assumptions regarding drug uptake and optical pro perties resulted in offsetting errors which caused the calculations of

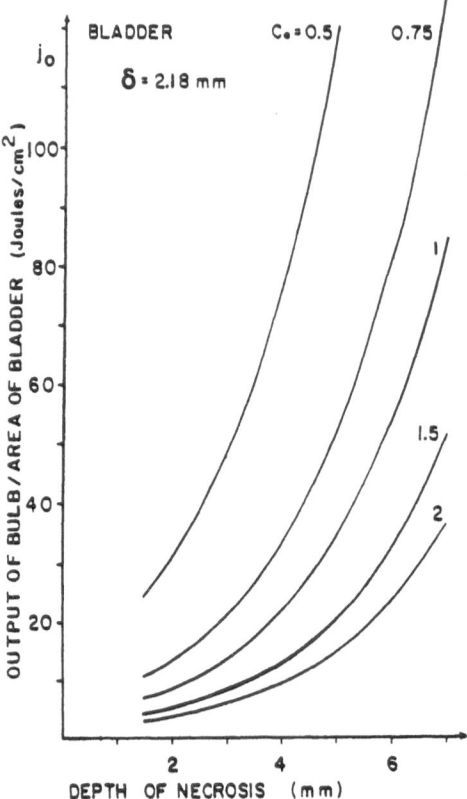

Fig. 4. J_O is the light dose to the surface of a water filled
bladder required to necrose tumor tissue to the depth (mm)
indicated by the horizontal axis. The total attenuation,
$\alpha = 4.58$ cm^{-1}, is the value measured in normal dog bladder
wall[5]. It is also from Reference 5 the injected dose
C_{inj} of Photofrin II.

D^*_{min} to be nearly identical for these otherwise dissimilar situations. If
in the future measured values of the actual tissue optical properties can
be collected as well as data on depth of tissue necrosis, it will be possible to determine the relative Photofrin II uptake under the implicit assumption that all tumors require the same photodynamic dose to produce
necrosis and that the theory correctly described the photodynamic dose in
the tissue.

Equation (12) predicts the point at which the PDT effect will fail
due to an inadequate level of Photofrin II. If

$$1-D^*_{min}\,(\beta*/C_{inj}) = 0 \tag{16}$$

then

$$1/\{\ln[1-D^*_{min}(\beta*/C_{inj})]\} = 0 \tag{17}$$

and

$$\alpha z_{max} = 0 \tag{18}$$

Using the extreme values for D^*_{min} ($D^*_{min} = 27.46$), the minimum therapeutic value of C_{inj} is in the range

$$0.34 \text{ mg/kg} < C_{inj} < 0.58 \text{ mg/kg}. \tag{19}$$

Thus the theory predicts that drug dose will become inadequate between 0.58 mg/kg and 0.34 mg/kg injected. The value 0.58 is clearly too high as patients treated with 0.5 mg/kg continue to respond. Continued reduction in the injected dose will determine if the lower limit of 0.34 is correct.

REFERENCES

1. C. Konaka and J. Ono, "Skin metastasis from breast cancer", in "Laser and Hematoporphyrin Derivative in Cancer" (Y. Hayata and T. J. Dougherty eds), pp.110-114, Igaku-Shoin (Tokyo, 1983)

2. W. R. Potter, T. S. Mang and T. J. Dougherty (1987) "The theory of photodynamic therapy dosimetry: consequences of Photodestruction of sensitizer", Photochem. and Photobiol., vol.46, 1:97-101 (1987)

3. W. R. Potter and T. S. Mang, "Photofrin 2 levels by in vivo fluorescence photometry", in "Porphyrin Localization and Treatment of Tumors" (C. J. Gomer and D. R. Doiron eds.), pp. 177-186, Alan R. Liss (New York, 1984)

4. A. E. Profio and D. R. Doiron, "Transport of light in tissue in photodynamic therapy", Photochem. and Photobiol. 46, 5:591-599 (1987)

5. W. M. Star, H. P. Marijinissen, H. Jansen, M. Kreuzer and M. J. C. van Gemert, "Light dosimetry for photodynamic therapy by whole bladder wall irradiation", Photochem. and Photobiol. 46, 5:619-624 (1987)

6. H. Barr, S. G. Bown and N. Krasner, "Photodynamic therapy for colorectal cancer" (Abstract), The British Society of Gastroenterology (1987).

LASER EXCITED FLUORESCENCE LIFETIMES OF HEMATOPORPHYRIN IX AS A FUNCTION

OF SOLUTION pH: IMPLICATION IN PHOTODYNAMIC THERAPY

P. Nadeau and R.H. Pottier(*);
A. Szabo(**);
D. Brault and C. Vever-Bizet(***)

I. INTRODUCTION

The combination of a harmless chemical photoactivated with safe visible electromagnetic radiation in order to eradicate tumors with a photo-sensitization reaction, offers much hope for the treatment of a large variety of neoplastic diseases. The origin for the selective biodistribution of photosensitizer toward tumor tissue is not yet clearly understood. The pH of many rapidly growing tumors is often found to be significantly lower than that of normal tissue in the same individual[1÷4]. Recent volumetric titration of hematoporphyrin IX (HP)[5] have yielded the distribution diagrams of five HP ionic species, having net charges of +2, +1, 0, -1 and -2. However, only three ionic species may be associated with the steady state absorbance and fluorescence spectra[6]. Further, fluorescence decay measurements carried out in aqueous solutions at physiological pH generally show two lifetimes; a 15 nsec component normally ascribed to a monomeric species and a 3.8 nsec component usually assigned to a dimeric species[7]. Since both monomeric and dimeric species, as well as different ionic species, may play a role in the selective biodistribution of drugs toward tumor tissue, it was deemed important to evaluate the fluorescence lifetimes of HP as a function of both pH and concentration. These measurements were carried out in aqueous solutions as well as in the presence of a surfactant, sodium dodecyl sulfate (SDS).

II. MATERIAL AND METHODS

Hematoporphyrin IX dihydrochloride was obtained from Porphyrin Products, Logan, UT, USA (Lot No. 111684) and was used without further purification. The neutral form of hematoporphyrin IX was obtained from the "Museum National d'Histoire Naturelle de Paris, Laboratoire de Biophysique" and was purified as reported by Vever-Bizet et al.[8] and Dellinger et al.[9]. Sodium dodecyl sulfate salt, Baker Analyzed Reagent grade lot No. 527220

(*)   Department of Chemistry and Chemical Engineering, The Royal Military
       College of Canada, Kingston, Ontario, Canada
(**)   Department of Biology, The National Research Council of Canada,
       Ottawa, Ontario, Canada
(***) Laboratoire de Biophysique, Museum National d'Histoire Naturelle,
       INSERM, Paris, France.

was purchased from J. T. Baker Chemical Co., Phillipsburg, N. J. Stock solution were prepared by dissolving 1 mg HP into a 25 ml volumetric flask, using 0.010 molar HCl as the solvent. Complete solubility could be achieved by placing the solution flask in an ultrasonic bath for few minutes. From this stock solution, 2 ml aliquots were diluted to a final volume of 5.00 ml. The dilution was done with buffered solutions adjusted to the desired pH. The HP concentration range studied was between 0.5 μM and 30 μM. The final SDS concentration was 0.4% W/V when present in solution. The solution pH was obtained with the use of an Orion Research Model 811 microprocessor pH/millivolt meter equipped with an Orion Ross type model 8103 combination electrode. A two-points calibration was carried out daily prior to the start of any measurement. The solutions were bubbled with nitrogen just prior to each run, in order to be oxygen free. All measurements were carried out on freshly prepared hematoporphyrin solutions. Fluorescence decay curves were obtained from two independent systems, which were both single photon counting instruments. On the first, a 575 nm laser excitation beam was used, the details of which have been reported previously by Szabo[10]. The fluorescence emission was detected at 610 and 660 nm. The second system consisted of a 400 nm pulsed excitation source having a 50 nm band pass, which has been previously described by Sun et al.[11]. The fluorescence emission was detected using a low-pass filter at 590 nm.

Steady state fluorescence spectra were measured using a 75 W xenon lamp as excitation source (PTI A-1010), stabilized with a PTI LPS200X feed back control unit. The 404 nm excitation wavelength was obtained using an interference filter that has a 17 nm band pass. The excitation radiation was directed on the sample at a 90° angle from the emission monochromator, a Jarrell-Ash, 0.27 meter monochromator (Monospec - 27,100 grooves/mm grating blazed at 600 nm). The entrance slit of the monochromator was 25 μm. The entire fluorescence spectrum, from 500 to 800 nm, was recorded via an intensified photodiode array having 1024 diodes (Princeton Instruments, Inc. Model No. DMCP 700G). The exposure time of the analyzing light was typically of the order of 6 sec, during which 200 scans of the detector were integrated.

Absorbance measurements were carried out on Shimadzu-UV 160 double beam recording spectrophotometer. All quartz cuvettes used were cleaned in 3M nitric acid overnight prior to any series of measurements. Before any spe- cific measurement was done, the cuvettes were further washed with ethanol and distilled water, and then rinsed with the solution to be analyzed.

III. RESULTS AND DISCUSSION

Except at very low pH values, three lifetimes were obtained in aqueous solution of HP. The solutions studied were adjusted to pH values that correspond to regions in which the steady state absorbance and fluorescence spectra reveals the presence of distinct monomeric species. A study between 0.5 μM and 30 μM in HP shows that the concentration of the solution does not affect markedly the monomeric lifetime values. Thus from the results in table 1A, the monomeric species of HP in aqueous solutions is characterized by a fluorescence lifetime of 5.8 nsec at pH 0.5, 7.6 nsec at pH 3 and 15.2 nsec at pH 8. In addition, a shorter lived (3.8 nsec fluorescent species is detected at high pH, probably reflecting the dimeric form of HP, as has been suggested in the literature[7,12]. A very short-lived species (≈0.2 nsec) can be detected at pH higher than 2.5, possibly arising form higher aggregates of Hp. The pH dependant relative proportions of the longer lived monomeric species are in good agreement with recent results obtained via volumetric titration[5].

Table 1. Fluorescence lifetimes (τ) in nsec and the relative amplitude (A) of 29 μM aqueous solutions of HP (1A) and 25 μM aqueous solutions of HP in the presence of 0.4% SDS (1B)

pH	τ_1	A_1	τ_2	A_2	τ_3	A_3
(1A)						
0.5	5.84	0.93	0.26	0.07	---	---
3.0	7.61	0.53	3.84	0.08	0.18	0.33
8.5	15.23	0.57	3.04	0.12	0.22	0.31
(1B)						
0.5	6.72	0.63	2.81	0.09	0.04	0.27
4.0	9.13	0.83	3.44	0.05	0.04	0.20
8.0	16.74	0.95	2.34	0.05	---	---

The surfactant SDS tends to disrupt the dimeric and aggregated species of HP. As revealed in table 1B, the long lived fluorescence species of HP in the presence of SDS are in parallel with those observed in aqueous solutions, except that all long lived lifetimes are increased by approximately 1nsec and the pH values in which these can be detected are slightly different. The disrupting effect of the surfactant is revealed by the large amplitude values of the long lived species. A 3.4 nsec lifetime component could be observed over all the pH region studied, having an amplitude of ≈ 0.05 at pH lower than 5. This amplitude increased to 0.10 at pH 6 (data not shown) and decreased again to 0.05 at pH 8. As in the case of aqueous solutions, a third species is detected having a very short lifetime of approximately 0.04 nsec. This species could be detected mainly at pH lower than 5. The nature of this species could possibly be due to the presence of HP-SDS complexes.

The three long lived fluorescent species 5.8 nsec, 7.6 nsec and 15.2 nsec in aqueous solutions (6.7 nsec, 9.1 nsec and 16.7 nsec in the presence of SDS) may be assigned respectively to the dicationic, monocationic and neutral species of HP, in conformity with the volumetric titrations[5]. It must be noted that this nomenclature refers to the charge on the porphyrin ring only (due to protonation of the imino nitrogens) and do not include possible charges on the peripheral carboxylic acids[5,6].

It is now well established that the pH of many rapidly growing tumors is often lower than that of normal tissue[1÷4]. In fact, it has been shown that both selective tumor accumulation and photosensitization of tumor cells are increased at lower pH[14]. Photoacoustic studies also reveal increased molecular association of boundary surfaces in aqueous solutions of HP at pH 6[14]. Before the exact role of both the monomeric ionic species and the dimeric/aggregated forms of photosensitizing porphyrins can be fully elucidated, the characterization of the photophysical parameters of all species present in solutions of photochemotherapeutic agents must be made. As revealed in this study, not only monomers and dimers/aggregates exist in porphyrin solutions, but one must also include the various ionic species. Thus, both selective biodistribution and phototoxic reactivity must include the role of these various components. For example it is well known that neutral species can cross cell membranes with greater ease than charged species[15,16]. Thus charged species would be expected to have different pharmacokinetic, and possibly different photochemical and photosensitizing behavior.

REFERENCES

1. W. J. Waddell and R. G. Bates, "Intercellular pH", Physiol. Rev., 49:285 (1969)
2. T. C. Ng, W. T. Evanochko, R. N. Hiramoto, V. K. Ghanta, M. B. Lilly, A. J. Lawson, T. H. Corbet, J. R. Durant and J. D. Glickson, "31P NMR spectroscopy of in vivo tumors", J. Magn. Reson., 49:271 (1982)
3. J. L. Wike-Hooley, J. Haveman and H. S. Reinhold, "The relevance of tumor pH to the treatment of malignant diseases", Radiother. Oncol. 2:366 (1984)
4. A. J. Thistlethewaite, D. B. Leeper, D. J. Moylar III and R. E. Nerlinger, "pH distribution in human tumors", Int. J. Radiat. Oncol. Biol. Phys., 11:1647 (1985)
5. A. J. Barrett, J. C. Kennedy, R. A. Jones, P. Nadeau and R. H. Pottier, "The effect of tissue and cellular pH on the selective biodistribution of porphyrin-type photochemotherapeutic agents: A volumetric titration study", J. Photochem. Photobiol. (B:Biology) 6:1 (1990)
6. R. H. Pottier, J. C. Kennedy, Y. F. A. Chow and F. Cheung, "The pK$_a$ values of hematoporphyrin IX as determined by absorbance and fluorescence spectroscopy", Can. J. Spectrosc. 33(3):57 (1988)
7. A. Andreoni, R. Cubeddu, S. De Silvestri, G. Jori, P. Laporta and E. Reddi, "Time-resolved fluorescence studies of hematoporphyrin in different solvent systems", Z. Naturforsch. 38c:83 (1983)
8. C. Vever-Bizet, O. Delgado and D. Brault, "The purification of hematoporphyrin IX and its acetylated derivatives", J. Chromatog., 283:157 (1984)
9. M. Dellinger and D. Brault, "Normal-phase high performance liquid chromatography of free acid dicarboxylic porphyrin and hematoporphyirin derivative on silica, J. Chromatog. 422:73 (1987)
10. M. Zuker, A. G. Szabo, L. Bramall, D. T. Krajcarsk and B. Sellinger, "Delta function convolution method (9DFCM) for fluorescence decay experiments", Rev. Sci. Instrum. 56(1):14 (1985)
11. J. S. Sun, M. Rougé, M. Delarue, T. Montenay-Garestier and C. Hélène, "Solvent Relaxation around Excited 2-Methoxy-6-chloro-9-amino-acridine in aqueous solvents", J. Phys. Chem. 94:968 (1990)
12. M. Yamashita, M. Nomura, S. Kobayashi, T. Sato and K. Aizawa, "Picosecond time-resolved fluorescence spectroscopy of hematoporphyrin derivative", IEEE J. Quant. Electr. QE-20 (12):1363 (1984)
13. J. P. Thomas and A. W. Girotti, "Glucose administration augments in vivo uptake and phototoxicity of the tumor-localizing fraction of hematoporphyrin derivative, Photochem. Photobiol. 49(3):241 (1989)
14. R. Pottier, A. Lachaine, M. Pierre and J. C. Kennedy, "A new electronic absorbance band in concentrated aqueous solutions of hematoporphyrin IX detected by photoacoustic spectroscopy", Photochem. Photobiol. 47(5):669 (1988)
15. D. Kessel, "Hematoporphyrin and HPD: photophysics, photochemistry and phototherapy", Photochem. Photobiol. 39:851 (1984)
16. P. Seeman and H. Kalant, "Drug solubility, K absorption, and movement across body membranes", in "Principles of Medical Pharmacology", H. Kalant, W. H. E. Roschian and E. M. Sellers (eds.), University of Toronto Press, Toronto, 4th edn., p. 11 (1985)

BIOLOGICAL APPLICATIONS OF TIME-GATED FLUORESCENCE SPECTROSCOPY

R. Cubeddu, R. Ramponi and P. Taroni

C.E.Q.S.E., C.N.R.
Polytechnic of Milan
Milan, Italy

I. INTRODUCTION

The diagnostic use of fluorescence emission from both endogenous and exogenous fluorophores has found wide application in biology and medicine[1÷4]. At present in most cases fluorescence diagnosis is based on continuous wave (cw) measurements. This technique is applicable when the fluorescent emission from the target is sufficiently intense to be detected, and can be separated from the background through an appropriate selection of either excitation or emission wavelength[5,6]. However in many cases quenching mechanisms strongly reduce the fluorescence, or the spectral overlapping of different fluorophors does not allow effective discrimination or localization of a target exogenous dye.

However, the fluorescence emission process is characterized not only by its spectral intensity but also by its time evolution. In fact different fluorophores, even when overlapped in spectra, usually differ appreciably in their fluorescence decays. Thus time-resolved techniques may increase both selectivity and sensitivity of fluorescence-based diagnosis[7÷9]. Moreover, if the time-resolution of the measuring system is sufficiently short, the detection of a dye is not affected by quenching mechanisms which leave unaltered the number of excited molecules. Among the different techniques, photon counting detection allows the measurement of ultra weak fluorescence; and time-correlated single photon timing provides both the fluorescence decay and, by a suitable choice of a time observation window, the possibility of discriminating the contribution of different fluorophores to the overall emission spectrum through their temporal behaviour[10,11].

In this paper we will describe a computer-controlled system for time-resolved fluorescence spectroscopy, which is able to directly measure gated fluorescence spectra with programmable gate position and width. This apparatus has been utilized in several fields of biomedicine such as photosensitization, photomovement and ocular pigments. In particular, we will discuss its application to the study of the fluorescence properties of Hematoporphyrin Derivative (HpD). This drug is used in the photodynamic therapy of cancer because it is retained preferentially in tumor tissues compared with normal surrounding ones and produces, once activated by suitable light, cytocidal effects. Thus the aim of this study is to evaluate the modification induced by the environment on the fluorescence properties of

HpD, in order to elucidate some incorporation mechanisms of the drug "in vitro" and "in vivo".

II. MATERIALS AND METHODS

II.1. Chemicals

HpD was prepared following the procedure described by Lypson et al.[12] and stored as a lyophilized powder, ready to be dissolved in buffer. It was tested by analytical HPLC and found to contain 60% of tumor localizing fraction (TLF). TLF was obtained as described by Keir et al.[13] and supplied as a powder. Both drugs were kindly provided by the Department of Chemistry of the Paisley College of Technology (Paisley, Scotland, UK).

For the experiments in solution, TLF powder was dissolved in buffer (pH 7, 0.2 M phosphate) at a concentration of 2.5 mg/ml. This was the stock solution for all experiments including those performed in cells and was chosen since it is the same as the injectable solution used in clinical photodynamic treatment.

The stock solution was then diluted in buffer to a concentration of 5 µg/ml with an appropriate amount of surfactant (CTAB from Carlo Erba, Italy) added below and above the critical micelle concentration (cmc).

II.2. Cell line and treatments

Hybrid (Balb/cxDBA/2 F_1, hereafter called CD2F$_1$) mice of both sexes, 68 weeks old, from Charled River Breeding Laboratories (Calco, Italy) were used. The chemically induced lymphoid leukemia L1210 (H-2^d) was maintained by weekly i.p. injection into CD2F1 mice. In the experiments, when ascitic tumors developed, the cells were collected and extensively washed.

Cells were diluted to 10^6/cc in Hank's balanced salt solution (HBSS), treated with th drug (TLF or HpD containing 60% TLF) at different concentrations (5 and 20 µg/ml), and incubated in the dark. Every 30 min (up to 120 min and, in several cases, to 180 min), 1 cc was centrifuged, and resuspended in PBS (pH 7.4) for the measurements. Cell viability was checked by the Trypan blue dye exclusion test.

In a second set of experiments the drugs were injected i.p. or i.v. into mice at the concentration of 25 mg/kg b.w. for HpD and 12.5 mg/kg b.w. for TLF. Every 2 hr and up to 22 or 30 hr after i.p. or i.v. injection respectively, two mice were sacrificed, and cells were drawn from the peritoneum, washed, centrifuged and resuspended in phosphate buffered solution at a concentration of 2×10^6/cc for the fluorescence measurements.

At various times, in all the experiments, control measurements were performed on peritoneal cells extracted from tumor-bearing mice not injected with any drug.

II.3. Experimental apparatus

Fluorescence decay waveforms, consisting of both time-integrated and time-gated fluorescence spectra, were collected by using the system described in[11]. A schematic drawing of the experimental apparatus is shown in Figure 1.

The excitation source was a mode-locked Ar$^+$ ion laser (Coherent CR-18) tuned to 364 nm, with a pulse duration of $\approx$ 150 ps. The repetition rate of $\approx$ 75 MHz was reduced to $\approx$ 750 kHz by means of an acousto-optic pulse-

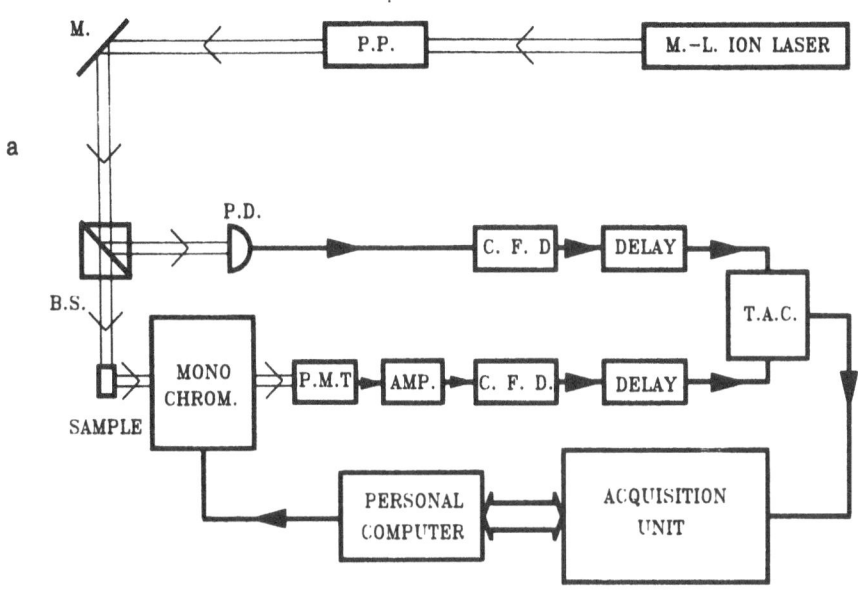

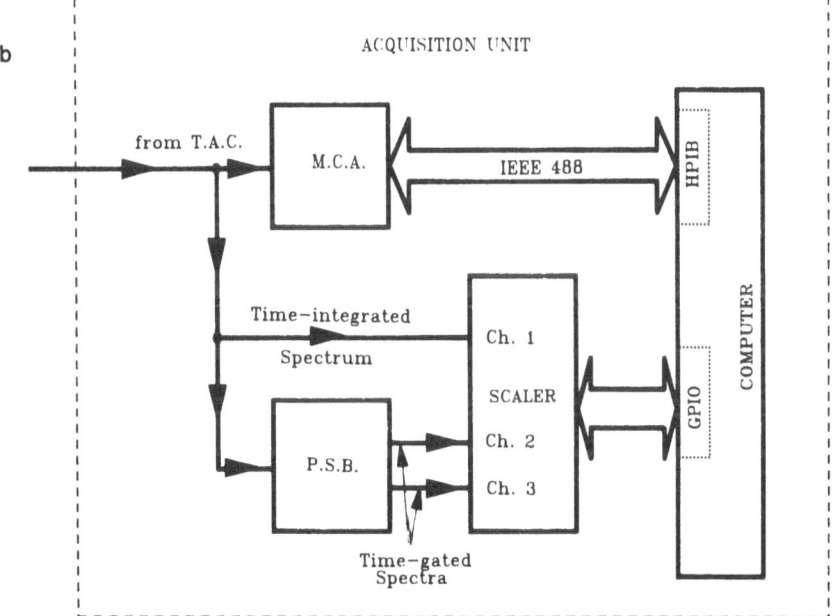

Fig. 1.   1.a: Block diagram of the system for time-resolved fluorescence spectroscopy with picosecond gating. Legend: P.P.:
pulse-picker; M.: mirror; B.S.: beam splitter; P.D.: p-i-n
photodiode; AMP.: signal preamplifier and amplifier circuitry;
C.F.D.: constant fraction discriminator; T.A.C.: time-to-
amplitude converter.
1.b: Details of the acquisition unit for decay waveforms,
and for time-integrated and time-gated spectra.

picker (Coherent 7200). The samples were contained in a 1 cm-pathway quartz cuvette. The emitted fluorescence was collected at 90° through a monochromator (Jarrell-Ash 82-410) for selection of the observation wavelength, and detected by a microchannel-plate photomultiplier (Hamamatsu R1564U-01). Cut-off filters (Kodak Wratten) were used to eliminate any residual excitation light. The overall time resolution of the system was ≈ 150 ps.

The experimental data were obtained using the time-correlated single photon counting technique[14]. The output pulses of the time to amplitude converter (TAC) were sent to a multichannel analyzer (Silena System BS 27), operated in the PHA mode, to measure the decay curves, and to a home-made acquisition unit. This unit allowed the simultaneous collection of the time-integrated emission spectrum, by accumulating all the TAC output pulses contributing to the decay curve, and two time-gated fluorescence spectra, by counting only the signals falling within preselected time windows. Therefore, a single measurement provided three fluorescence spectra and the full-spectrum decay curve, obtained by setting the multi-channel analyzer to the acquisition mode during the whole time interval of the collection of the spectra.

Experimental data were sent to a personal computer for storage and analysis. In particular, fluorescence decay waveforms were analyzed by means of a non-linear least squares fitting method, and weighted residuals and their autocorrelation function were used to judge the quality of the fitting. Wavelength scanning and setting of the gate parameters were computer controlled.

III. RESULTS AND DISCUSSION

HpD as used in photodynamic therapy is a complex mixture of porphyrins where only a fraction is tumor-localizer (TLF). Although its composition is still not completely understood, it seems likely that it is made by polymeric chains of porphyrin with an average number of 8 porphyrin rings[13]. The TLF is strongly aggregated in aqueous solution, so that its fluorescence and singlet oxygen formation yields are very low. Thus, following therapeutic injection the TLF must undergo photophysical modification to become photodynamically active. In order to evaluate this effect time-gated fluorescence spectroscopy was performed on both HpD and TLF in different environments.

The simpler model utilized was that of surfactants which can provide a hydrophobic microenvironment similar to the cell membrane. Since porphyrins are negatively charged, major effects were observed, as expected, using the positively-charged surfactant CTAB. Figure 2 shows the emission spectra of 5 μg/ml TLF in buffer at different CTAB concentrations below and above the cmc. As reported in several studies[15,16], there is substantial increase in the fluorescence intensity and a red-shift (the peak moves from 615 to 626 nm) as a result of "solubilizing" the TLF in detergent micelles. However the most marked effect arises at CTAB concentration below the cmc, where a substantial modification of the spectrum is observed together with a reduction in fluorescence intensity on adding CTAB at a concentration of 0.05 mM. The fluorescence lifetime measurements evidenced the presence of three emitting molecular species with decay time values of ≈ 0.7, ≈ 3 and ≈ 15 ns respectively, and relative abundance dependent on the CTAB concentration as shown in Table 1.

The slow component, which is predominant in buffer and in micelles, has been attributed to residual monomers and Hematoporphyrin (Hp) moieties of unfolded TLF oligomers. This species, which has an emission peak at

Table 1. Fluorescence decay time constants and relative amplitudes
of 5 µg/ml TLF in buffer at different CTAB concentrations.

CTAB (mM)	τ_1 (ns)	A_1 (%)	τ_2 (ns)	A_2 (%)	τ_3 (ns)	A_3 (%)
0	14.56	56	3.13	22	0.48	22
0.05	11.61	4	3.04	43	1.11	53
1	15.00	69	3.54	31	--	--

enhancement at about 660 nm in the undelayed spectrum. Thus the fluorescence spectra are given by the superposition of the emission spectra of
different molecular species present in the solution and no appreciable interaction among the chromophores is observed.

Different observations can be made on the spectra of the TLF in the
with aggregated or polymeric material. Since their predominance respect to
the slow decay at 0.05 mM CTAB concentration could not be explained in
terms of a disaggregation process, time-gated fluorescence spectroscopy was
utilized to elucidate this result.

Taking into account the measured values of the decay times, three sets
of gates were used, having a delay with respect to the peak of 0, 5 and 18
ns and widths of 0.2, 3 and 6 ns respectively. Assuming that each molecular species fluoresces independently of the presence of the others, the
spectrum delayed by 18 ns is expected to be mainly related to the slow component while that at 0 delay should provide information on the excited molecules independently of their fluorescence decay. Figures 3-5 show the
fluorescence emission spectra of TLF in the three solution environments
considered. According to the above considerations for buffer and micelle
solution, in both situations the time-integrated spectrum is quite similar
to the long-delayed one, indicating a predominance of the 15 ns component,
while the contribution of the other two species is evidenced by the slight

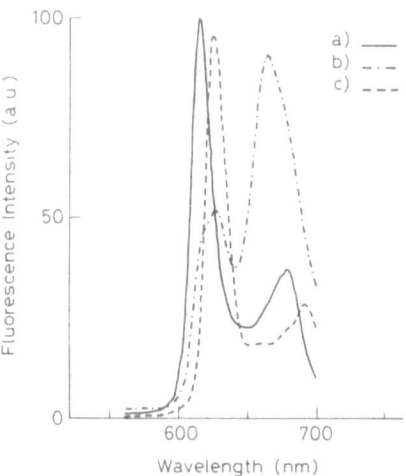

Fig. 2. Emission spectra of 5 µg/ml TLF in buffer (pH 7,
µ = 0.2) at different CTAB concentration. All
curves were obtained under excitation at 364 nm
and normalized to the peak value. (a) No CTAB,
emission range = 88.7. (b) 0.05 mM CTAB, emission
range = 68.1. (c) 1 mM CTAB, emission range = 502.

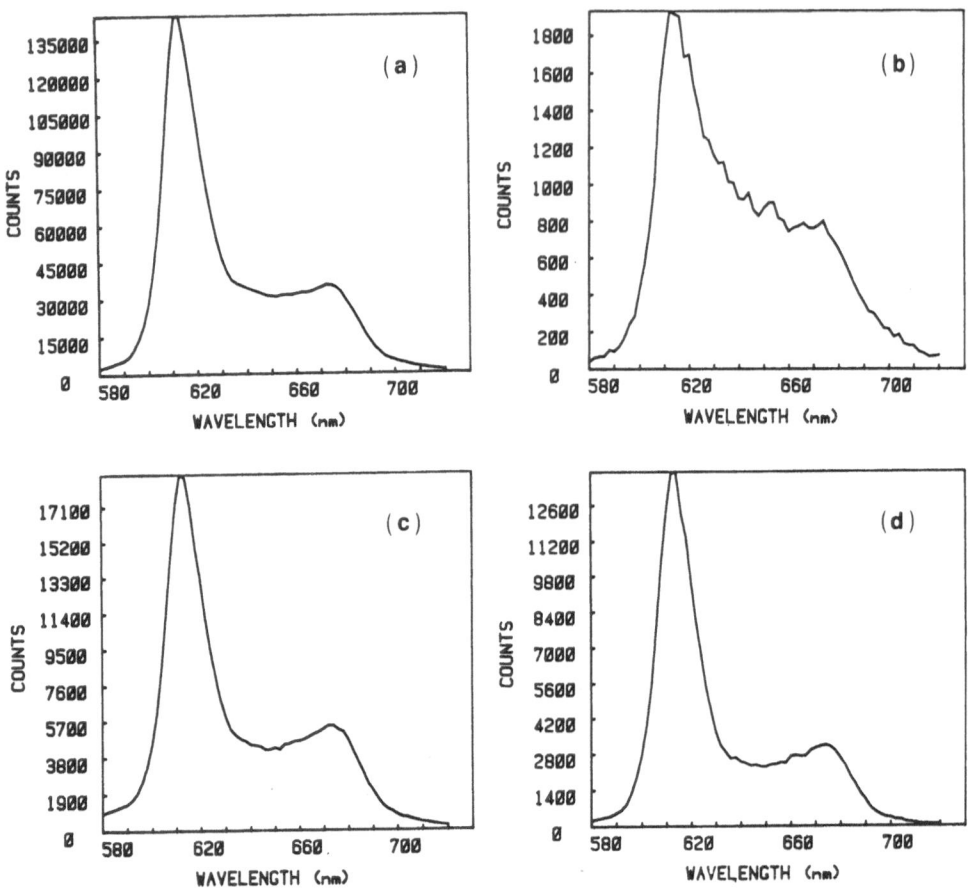

Fig. 3. Emission spectra of 5 μg/ml TLF in buffer: (a) time-integrated
spectrum. (b) gated spectrum, 0 delay, 0.2 ns width. (c) gated
spectrum, 5 ns delay, 3 ns width. (d) gated spectrum, 18 ns
delay, 6 ns width.

≈ 615 nm in an aqueous environment and at ≈ 626 nm in a hydrophobic environment, can account for the increase in emission intensity through a progressive unfolding induced by the micelles. The other two components are
attributed to aggregated polymers (leading to the fastest decay) and to
"free chromophore" moieties, probably associated with CTAB, interacting
presence of CTAB below the cmc (Fig. 5). The 660-670 nm emission band
becomes predominant in the time-integrated spectrum, and both the 615 and
626 nm peaks, respectively related to non-interacting and CTAB-interacting
long-living species, are present consistently with the data reported in
Table 1. However, the analysis of the gated spectra indicates that the results obtained cannot be explained simply in terms of relative abundance of
the molecular species. In fact, the undelayed spectrum shows no evidence
of the 615 nm long-living species, whereas the 626 nm peak is more enhanced. On the contrary, in the long-gated spectrum the 615 nm peak is
predominant and almost completely masks that at 626 nm, barely observable
as a small shoulder. It must also be noted that even after 18 ns the

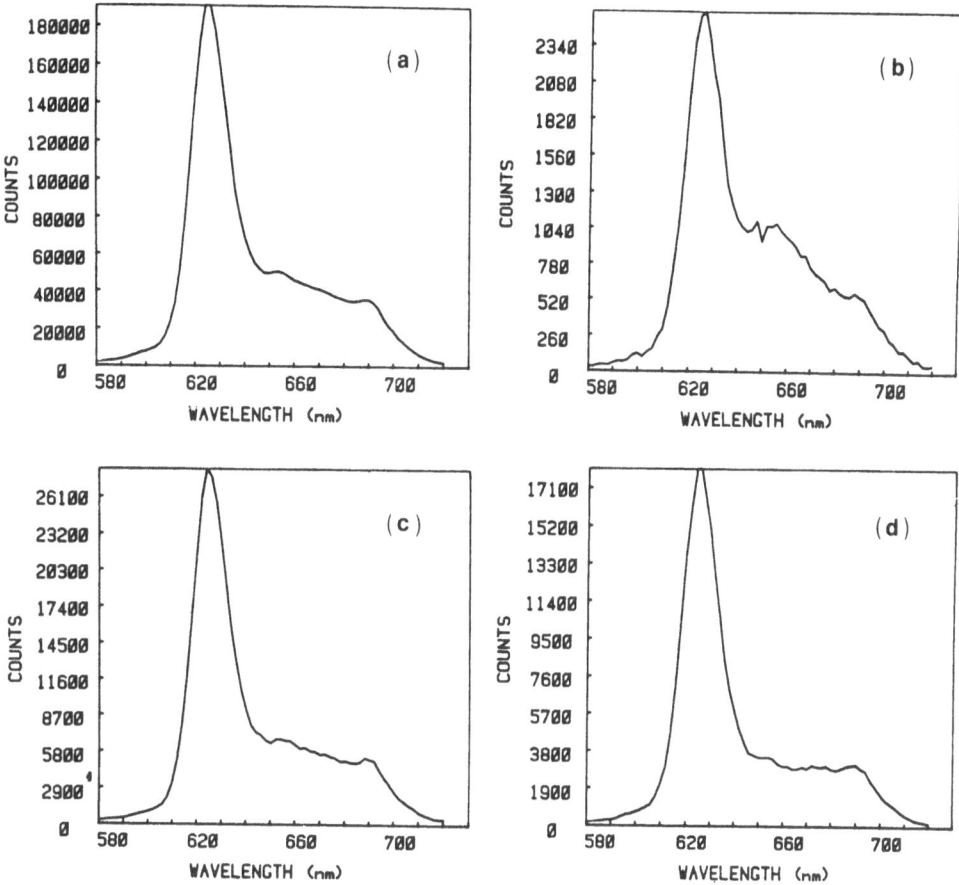

Fig. 4. Emission spectra of 5 µg/ml TLF in buffer with 0.05 mM CTAB:
(a) time-integrated spectrum. (b) gated spectrum, 0 delay,
0.2 ns width. (c) gated spectrum, 5 ns delay, 3 ns width.
(d) gated spectrum, 18 ns delay, 6 ns width.

660-670 nm emission band is still present in an appreciable amount, in contrast with the behaviour expected for the value of its decay constant.

These experimental results thus imply the existence of an energy transfer mechanism among the molecular species. In particular, a conversion from the molecules emitting at 626 nm to those fluorescing at 660-670 nm seems to be the most significant. On the contrary, the molecules emitting at 615 nm behave independently. Thus the energy transfer seems to be mediated by the presence of CTAB, and possibly to occur between molecules belonging to the same polymeric chain interacting in different ways with the detergent in the pre-micellar range.

As shown above, polar hydrophobic structures strongly interact with the TLF or HpD. Thus similar behaviour is expected when these drugs are incorporated into cells, where cell and cytoplasm organelle membranes create a highly hydrophobic environment. This aspect was evaluated on cells collected from ascitic tumors implanted in mice (L1210 cells), and

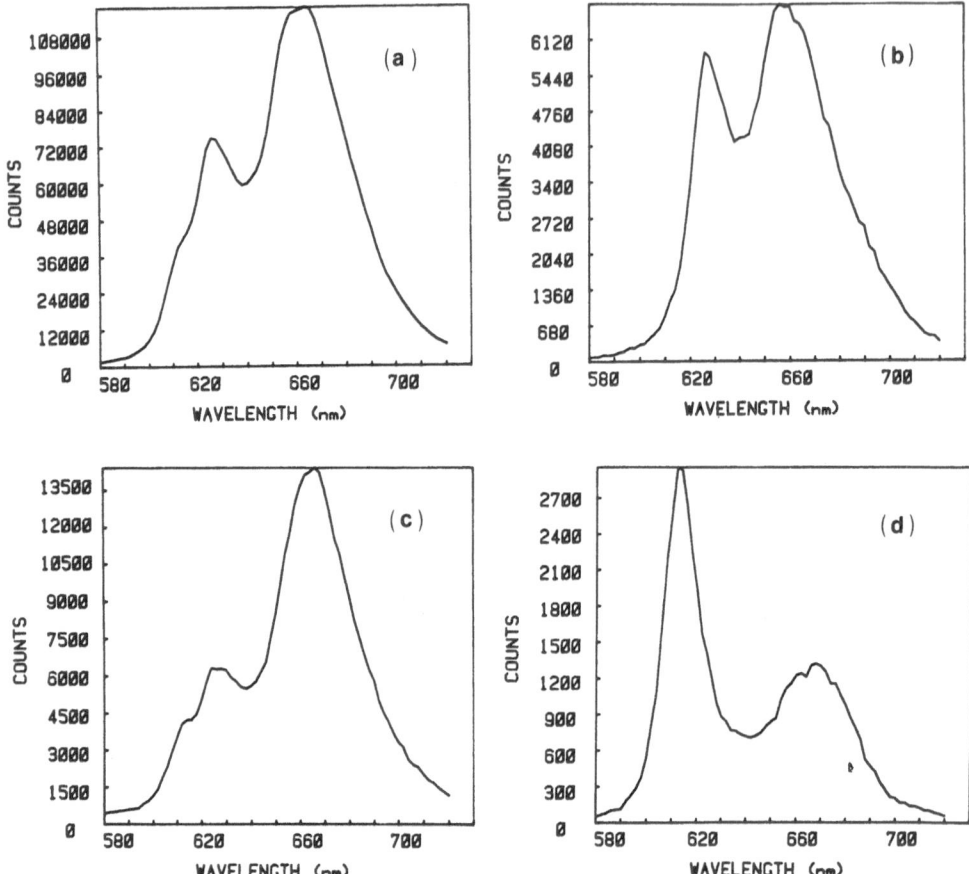

Fig. 5. Emission spectra of 5 µg/ml TLF in buffer with 0.05 mM CTAB:
(a) time-integrated spectrum. (b) gated spectrum, 0 delay,
0.2 ns width. (c) gated spectrum, 5 ns delay, 3 ns width.
(d) gated spectrum, 18 ns delay, 6 ns width.

time-gated fluorescence spectroscopy was utilized to investigate the photophysical properties of the TLF, and HpD containing 60% TLF, incorporated
into cells and their dependence on the uptake time.

Typical fluorescence spectra of L1210 cells treated with 20 µg/ml of
TLF for different incubation times are shown in Figure 6. The time-integrated spectrum collected after 30 min is characterized by the presence of
two peaks at 615 nm and 630 nm respectively, and by a shoulder at ≈ 665 nm.
According to the results obtained for the TLF in solution, three molecular
species are present in the fluorescence decay: the fast one related to aggregates or folded polymeric chains (τ_3 =0.92 ns, A_3 =37%), the intermediate
to porphyrin moieties interacting with cellular compponents and/or porphyrin aggregates (τ_2 =4.7 ns, A_2 =30%), and the slow one to monomers and
end-rings of unfolded polymeric chains (τ_1 =14.69 ns, A_1 =33%). No significant changes were observed for the values of the fitting parameters as a
function of the uptake time, indicating time equilibrium among the fluorescent molecular species. As evidenced by the values of their relative am-

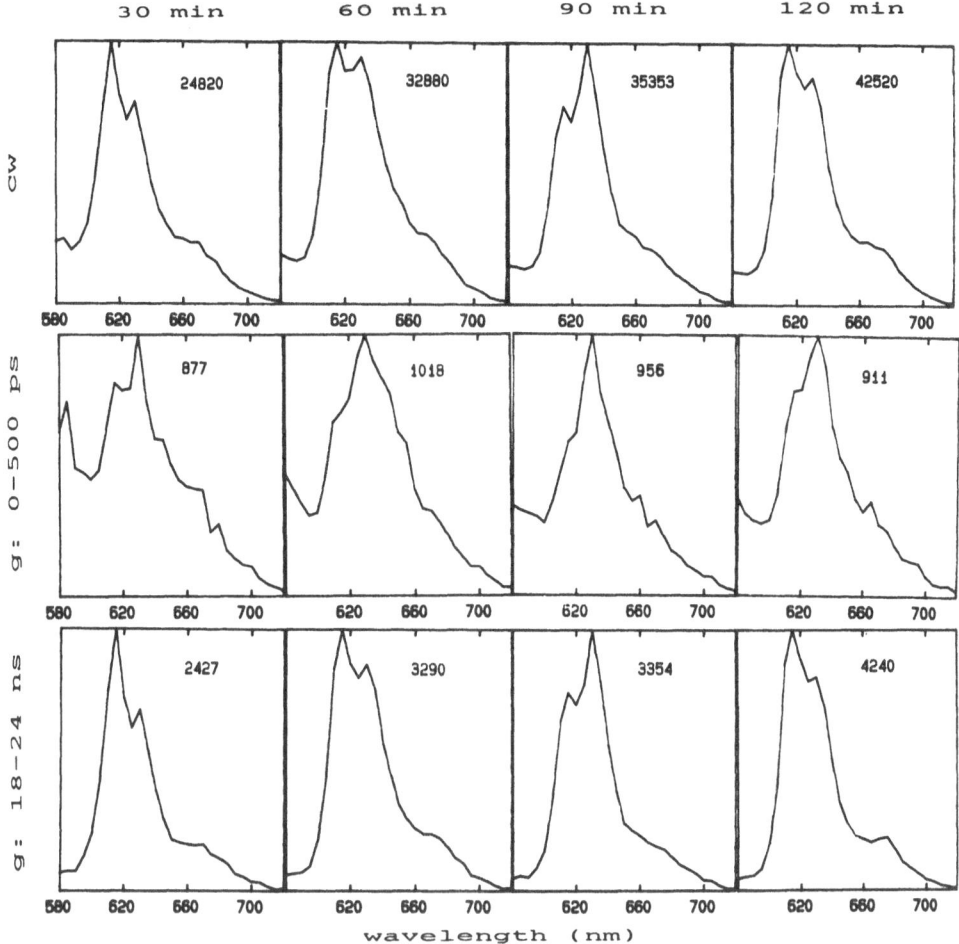

Fig. 6. Fluorescence spectra of L1210 cells treated with 20 µg/ml of TLF for different incubation times. Top line: time-integrated spectrum; center line: time-gated spectrum, 0 delay, 500 ps width; bottom line: time-gated spectrum, 18 ns delay, 6 ns width. Successive plots from left to right show emission spectra taken after different uptake times: 30 min, 60 min, 90 min, and 120 min.

plitudes, the long-living component provides the major contribution to the time-integrated spectrum. The appearance of two main peaks indicates that the porphyrin molecules are in the presence of either an aqueous (615 nm) or hydrophobic (630 nm) microenvironment. The attribution of the long-living component is also confirmed by the shape of the delayed spectrum, where the two peaks are still dominant. As expected, the fast and intermediate components become more apparent in the undelayed spectrum, where an enhancement in the emission between 630 and 680 nm is observed.

When the uptake time is increased, the cells seem to incorporate the porphyrin molecules to a larger extent, as shown by the observed increase of the emission peak at 630 nm and the similarity among the gated spectra.

After 120 min of incubation, however, the peak at 615 nm is again apparent and the shoulder at 630-680 nm is observable in the undelayed spectrum. The modifications observed in the spectra can be explained in terms of metabolic effects on the TLF. It appears that after incorporation the TLF is reduced, possibly by the breaking of large polymeric chains that leads to a consequent partial drug release or weaker binding at cellular sites (615 nm peak). This seems to be consistent with the hypothesis that the uptake mechanism of the TLF aggregates is phagocytosis[17].

Since the TLF is derived from Hematoporphyrin Derivative via a purification technique, as described in Ref. 13, the whole set of measurements was repeated using HpD with a content of 60% of TLF. The effect of monomeric material was evident for all the spectra in a predominance of the emission at 615 nm and in an increase in the relative amplitude of the long-life component (A_l=78%). Thus, the presence of the monomers apparently modifies the uptake process by saturating the binding sites, possibly on the outer cell membrane, and thereby reduces the amount of material incorporated in hydrophobic environment.

The results obtained seem therefore to indicate that both the drug composition and the metabolic properties of the biological environment strongly influence the uptake process and the fluorescence behaviour of the incorporate sensitizer.

The influence of systemic administration at the cellular level was studied in a second set of experiments by collecting time-gated fluorescence data after intraperitoneal or intravenous administration of either TLF or HpD.

The shape of the spectra for both HpD and TLF are, as previously, dominated by the long-life components. However in this case the 630 nm peak is constantly dominant at all times for both drugs and increases in intensity until 14 hr after the injection, with no meaningful alteration thereafter. The 615 nm emission is visible as a shoulder only in the first hours after HpD administration and is weaker after i.v. than after i.p. injection.

Thus, systemic administration leads to a more continuous trend in the fluorescence behaviour and presumably in the uptake process, and to a possible selectivity in carrying the drug to the peritoneum. In fact, the intensity ratio between the 630 nm emission and the 615 nm one is always higher not only for i.v. injection but also for TLF with respect to HpD.

The monomers contained in HpD seem therefore to be partially eliminated before reaching the peritoneum and to bind weakly at the available binding sites, possibly on the outer surface of the cell membrane. The weakness of this binding favours a subsequent fast release, as evidenced by the disappearance with time of the 615 nm fluorescence. These results also confirm that TLF is the porphyrin fraction most efficiently delivered and retained by neoplastic cells and, therefore, the one responsible for the therapeutic effect in the treatment of tumors.

In conclusion, the experimental results reported in this paper have shown the potential utility of picosecond time-gated fluorescence spectroscopy due to its sensitivity, its time-resolution and the large amount of information that can be collected during a single measurement through gated spectra. Work is in progress to extend this technique for diagnostic applications "in vivo".

REFERENCES

1. J. R. Lakowicz, "Principles of Fluorescence Spectroscopy", Plenum Press, New York (1983)

2. C. S. Hoyt, R. R. Richards-Kortum, B. Costello, B. A. Sacks, C. Kittrell, N. B. Ratliff, J. R. Kramer and M. S. Feld, "Remote biomedical spectroscopic imaging of human artery wall", Lasers Surg. Med., vol.8, pp. 1-19 (1988)

3. Y. Hayata, H. Kato, J. Ono, Y. Matsushima, N. Hayasi, T. Saito and N. Kawate, "Fluorescence fiberoptic bronchoscopy in the diagnosis of early stage lung cancer", Recent Results Cancer Res., vol. 82, pp.121-130 (1982)

4. A. E. Profio, "Laser-excited diagnosis of hematoporphyrin derivative for the diagnosis of cancer", IEEE J. Quant. Electron., vol. QE-20, pp.1502-1506 (1984)

5. R. Baumgartner, H. Fisslinger, D. Jocham, H. Lenz, L. Ruprecht, H. Stepp and E. Unsoeld, "A fluorescence imaging device for endoscopic detection of early stage cancer - instrumentational and experimental studies", Photochem. Photobiol., vol. 6, pp. 759-763 (1987)

6. S. Montan, K. Svanberg and S. Svanberg, "Multicolor imaging and contrast enhancement in cancer-tumor localization using laser-induced fluorescence in hematoporphyrin-derivative bearing tissue", Opt. Lett., vol. 10, pp. 56-58 (1985)

7. F. Docchio, R. Ramponi, C. A. Sacchi, G. Bottiroli and I. Freitas, "An automatic pulsed laser microfluorometer with high spatial and temporal resolution", J. Microsc., vol. 134, pp. 151-160 (1984)

8. S. Kinoshita, H. Ohta and T. Kushida, "Subnanosecond fluorescence life-time measuring system using single photon counting method with mode-locked laser excitation", Rev. Sci. Instrum., vol.52, pp. 572-575 (1981)

9. H. Schneckenburger, H. K. Seidlitz and J. Eberz, "New trends in photobiology: time-resolved fluorescence in photobiology", J. Photochem. Photobiol. B: Biology, vol.2, pp. 1-19 (1988)

10. D. V. O'Connor and D. Phillips, "Time-correlated single photon counting", Academic Press, London, New York, Ch. 7, pp. 211-251 (1984)

11. R. Cubeddu, F. Docchio, W. Q. Liu, R. Ramponi and P. Taroni, "A system for time-resolved laser fluorescence spectroscopy with multiple picosecond gating", Rev. Sci. Instrum., vol. 59, pp. 2254-2259 (1988)

12. R. L. Lipson, E. J. Baldes and A. M. Olsen, "The use of a derivative of hematoporhyrin in tumor detection", J. Natl. Cancer Inst., vol. 26, pp. 1-8 (1961)

13. W. F. Keir, E. J. Land, A. H. MacLennan, D. J. McGarvey and T. G. Truscott, "Pulsed radiation studies of photodynamic sensitizers: the nature of DHE", Photochem. Photobiol. 46: 587-589 (1987)

14. S. Cova, A. Longoni, A. Andreoni and R. Cubeddu, "A semiconductor detector for measuring ultraweak fluorescence decays with 70 ps FWHM resolution", IEEE J. Quantum Electron. QE-19: 630-634 (1983)

15. R. Cubeddu, R. Ramponi and G. Bottiroli, "Time-resolved fluorescence spectroscopy of hematoporphyrin derivative in micelles", Chem. Phys. Lett. 128: 439-442 (1986)

16. R. Redmond, E. J. Land and T. G. Truscott, "A comparison of the photophysical properties of porphyrins used in cancer phototherapy", in: "Primary Photoprocesses in Biology and Medicine", edited by R. V. Bensasson, G. Jori, E. J. Land and T. G. Truscott, Plenum Press, New York, pp. 335-339 (1985)

17. T. J. Dougherty, "Photosensitizers: therapy and detection of malignant tumors", Photochem. Photobiol. 45: 879-889 (1987).

INSTRUMENTATION FOR PHOTODYNAMIC THERAPY

D. R. Doiron

Laser Therapeutics, Inc.
Buellton, California, USA

Photodynamic Therapy (PDT) is a modality that includes pharmaceuticals, devices and techniques. Without the proper instrumentation and their proper use, PDT is ineffective. Instrumentation for PDT includes: light sources, delivery systems, fluorescence detection systems and dosimetry devices. All of these devices are important in optimizing PDT's potential as both a diagnostic and therapeutic modality.

PDT light source requirements are that the output spectrum of the source must provide ample energy in the absorption band of the photosensitizer used. In addition, the delivered light should not contain a significant amount of energy outside of the absorption band of the photosensitizer to minimize overheating or the generation of undesirable photochemical reactions. Another major factor in the choice of the light source is its deliverability to the treatment site. The two major categories of light source available for PDT include laser and non-laser sources. The laser is a good PDT light source, if its output wavelength matches the drug absorption due to their monochromatic nature, but also due to their efficient delivery through small flexible fiber optics. This deliverability gives raise to the possibility of efficient use for PDT through standard flexible endoscopes and for direct interstitial applications. The price of this deliverability, though, is high in terms of complexity, reliability and money. Solid state lasers hold promise in providing better PDT laser systems than the present argon pumped dye systems in use.

Non-laser light sources for PDT hold significant promise for certain PDT applications. These applications tend to be the conditions readily accessible without the use of endoscopy and which are superficial in nature. Non-laser sources suitable for limited PDT applications include lamps and light emitting diodes (LED's). In lamp sources the arc systems, xenon or mixed gas, hold the most promise due to their relatively high brightness. Addition of additives to these lamps can significantly enhance their output in the red spectrum for use in PDT while newer lamp designs and coupling methods have recently led to significantly better deliverable power levels through flexible deliverable systems. At present it is possible to deliver over 400 mW in the (630 ± 20) nm band with such a system. The arc lamps, as well as the incandescent lamps require significant filtering to remove undesirable wavelengths such as UV and infrared. LED's are highly efficient light sources with relatively narrow emission bands, typically 40 nm. LED outputs are limited as individual units, but are effective sources for large areas is used in two dimensional arrays. Such sources do offer significant advantages in cost, reliability and easy of use compared to most lasers applicable to PDT. The deliverability of such sources make them applicable only for non-endoscopic and non-interstitial applications. In order to compare one light source to another on a per watt or joule basis, it is necessary to convolute the normalized source spectrum to the absorption of the photosensitizer used in-vivo. Even though non-laser sources for PDT are limited in their applications of PDT, they do hold significant promise for certain applications.

Delivery systems for PDT are unique compared to surgical lasers. In most surgical applications the desire is to concentrate the light out of the delivery system to obtain high power densities to cut, coagulate or induce photoacoustical effects. In PDT the desire is to disperse the light uniformly either over a large area or into a significant volume of tissue in order to activate the photochemical process without inducing unwanted effects. To disperse the light, a number of lenses and diffusing delivery systems have been developed for PDT. The lens system provides uniform circular illumination fields, while the diffusing fiber provides for illumination in spherical or cylindrical type cavities or for interstitial use. In addition, balloon diffusers have also been found useful for defining the illumination cavities and protecting the delivery fiber in certain applications. For lasers the delivery systems are generally small (<600 micron core, single optical fibers); while for non-laser sources, they are either mirror/lens systems or light guides for arc lamps.

Fluorescence detection equipment in PDT consists of both non-imaging and imaging types of detection. It has been found that the limiting factor in fluorescence localization in PDT is the inherent background (autofluorescence) of the tissue. This background masks the fluorescence of the fluorescence marker leading to excessive false negative findings. In addition, the rapid variation of the fluorescence signal with distance between the tissue and the detection system can make it very difficult to determine what is true positive fluorescence and what is simply distance variations. Methods are presently under development to minimize these problems by using various methods of background subtraction and distance correction. These corrections are straight forward for non-imaging detection systems but are complicated for imaging based detection methods. With these corrections, fluorescence localization holds a significant potential in the application of PDT to early localized lesions and possibly as an aid to dosimetry.

PDT dosimetry devices today consist of systems to measure the delivered power to the tissue prior to treatment only. Such methods do not account for variations of the delivered light during the treatment, or do they account for the physical parameters that might impact the effectiveness of the PDT treatment. These parameters include the photosensitizer concentration in the tumor and surrounding normal tissue, the optical characteristics of the tissue and the physiological factors that might limit the treatment effectiveness. To overcome this and better optimize the PDT treatment, a number of devices and methods are under development to permit measurement and possible real time monitoring of the important parameters. These devices include systems that are both non-invasive and invasive. A review of these various techniques and their application will be presented. Each of these devices need to be tested to determine the correlation between the measured parameter and the biological response.

Overall, PDT requires a significant variation of devices to be effective. The development of these devices and their availability will directly impact the effectiveness and ultimate acceptance of PDT as a new therapeutic modality in the battle against cancer.

230

THE USE OF OPTICAL FIBERS IN BIOMEDICAL SENSING

A. G. Mignani and A. M. Scheggi

IROE - CNR
Firenze, Italy

I. INTRODUCTION

The use of optical fibers in biomedical instrumentation goes back to the 1960s, when thousands of thin glass fibers were successfully assembled in flexible bundles, for imaging hidden areas of the human body. Some fibroscopes were further provided with an ancillary channel, so as to make possible power-laser cavitational-surgery and therapy using large-core optical fibers as a delivery system. In addition to this fiber-based instrumentation, attention has also been given to sensors for monitoring biological functions by means of optical fibers[1,2,3].

Fiberoptic biosensors (FOBS) are very satisfactory for continuous internal monitoring of the patient, owing to the high degree of integration of the probes, the biocompatible material of which the fibers are made, and the safety offered to the patient by the absence of electrical connections.

FOBS are now sensitive and fast and since their cost is destined to decrease, they are particularly suited for disposable probes. As conventional sensors, fiber-based sensors are easy to handle by medical staff that is not particularly skilled in optics: the entire optics and electronics are housed in a compact unit, and the fiberoptic link feeding the probe is joined to the electro-optic unit by a simple threaded connector. Even when a start-up calibration is necessary, very simple procedures are involved.

II. THE BASIC WORKING PRINCIPLE

The basic architecture of a FOBS is a fiberoptic-link which connects the electro-optic module to the sensing tip, as shown in Figure 1. The physical or chemical parameter being tested modifies one optical characteristic of the guided light: either intensity, or wavelength or polarization state. The sensor output is obtained by intensity modulation, directly from the fiber or through suitable filters or polarizers.

Depending on the sensing tip, FOBS are usually classified as spectrometric, transducer and all-fiber sensors.

Spectrometric sensors perform a photometric analysis "on the patient",

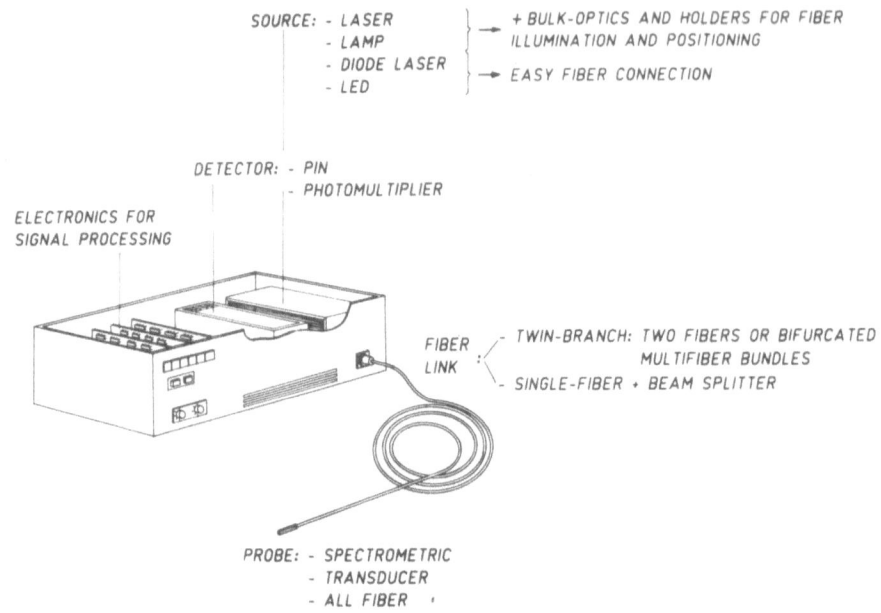

SOURCE: - LASER } + BULK-OPTICS AND HOLDERS FOR FIBER
- LAMP ILLUMINATION AND POSITIONING
- DIODE LASER } → EASY FIBER CONNECTION
- LED

DETECTOR: - PIN
- PHOTOMULTIPLIER

ELECTRONICS FOR
SIGNAL PROCESSING

FIBER - TWIN-BRANCH: TWO FIBERS OR BIFURCATED
LINK : MULTIFIBER BUNDLES
- SINGLE-FIBER + BEAM SPLITTER

PROBE: - SPECTROMETRIC
- TRANSDUCER
- ALL FIBER

Fig. 1. Basic architecture of a fiberoptic bio-sensor.

patient", thus allowing localized measurements and obviating sample-taking
Transducer sensors are based on an additional sensing element, fixed at the
fiber end, which induces the optical modulation of the guided light. All-
fiber sensors use the fiber itself as transducer, generally through a modi·
fied portion of the core or of the cladding.

Optical fibers are generally multimode-type, all-silica as well as PCS
or plastic type. Separate fibers or bi-/trifurcated bundles can be used,
each arm coupled to the source and to the detector, respectively, and
joined together at the probe. A more compact arrangement and a size reduc·
tion of the probe are obtained by using a single fiber, which uses a split·
ting device for separating the illumination from the return signal. Different sources can be used, depending on the working wavelength of the
sensing tip. Diode lasers and LEDs, housed in proper receptacles, allow
easy connection with the fiber, but sometimes lasers or broad-band white
light lamps provided with filters must be used. The detectors are generally photomultipliers or photodiodes, equipped also with filters in case a
restricted spectral response is required.

The optical system presents an "all-fiber" or "hybrid" architecture.
The first configuration is the more compact and reliable, since all the optical components such as X, Y, and star couplers and wavelength division
multiplayers (WDM) are fiberoptic-type, and are easily joined together by
commercially available connectors. The hybrid archiitecture, on the contrary, uses conventional optical components miniaturized for fiberoptic applications, and custom holders and alignment devices.

The sensors based on light intensity modulation suffer heavily from a
change in intensity induced by the electro-optic system, caused for example
by fiber connections and bends or by source fluctuations. This drawback
must be overcome by a referencing technique, which naturally increases the
complexity and the cost of the system. For example, a down-lead insensitive output isobtained by the dual-wavelength technique, which processes
the intensity of light, at two wavelengths, one modulated by the measurand,
and the other, insensitive. Even if extra processing of the output signal

234

is not complicated, a computer-based system or a suitably programmed micro-processor is advisable for management of the electro-optic module.

III. SPECTROMETRIC SENSORS

A conventional spectrometer, with an illuminating and detecting unit suitably coupled to fibers for remote operation, is the basic configuration of spectrometric FOBS for general-purpose application. Measurements of absorbance and fluorescence need a measuring cell at the fiber end, both for determining the optical path inside the sample and for providing light backtransmission. Measurements of reflectance, on the contrary, are performed by a single fiber or a bundle of bare fibers, the numerical aperture of which determines the inspected viewing angle. On this basis, custom devices have been developed for the measurement of specific physiological parameters, using simplified electro-optic units ad custom-made probes, as shown in Figure 2.

III.1. Reflectance-Based Sensors

The monitoring of the oxygen saturation level (SaO_2) of blood was one of the first analyses performed by optical fibers, by means of a reflectance-based FOBS (Figure 2a). Long-term monitoring of blood oxygenation is important in cardiac surgery and in neonatal therapy. These stressed patients are safely monitored by fibers inserted in a catheter, without the necessity of actually taking many blood samples. The reflectivity of hemoglobin, the oxygen vector, is analyzed at the isobestic point (≈ 900 nm) where both fully oxygenated and reduced hemoglobin have the same reflectivity, and also at ≈ 650 nm where the highest difference between the reflectivities occurs. SaO_2 is linearly related to the ratio between the reflectivities at the two wavelengths[4,5].

Another analysis concerning blood, performed by bare-fiber reflectometry, is blood perfusion monitoring by laser-Doppler velocimetry (LDV),

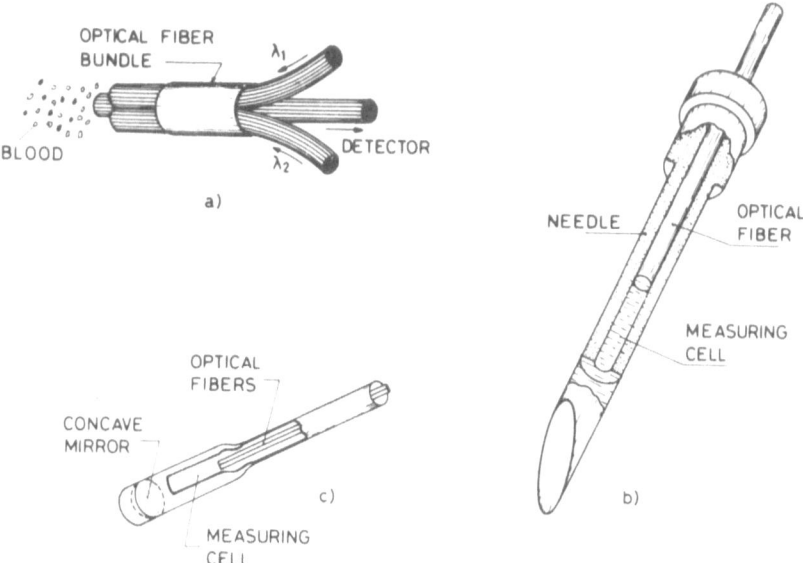

Fig. 2. Spectrometric probes for: a) blood oximetry; b) drug dosimetry; c) bile analysis.

which is of interest in cardiovascular clinic physiopathology, in translu-
minar angioplasty, and in other diagnostic related to vascular and plastic
surgery and circulation monitoring. Miniaturization and flexibility of optical fibers are essential in these applications, where measurements within
thin vessels are performed[6]. A single fiber guiding the laser light illuminates the blood and also collects back-scatterered light from the circulating red blood cells. The rate of blood flow is derived by spectral analysis of the back-scattered signal, which presents a flow-dependent
Doppler-shifted frequency.

III.2. Fluorescence-Based Sensors

Fluorescence-based FOBS have been developed for drug dosimetry in
tissues and body fluids. The antitumoral drug doxorubicin is detected in
tissues by fluorescence spectrometry, by using a microcuvette at the fiber-
end inserted in an hypodermic needle (Figure 2b). A single fiber is used
either for illuminating the sample at 488 nm and for collecting the fluorescence signal at 360 nm. The input of the fiber is coupled to a dichroic
filter, which separates the excitation from the fluorescent signal. This
technique allows one to analyze extremely small volumes (200 nl) of interstitial soft tissues, which cannot be sampled in sufficient amounts for
conventional "in vitro" spectroscopy[7].

Another drug dosimetry performed by fiber-based fluorescence spectroscopy involves the hematoporphyrin derivative (HPD). This drug is suitable
both for the localization of tumors and for their destruction by laser-
power radiation. The sensor, which has been developed for in vitro measurements, uses two bare fibers without any additional measuring cells. One
fiber illuminates the sample at 488 nm, and the other one collects the
fluorescence of the drug at 614 nm and the Raman scattering of the water
molecules at 580 nm. Both signals are due to the same exciting radiation
and are relative to the same sample volume. Hence, the ratio of two signals measures the HPD concentration, normalized to the incident radiation
and sample volume, and down-lead insensitive as well[8].

III.3. Absorption-Based Sensors

An absorption-based FOBS has recently been developed for the continuous monitoring of entero-gastric and non-acid esophageal refluxes. Conventional techniques are based mainly on the detection of pH change caused
by the refluxes, but they are not very satisfactory, because of differing
pH conditions not strictly due to the refluxes . Taking into account that
the reflux primarily consists of bile secreted by the liver, the working
principle of the sensor is based on the spectrophotometric properties of
bilirubin, which is the main biliary pigment. Absorption measurements are
performed at two wavelengths, 460 nm and 570 nm, relative to the absorption
peak of bilirubin and a reference respectively, and processed by the ratioing technique.

Two PCS fibers 200 μm core diameter are used, one to illuminate and
the other to collect the sensing and reference signals. A particular measuring cell has been used at the fiber tip, so to perform absorption measurements (Figure 2c). A stainless steel capillary (0.7 mm o.d.), having
two open windows on the lateral surfaces and a gold concave mirror as cap,
is coupled to the fibers, which are enclosed inside a catheter. The windows allow bilirubin penetration and the concave mirror ensures a better
coupling of the light between input and output fibers. The probe is e-
quipped with a proper washing system, able to clean the fiber-end surfaces
and to remove occlusion or pollution of the probe. A prototype of the instrument was made, and "in vivo" measurements have been carried out, continuously during several hours. The results are satisfactory and many

tests are still in progress, to compare fiber-sensor performances with those obtained by conventional techniques.

IV. TRANSDUCER SENSORS

Transducer-based sensors are the most developed FOBS, since many transducers can be found which change their own optical properties according to the physical or chemical parameter being tested. Physical parameters of interest in medicine are mainly temperature and pressure, while chemical parameters are blood gases (pH, pO_2, pCO_2), glucose, and immunological chemicals.

Because of the presence of an external element, the probe tip must be carefully designed to insure miniaturization, biocompatibility and proper packaging. Probe design also involves problems such as transducer stability and sealing. In addition, because of thrombogenic problems, probe surfaces must be smooth, and provide few sites for fibrin buildup.

IV.1. Physical Sensors

Optical fibers are used for thermometry when electrical insulation and immunity from electromagnetic interference are necessary. Typical applications involve tissue-heating control dring hyperthermia therapy for cancer, the mapping of thermal distribution in cancer phototherapy, patient monitoring during magnetic resonance imaging, and cardiac-output monitoring by means of the thermodilution technique[10]. A restricted sensing range of 35°-50°C is sufficient, with a resolution at least of 0.1°C. Thermometers such as thermistors and thermocouples are not suitable for these applications, since they can produce localized efffects or perturb the incident field. In addition, the metallic wires are an electrical path for dangerous shocks to the heart.

Many fiber-based thermometers have been developed, since many transducers have relevant sensitivity to temperature; some examples are given in Figure 3. The optical properties which are modified by temperature are generally the intensity, through absorption or reflectance, the wavelength, and the temporal distribution.

The fiberoptic thermometer based in the reflectance of a liquid-crystal mixture was one of the first devices developed to a prototype stage[11]. It exhibits adequate sensitivity, but since little hysteresis and aging are present, the sensor is an interesting candidate for a disposable probe. The probe has been also housed in a catheter with a fiber-optic oximeter, for combined measurements in the coronary sinus[12].

The absorbance of a thermochromic solution is exploited for thermometric purposes, since a band of the absorption spectrum is strongly modulated by temperature (≈ 660 nm) and another band is temperature insensitive (> 800 nm)[13]. Hence the dual-wavelength ratioing technique is used to obtain a down-lead insensitive temperature measurement. The probe has an original design, using a single optical fiber which is coupled to a microoptic system (Figure 3b). Two facing quarter-pitch gradient refractive index (GRIN) microlenses, inserted in a glass microcapillary, confine the thermochromic solution. The first GRIN is cemented by an index-matching glue to the optical fiber, and the other has a reflecting end surface. At the probe bottom an air bubble expands and compresses, compensating for solution volume change due to temperature variations. Such a probe is a "nearly autofocusing" cavity, and exhibits very high coupling efficiency between the fiber and the transducer ($\approx 63\%$)[14]. A thermometric device utilizing 19 thermochromic probes has been also developed; the sensors are

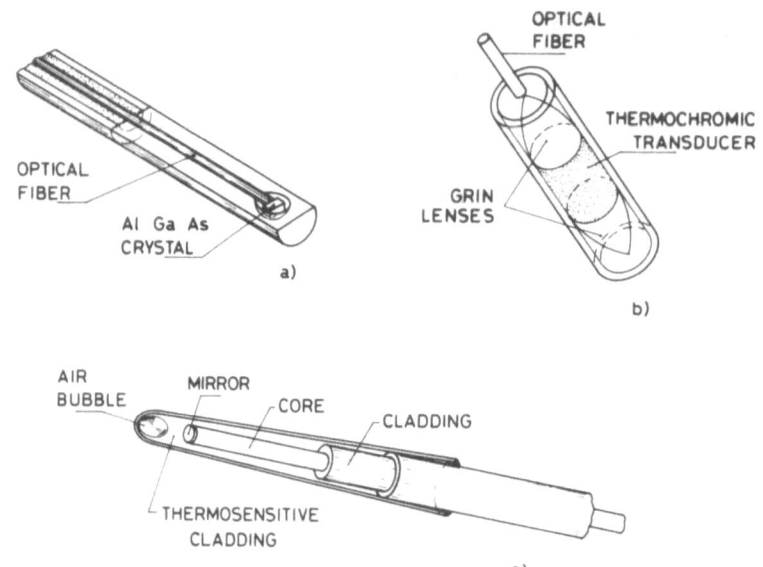

Fig. 3. Transducers for fiberoptic thermometers: a) semi-conductor crystal; b) thermochromic solution; c) thermosensitive clad (all-fiber sensor).

managed by a single electro-optic unit and are addressed by an electro-
mechanical optical scanner[15].

Another thermometer based on light intensity modulation utilizes as
transducer a phosphor mixture, whose fluorescence is quenched by temperature. Two fluorescence lines are monitored, the intensities of which are
ratioed so as to obtain a source-insensitive output. A more recent version
of this sensor is based on the fluorescence decay time of a magnesium fluo-
rogermanate phosphor, and presents the advantage of being intrinsically
down-lead insensitive[16]. An eight-channel system is now commercially a-
vailable, which is provided either with linear-array probes or with single-
sensor probes.

A well-established technique is based on wavelength modulation, and
often the photoluminescence of semiconductors is exploited (Figure 3a).
Down-lead insensitivity is achieved by ratioing the outputs of two spec-
trally matched photodiodes, so as to convert wavelength modulation into an
intensity modulation[17].

A Fabry-Perot cavity at fiber-end (FFP) is another type of temperature
transducer based on wavelength encoding[18]. In this case, the thermometric
technique is based on shift compensation of the FFP-resonator output of by
means of a suitable wavelength change. The same technique has also been
used as a strain gauge for the detection of contractions and bending motions of the upper gastrointestinal tract[19]. The working principle is the
FFP output modulation induced by peristaltic motility. Excursions in the
0.01 to 10 mm range are detected, independently of non-peristaltic motion
such as respiration.

Pressure measurement is important in medicine for describing the functions of the body, especially the cardiovascular and urological ones. Conventional devices are based on electric micro-tip transducers, generally
capacitive or piezoresistive. Their drawbacks are long-term drift and e-
lectrical shock hazard. In addition, these devices are very fragile and

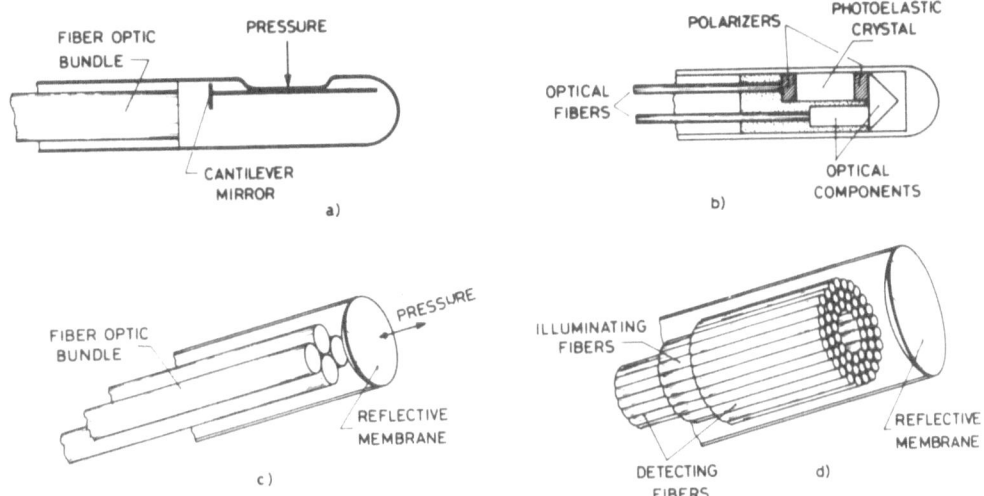

Fig. 4. Examples of transducers for fiberoptic pressure sensors:
a) cantilever mirror; b) photoelastic crystal; c) flat
diaphragm; d) curve diaphragm.

must be carefully packaged, since they can be damaged by body fluids;
hence, they are expensive and not suitable for disposable probes. A fiberoptic pressure sensor is often a good choice, especially if it matches the
dimensions of conventional catheters. The principal measurement requirements are an accuracy of at least 1% in a working range up to 300 mmHg.

Many fiberoptic pressure sensors have been proposed based on mechanical reflective transducers which are displaced by pressure at the end of
the fiber-link, giving rise to an intensity modulation as shown in Figure 4.

A fiberoptic pressure sensor designed for intravascular measurements
uses as transducer a flat and thin metal membrane (10 m thik), which is
mounted at the common end of a bifurcated fiber bundle (Figure 4c). Hysteresis and non-linearity in the range of interest are within 1% and the
allowed overload is greater than 3000 mmHg[20].

Another pressure sensor utilizes a fiber optic bundle inserted in a
catheter with a side hole at the tip. A pressure-sensitive watertight membrane covers the hole and inside the membrane a cantilever is fixed, termi-
nating with a small mirror (Fig. 4a). If no pressure is applied, the
mirror plate is parallel to the cross section of the fiber optic bundle[21].
Any pressure on the membrane presses down the cantilever, thus changing the
position of the mirror and hence the backtransmitted light intensity.

A more recent fiberoptic pressure sensor is based on curvature of a
diaphragm positioned at the end of three concentric circles of fibers (Figure 4d). Down-lead insensitive pressure measurement is obtained by the
ratio between the intensity of the light collected by the outside and inside fibers[22].

The most recent pressure sensor has been developed for the measurement
of intravascular pressure and intracardiac sounds and murmurs. It is based
on the stress-induced birefrangence of a photoelastic material, which si-
nusoidally modulates the intensity of elliptically polarized light[23]. A
particular polyurethane material allows one to obtain a good linear response (a ramp of the sinusoidal response) in the range of physiological interest. The probe is a micro-optic system, containing a GRIN-rod for input

beam collimation, an elliptical polarizer, the active birefringent element, and an analyzer connected to the output fiber (Figure 4b). In the entire physiologic range, the combined non-linearity and hysteresis is better than 3%, but is better than 2% in the smaller pressure range up to 180 mmHg. System improvements are still in progress, mainly to compensate the sensitivity to temperature variations and to fiber bending.

IV.2. Chemical Sensors

Chemical sensors involve a reagent as transducer, and the fiber tip is commonly called an "optrode"; the term is suggested by the combination of "optical" and "electrode". The optrode is a sort of miniaturized bio-chemical laboratory, where chemical reactions are induced and detected by optical fibers. Optrodes offer significant cost advantages over electrodes, but have as drawback limited dynamic range, long-term stability, and increased response time. In particular, they are promising candidate for disposable probes.

Two different types of optrode have been defined: extrinsic and intrinsic. Extrinsic optrodes use a reagent immobilized on a polymeric support and attached at the fiber end; the trip isoften enclosed in a membrane which acts both as selective and protective coating. Intrinsic optrodes use the reagent directly fixed at the end of the fiber, by absorption or by chemical bond, and allow reduced probe dimensions. In addition, when the membrane is unnecessary, the response time decreases. Intrinsic optrodes can be considered all-fiber sensors[24÷27].

In the area of chemical sensing, pH is the most frequently investigated parameter; it is generally measured by an extrinsic optrode based on absorption or fluorescence of different reagents, as shown in Figure 5.

The first fiberoptic pH sensor was designed for tissue and blood measurements in the physiological range of pH 7.0 to 7.4, and presented an accuracy of 0.01 pH units (Figure 5a). It is based on the absorption of

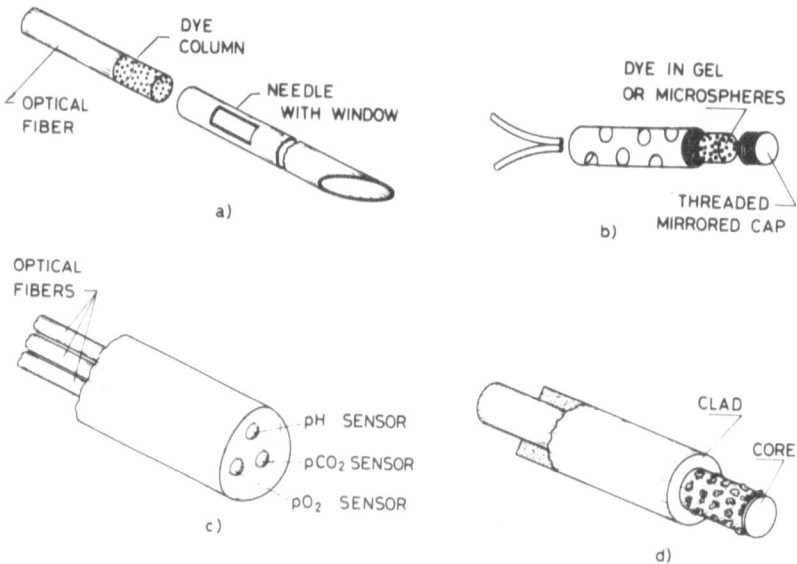

Fig. 5. Examples of transducers for fiberoptic pH sensors:
a) dye trapped in microspheres; b) dye trapped in gel;
c) fluorescent reagents; d) intrinsic optrode (all-
fiber sensor).

240

phenol red dye, packaged in polyacrylamide microspheres containing small polystyrene microspheres for light scattering, and inserted in a cellulosic dialysis tube [28].

Phenol red dye, immobilized on XAD-2 and fixed inside a hydrophilic gel, has been used as the pH transducer of another fiberoptic sensor. The probe consists of stainless steel capillary with holes on the lateral surface, so to allow the flow of the sample under test. The probe cap is mirrored and threaded, so to obtain the best coupling position of the cylindrical gel-optrode with the two PCS fibers (Figure 5b). The working range is 5.9 to 11.2 pH units, and the sensitivity 0.06 pH units [29].

The principal drawbacks of these absorption-based sensors are the low signal-to-noise ratio and the limited dynamic range. Fluorescence-based sensors overcome these drawbacks, even if other problems arise, such as the absorption of both excitation and emitted radiation caused by the polymer host of the fluorophore, and the sometimes difficult identification of an analyte-specific fluorophore.

A very interesting example of chemical FOBS based on fluorimetry has been set up for intravascular blood gas monitoring: pH, pO_2, pCO_2 [30,31]. The probe is made of three individual fluorescence-based sensors, each fabricated on a single fiberoptic tip, and together integrated within a polymeric structure (Figure 5c). Each dye mixture provides a sensing and a reference signal, and a down-lead insensitive output is obtained. The pH sensitive dye is hydroxy-pyrene trisulphonic acid, covalently bonded to a cellulose matrix. The acid and base forms of the dye, excited by different spectra, show emission in the same band, however differently influenced by pH in the range 6.8 to 7.8 pH units. The pCO_2 sensor uses the same pH sensitive dye to measure pH variation, induced by changes in CO_2 concentration, on an isolated bicarbonate buffer. A hydrophobic gas-permeable silicon matrix encapsulates the buffer, the concentration of which is chosen to give sufficient pH variation for obtaining good pCO_2 sensitivity in the physiological range 10 to 100 mmHg. The pO_2 sensor uses a dye having emission quenched by O_2 in the physiological range 20 to 300 mmHg, mixed with a dye unaffected by oxygen; the dyes are hosted in a solid polymer. An opaque cellulose overcoat is applied to each sensor for optical isolation of the sensor chemistry from environmental influences. The electro-optic instrumentation, optimized for the very low backtransmitted fluorescence signals, is divided into three parts: an analyzer, a patient module interface and the probe. The probe is disposable and can be inserted in a radial artery catheter. The sensor has a compact design and can be placed near the patient, with the module interface clamped at beside.

The measurement of fluorescence quenching induced by oxygen can also be used to detect other chemicals which give rise to reactions with oxygen production or consumption. For esample, a model of a glucose sensor has been developed, which uses both a fluorophore and the glucose oxidase enzyme trapped within a hydrophilic polymeric matrix. Glucose spreading into the matrix lowers oxygen concentration within the polymer and causes increased fluorescence intensity, thus producing a glucose-modified oxygen signal [32]. A comparison between the glucose-modified oxygen signal and a reference signal from a polymer containing only the fluorophore provides a quantitative measure of glucose concentration. Further sensor development is desired since an optical fiber sensor for continuous glucose monitoring would be the basic element of a closed-loop insulin delivery system.

V. ALL-FIBER SENSORS

These FOBS make use of modified tips, which are sensitive because of a doped core or cladding, or because of chemical modification of the fiber-end. All-fiber sensors have been developed especially for temperature, for pH measurements and for immunological assay.

Temperature has been measured by using a liquid cladding with a thermosensitive refractive index, so that the numerical aperture of the fiber tip is temperature modulated (Figure 3c). A single-branch fiberlink is utilized, and the fiber end is mirrored for providing light backtransmission[33]. Two different enclosures of liquid cladding have been tested: a conically-shaped glass microcapillary which also entraps an air bubble for liquid expansion compensation, and a platinum or gold microcapillary, which allows very fast response time ($\approx$ 300 msec)[34].

Very recently, several pH sensors with an intrinsic optrode have been proposed, using a chemically-modified fiber tp with a fluorescent species attached (Figure 5d). These sensors present a faster response time compared to the extrinsic pH optrodes.

A new and original all-fiber pH sensor uses both a dye and a fluorophore, but limits the disadvantages of similar absorption- and fluorescence-based sensors. The pH-insensitive eosin fluorophore is coimmobilized with the pH-sensitive phenol red dye on the silanized fiber tip[35]. The phenol red absorbance spectrum, modulated by pH, is in the same region as the eosin emission spectrum, so that a non-radiative energy transfer modulated by pH occurs from eosin to phenol red. In practice, the fluorescence signal is modulated by pH though the absorber dye. The relative response time is $\approx$ 4 secs.

Other chemical parameters of interest in biomedicine and which can be investigated by all-fiber sensors are the immunological parameters[36]. Immunoassay is performed by intrinsic optrodes, using a fiber-clad modification, so that the core-cladding interface acts as a sensing element. This technique is very promising for measuring very low concentrations of chemicals in complex solutions. Conventional techniques are based on multi-steps analysis, which is more expensive and time-consuming, and which requires specialized personnel. The fiberoptic technique is based on internal reflection or fluorescence spectroscopy of a thin layer of suitable fiber-cladding, which is able to trap specific chemicals in a complex solution. The fiber-clad interface is a layer of proteins with specific binding properties, called antibodies, which are able to capture the correspondent antigens, so that the evanescent-wave component of the light guided by the fiber is modified by the antibody-antigen binders.

The main problems of this futuristic technique are the choice of suitable antibodies, their immobilization on the silica substrate, their stability, and their reversibility. In spite of the many problems still open, fiber-optic immunoassay does not have serious competitors, and it is of interest not only for immunological chemicals, but also for measuring the concentrations of molecular species inside multicomponent media.

VI. CONCLUSIONS

Optical fibers have proved useful in several medical applications, and interest is still growing. Many fiber-sensors have been developed for monitoring biological functions, even if most of them are only experimental prototypes and need further improvement. Very few of these are available on the market or are under industrial development and engineering, especially if compared with fiber-sensors for industrial applications.

Market studies on FOBS confirm the interest in fiber-biosensors, but show that the main barrier to further market growth is the cost, which is still too high in comparison with that of conventional sensors[37,38]. Cost reduction will come not only with the growth in the optoelectronic market foreseen for sensor components, but also with improvements in probe technology. In fact, while electro-optic components including those for sensors are now mass-produced, probe manufacturing is still a problem because it is rarely automatized and needs skilled personnel.

In spite of these problems, many ongoing efforts are dedicated to improving existing sensors and to developing new devices intrinsically more reliable, specifically for biomedical applications. Attention is being given to improvements in chemical technology, for the monitoring of many other analyses of interest in the biomedical field, such as urea, lactate drugs, metallic ions and immunological parameters.

Some companies are researching and developing fiber-based blood-test kits, with simplified electro-optic units using LEDs and diode lasers, and custom-made components. In fact the market for multitest fiberoptic packages is estimated to be high, since on-line continuous testing allows individual pharmacological therapy, based on the patient's own metabolism.

A further goal of optical fiber biosensing is the integration in a single catheter of many different sensors, managed by a single electro-optic unit, serving as a portable laboratory. This is ideally possible, since fibers are crosstalk-free, and many optical components for fast switching and beam dividing are now commercially available.

REFERENCES

1. J. I. Peterson and G. G. Vurek, "Fiber-optic Sensors for Biomedical Applications", Science 224, 123-127 (1984)
2. A. M. Scheggi, "Optical Fibres Sensing in Medicine", Int. J. Opt. Sens. 1(1), 5-25 (1986)
3. F. P. Milanovich, T. B. Hirschfeld, F. T. Wang, S. M. Klainer and D. Walt, "Clinical Measurements Using Fiber Optics and Optrodes", Proc. SPIE 494, 18-24 (1984)
4. R. J. Volz and D. A. Christensen, "A Neonatal Fiberoptic Probe for Oximetry and Dye Curves", IEEE Trans. Biom. Eng. BME-26(7), 416-421 (1979)
5. A. W. Domanski, S. Kostrzewa and T. R. Wolinski, "Fiber-optic Absorptive Oximeter", Proc. SPIE 1085, 450-456 (1988)
6. H. Nishihara, J. Koyama, N. Hoki, F. Kajiya, M. Hironaga and M. Kano, "Optical-fiber Laser Doppler Velocimeter for High-resolution Measurement of Pulsatile Blood Flows", Appl. Opt. 21(10), 1785-1790 (1982)
7. M. J. Sepaniak, B. J. Tromberg and J. F. Eastham, "Optical Fiber Fluoroprobes in Clinical Analysis", Clin. Chem. 29(9), 1678-1682 (1983)
8. P. R. King, J. B.Dawson, I. Driver, D. J. Ellis and J. W. Feather, "Design of a Fibre Optic Probe for the In-Vivo Determination of Photosensitizing Drugs", Proc. SPIE 798, 214-217 (1987)
9. R. Falciai, F. Baldini, G. Conforti, F. Cosi and A. M. Scheggi, "A fiber Optic System for the Detextion of Entero-Gastric Reflux", Proc. SPIE 990, 18-21 (1988)
10. D. A. Christensen, "Fiberoptic Temperature Sensing for Biomedical Applications", Proc. SPIE 906, 108-113 (1988)
11. T. C. Rozzel, C. C. Johnson, C. H. Durney, J. L. Lords and R. G. Olsen, "A Nonperturbing Temperature Sensor for Measurements in Electro-magnetic Fields", J. Microw Power, 9(3), 241-249 (1974)
12. D. De Rossi, A. Benassi, A. L'Abbate and P. Dario, "A new Fibre-Optic

Liquid Crystal Catheter for Oxygen Saturation and Blood Flow Measurements in the Coronary Sinus", J. Biom. Eng. 2(10), 257-264 (1980)

13. A. M. Scheggi, M. Bacci, M. Brenci, G. Conforti, R. Falciai and A. G. Mignani, "Thermometry by Optical Fibers and a Thermochromic Transducer", Opt. Eng. 26(6), 534-537 (1987)

14. G. Conforti, M. Brenci, A. Mencaglia and A. G. Mignani, "Fiber-Optic Thermometric Probe Utilizing GRIN Lenses", Appl. Opt. 28(3), 577-580 (1989)

15. G. Conforti, M. Brenci, A. Mencaglia, A. G. Mignani and A. M. Scheggi, "Multiplexed Fiber Optic Thermometer", Proc. SPIE 985, 125-129 (1989)

16. K. A. Wickersheim, "A new Fiberoptic Thermometry System for Use in Medical Hyperthermia", Proc. SPIE 713, 150-157 (1987)

17. C. Ovrén, M. Sdolfsson and B. Hök, "Fiber-Optic Systems for Temperature and Vibration Measurements in Industrial Applications", Proc. Int. Conf. Optical Techniques in Process Control (The Hague, June 14-16, 1983), 6781 (1983)

18. R. Kist, S. Drope and H. Wolfelschneider, "Fiber Fabry-Perot (FFP) Thermometer for Medical Applications", Proc. 2nd International Conference on Optical Fiber Sensors, OFS-2, Stuttgart September 5-7, 1984, 165170 (1984)

19. H. Wolfelschneider, R. Kist, J. Schneider and H. Modler, "A Fiber Fabry Perot Mobility Sensor for the Measurement of Peristaltic Motions in the Upper Gastrointestinal Tract", Proc. 5th International Conference on Optical Fiber Sensors, OFS-5, New Orleans January 27-29, 1988), 357-360 (1988)

20. T. E. Hansen, "A Fiberoptic Micro-Tip Pressure Transducer for Medical Applications", Sensors and Actuators 4, 545-554 (1983)

21. H. Matsumoto, M. Saegusa, K. Saito and K. Mizoi, "The Development of a Fibre Optic Catheter Tip Pressure Transducer", J. Med. Eng. Tech. 2(5), 239-242, (1978)

22. C. M. Lawson and V. J. Tekippe, "Fiber-Optic Diaphragm-Curvature Pressure Transducer", Opt. Lett. 8(5), 286-288 (1983)

23. P. Dario, D. Fermi and F. Vivaldi, "Fiber-Optic Catheter-Tip Sensor Based on the Photoelastic Effect", Sensors and Actuators, 12, 35-47 (1987)

24. G. G. Vurek, "In Vivo Optical Chemical Sensors", Proc. SPIE 494, 2-6 (1984)

25. A. J. Guthrie, R. Narayanaswamy and D. A. Russel, "Optical Fibres in Chemical Sensing: a Review", Trans. Inst. M. C. 9(2), 71-80 (1987)

26. A. M. Scheggi and G. Conforti, "Review: Optical Fibre Sensors and Applications with Particular Reference to Chemical Sensing", Proc. SPIE 949, 146151 (1988)

27. R. Narayanaswamy and F. Sevilla, "Optical Fibre Sensors for Chemical Species", J. Phys. E: Scien. Instrum. 21, 10-17 (1988)

28. J. I. Peterson, S. R. Goldstein and R. V. Fitzgerald, "Fiber Optic pH Probe for Physiological Use", Anal. Chem. 52(6), 864-869 (1980)

29. M. Bacci, F. Baldini, F. Cosi, G. Conforti and A. M. Scheggi, "Probe Performance Optimization for pH Continuous Monitoring", Proc. 6th International Conference on Optical Fiber Sensors, OFS-6, Paris September 1820 1989), 425-430 (1989)

30. J. L. Gehrich, D. W. Lubbers, N. Opitz, D. R. Hansmann, W. W. Miller, J. K. Tusa and M. Yafuso, "Optical Fluorescence and its Application to an Intravascular Blood Gas Monitoring System", IEEE Trans. Biom. Eng. BME-33(2), 117-132 (1986)

31. J. Tusa, T. Hacker, D. R. Hansmann, T. M. Kaput and T. P. Maxwell, "Fiber Optic Microsensor for Continuous In-Vivo Measurement of Blood Gases" proc. SPIE 713, 137-143 (1986)

32. J. W. Parker and M. E. Cox, "Glucose/Oxygen Sensor", Proc. SPIE 713, 113-120 (1986)

33. M. Brenci, G. Conforti, R. Falciai, A. G. Mignani and A. M. Scheggi, "All-Fibre Temperature Sensor", Int. J. Opt. Sens. 1(2), 163-169 (1986)

34. M. Brenci, G. Conforti, R. Falciai, A. G. Mignani and G. Gironi, "Assembling Technique of Microcapsules for Optical Fibre Sensors", Int. J. Opt. Sens. 2(5), 357-362 (1987)

35. D. M. Jordan, D. R. Walt and F. P. Milanovich, "Physiological pH Fiber Optic Chemical Sensor Based on Energy Transfer", Anal. Chem. 59(3), 437439 (1987)

36. J. D. Andrade, R. A. Vanwagenen, D. E. Gregonis, K. Newby and J. N. Lin, "Remote Fiber-Optic Biosensors Based on Evanescent-excited Fluoro-immunoassay: Concepts and Progress", IEEE Trans. Electron. Dev. ED-32, 1175-1179 (1985)

37. J. D. Montgomery, "Fiber Optic Sensor Long-Range Market Forecast", Proc. SPIE 586, 2-13 (1985)

38. P. McGeehin, "Current Impact of Fibre Optic Sensors", Proc. SPIE 798, 16-23 (1987)

LASER HOLOGRAPHY AS A TECHNIQUE

IN EXPERIMENTAL MEDICINE

H. Podbielska*

Department of Electronics
The Weizmann Institute of Science
Rehovot, Israel

I. PRINCIPLES OF HOLOGRAPHY

Holography is a technique for recording and reconstructing light waves. Although a hologram is recorded on a flat surface, it produces a three-dimensional image. A conventional photograph records the real, two-dimensional image formed by a lens or a more complicated optical system. Optical detectors, like photographic film, respond only to irradiance, so only the distribution of real amplitude can be recorded and information about phase is lost. A hologram, however, records the intensity distribution that results from interference of the light scattered by an object and an additional wave coming directly from the coherent light source. When a holographic plate is developed and illuminated properly, it produces a three-dimensional image of the recorded object, thus information about both the amplitude and the phase of the scattered light is reconstructed.

The light beam from a laser source is expanded by means of a beam expander and then is divided by a beam-splitter into two beams. One of them, called the reference beam, directly illuminates the holographic film. The other beam is scattered from the surface of the object. A portion of the scattered light interferes with the reference beam at the film plane. Then the hologram is developed and placed in its previous position. When it is illuminated by the reference wave, called in this step the reconstruction wave, it produces an image which is located in space exactly where the object was during the recording step.

II. HOLOGRAPHIC INTERFEROMETRY

Holographic interferometry allows a non-contact, high resolution full-field deformation measurement of the object under examination. The technique is used extensively in industrial applications[1÷5] and can be used for biomedical research as well[6÷10]. It is not restricted to investigations on models, as when using photoelastic techniques, or to pointwise analysis, as when using extensometers or strain gauges.

(*) On leave from Institute of Physics, Technical University of Wroclaw and
    Medical Academy of Wroclaw, Poland; recipient of the Edmond I. and
    Lilian S. Kaufmann Fellowship at the Weizmann Institute of Science.

Laser Systems for Photobiology and Photomedicine
Edited by A. N. Chester *et al.*, Plenum Press, New York, 1991

Of the variety of holographic measuring methods, the most popular for biomedical applications is <u>double-exposure holographic interferometry</u>. In this method the object is displaced or deformed between exposures, so that two positions of the object in two consecutive states are recorded in the same recording medium. In this manner, by reconstructing the double-exposure hologram, one obtains two holographic images. Since both are coherent and exist in approximately the same place in space (if the deformation is small), they interfere with each other, producing an interference pattern. All information concerning the changes in object between exposures can be determined from this fringe pattern.

The most widely used application of double-exposure holographic interferometry is deformation analysis in experimental orthopaedics[11]. In particular, holographic deformation analysis is of special interest in the fields of osteosynthesis and endoprosthesis research[12]. Different types of surgical fixing devices have been studied: osteosynthesis plates of the Mittelmeier type and femur hip endoprostheses[13], mounted on tibial shaft AO plates under axial load[14] or in bending and torsion[15]. Double-exposure holographic examination of the tibia has been performed for bones loaded axially[16] by an external fixator. The same fixator was studied in torsion and bending[17]. The quality test of functioning of the joints connecting transfixing pins and side bars of a Hoffman-Vidal external fixator is reported by Jacquot et al[18]. The potential use of holographic measurements can augment results obtained by contact methods[19,20].

There are several review articles describing dental applications of holography[21,22]. Double-exposure holographic interferometry has been used in experimental dentistry to examine deformation of teeth, the jaw, prosthodontic appliances and the skull[23-25].

A further application of double-exposure holographic interferometry relates to the study of elastomechanical properties of bones. In this way the function of the tibio-fibular complex was examined[26], as well as the deformations of human pelvis[27] and human vertebrae[28,29]. The human femur under bending loading was tested by this method[30]. The rigidity of the human tibia was calculated from double-exposure holographic interferograms[31]. This method has also been used to measure the demineralization of bones[32].

Using a Q-switched ruby laser, vibration and non-periodic fast processes can be studied. An example of this is an in vitro experiment on the tympanic membrane of a guinea pig[33]. The first hologram of the membrane was made and, after some acoustic event, the second hologram was recorded on the same plate. The same technique has been applied with an intent to use it in clinical diagnostics for audiology[34]. Human chest motion has also been investigated in vivo to assist lung diagnosis[35]. The motion patterns originating from heart action were visualised by a double-pulse technique[36]. Vibration analysis of the human vocal organ was carried out in vivo on the frontal part of the human neck[37].

Another interesting application of double-exposure holographic interferometry relates to the examination of embryonic behaviour in chicks[38]. The prenatal motility cycles were studied in order to determinate different types of movement, their onset, duration and frequency of repetition.

Another method of holographic interferometry is <u>real-time holographic interferometry</u>. In this method, the hologram is replaced (after processing) in exactly the same position in which it was recorded. The problem of exact positioning is solved by using in situ development or a thermoplastic camera. When looking at the object through a recorded and processed hologram, changes in the object can be seen in real-time. Biomedical applica-

248

tions of this technique are complicated due to uncontrollable changes in biological specimens. Therefore, these experiments are done mostly on models or rigid objects, such as macerated bone. Real-time holographic interferometry has also been applied to study the human tibia after fracture and fixation by a compression plate[39], to study the function of the ankle joint and the leg-foot complex[40], and to optimize hip joint prostheses[41]. The deformation of the human calvaria[42], as well as the thermal expansion of human teeth and dental materials, were examined using this method[43]. Oscillation of tympanic membrane models were analyzed by real-time holography combined with synchronized stroboscopic illumination[44].

Time-averaged holographic interferometry is another method for holographic interferometry which can be exploited in experimental medicine. This technique is especially useful if the object moves periodically. Therefore, it has been used for the analysis of the vibration pattern of tympanic membranes. In this way the role of the tympanic membrane in sound transmission through middle ear was studied[45,46]. By using the time-averaged method the human ossicular chain[47] and the vibration of the round window in cats were tested[48].

III. METHODS RELATED TO HOLOGRAPHIC INTERFEROMETRY

Holographic contouring is an extension of holographic interferometry, closely related to it in concept and practice. Contour generation is the formation of an image of an object on which contours of constant elevation with respect to some plane are superimposed. In this way, full-field display of three-dimensional surface shape can be obtained.

There are three basic methods for holographic contour generation. One is dual source contouring, when the object illumination source is moved slightly between the two exposures. Two-refractive index contouring is carried out by immersion of the object in transparent liquid and changing the refractive index between exposures. Two-wavelength contouring is the recording of a double-exposure hologram using different wavelength for each exposure. The contour can be also generated by a reflection hologram or by a photogrammetric method.

Holographic contouring has been suggested to determine the curvature of the sclera or frontal corneal surface by generating depth contour lines of the eye[49,50]. The combination of contour mapping with the real-time holography was used to measure the wear of knee prostheses[51], acetabular cups and in vivo a worn mitral heart valve[52]. The rates of wear of dental materials can also be studied by this technique[53,54].

Another method related to holographic interferometry is electronic speckle pattern interferometry[55]. It uses on-line holography where a CCD detector array is used in place of a recording plate. This enables a quasi real-time display of speckle interferograms. In medicine it has found applications in vibration analysis of the human tympanic membrane and basilar membrane[56,57], as well as the ossicular chain[58].

IV. HOLOGRAPHIC ENDOSCOPY

By combining holographic interferometry with endoscopic imaging it is possible to obtain holographic imaging inside natural cavities of the body, thus making possible intracavitary measurements of size, shape or deformation of the objects under study. There are two approaches to record endoscopic holograms. Holograms can be recorded at the internal end of an endoscopic device (internal recording), or at the external end, where the endoscopic image is recorded outside the instrument (external recording).

The recording of holograms at the internal end of the endoscope was first suggested by Hadbawnik [59]. Internal recording gives the possibility of three-dimensional imaging with large depth of field, providing some intracavitary details from different perspectives. Holographic film can be inserted through the instrumentation channel of a commercially available endoscopic instrument and then be unfolded at the end of the endoscope. The illumination beam, guided via an optical fiber bundle, illuminates the object and simultaneously serves as the reference beam. This type of recording (reflection or Denisyuk's type hologram), however, requires the film to be as close as possible to the tissue. After recording, the film must be folded and retrieved for processing. The resulting hologram can be viewed in white light. A similar technique of internal holographic recording was proposed by Raviv [60,61]. He designed a holo-endoscope with object and reference beams guided separately by fiber optics, so as to allow for adjustment of the object/reference beam intensity ratio. Using a Q-switched ruby laser, the three-dimensional image of an internal part of the larynx of anaestetized dog was obtained, revealing details invisible in normal laryngoscopy.

Internal holographic recording offers the advantage of three-dimensional imagery, while reducing the problems associated with coherence length of the laser source. However, specially designed endoscopic instruments are required. For some applications, especially for metrology, these advantages are not of primary importance. For these applications external holographic recording is more practical.

With external recording, the holographic film is located outside the instrument. This technique allows for the use of a conventional endoscope. The reference wave does not propagate through the endoscope, but can be guided through a separate optical fiber to give greater flexibility in adjusting the holographic set-up. The external recording, however, leads to a loss of parallax and introduce speckle noise due to the small entrance pupil. For some medical applications the endoscope has to have a very small diameter (otoscopy, salpingoscopy), resulting in the small apertures for the imaging system. This can decrease the signal-to-noise ratio due to speckles. To overcome the problem of poor signal-to-noise ratio, gradient index optics can be used both for imaging and illumination [62, 63].

While there are some publications on technical and industrial applications of holographic endoscopy [64÷67], medical examinations using this technology are still rather rare. Endoscopic metrology by holographic interferometry was proposed to study the elastomechanical features of the urinary bladder in order to find eventual cancerous induration in the wall tissue [67]. In vitro study of the mechanical properties of the inner part of human skull bones were performed using a standard laryngoscope and real-time holographic interferometry [42]. Holographic endoscopic measurements of the human tympanic membrane for otoscopic studies on the middle ear disease have also been performed [68].

V. OTHER HOLOGRAPHIC TECHNIQUES

The possibility of three-dimensional imaging with a large depth of field is one of basic advantages of holography. This possibility can be used for some teaching purposes [69]. By placing holographic film inside the body cavities, three-dimensional imaging of the internal part of a living body can be obtained (see the section on Holographic endoscopy). Another field of medical research where the use of holography is potentially important is in ophthalmology. It is possible by means of holography to obtain more precise information than by using ordinary ophthalmic instruments[70]. Three-dimensional images of aerosols, sprays and other distributions of small particles[71] [72] can be formed on the basis of Gabor holograms[73]. Three-dimensional contour maps of the heart's electric field can be realized by white light holograms [74].

Holographic multiplexing can be also applied for three-dimensional imaging in medicine. This is especially useful in radiology, if one is interested in a three-dimensional display of a series of tomograms[75, 76]. Computer aided hologram synthesis enables good reconstruction with correction of distortion, as was demonstrated by Tsujuchi[77]. Recently a technique has been proposed for image plane integral holography, which can produce autostereoscopic three-dimensional images from conventional cine-angiography[78,79].

Holographic microscopy [80, 81] uses with advantage some properties of holography, particularly the large depth of field and the two-step principle of recording and reconstructing. Microscopic investigation can be carried out without preparing the section[82], and without focusing to a certain layer[83]. It can be also applied as the method of image transformation during reconstruction a posteriori[84]. Holographic microscopy is well suited to the study of cellular motion, such as cytoplasmic streaming or cell division[85]. Microscopic investigation of living marine plankton organisms, in stroboscopic illumination by a pulsed Argon laser, synchronized to the recording sequence of a camera, showed free movement of the objects in space[86]. Cineholomicroscopy was also used to study decompression sickness[87].

Using holographic spatial filtering, recognition of patterns for biomedical purposes can be accomplished. This is useful in cell identification[88,89], and for inspection of human vision[90,91]. Holographic spatial filters also give the possibility of a posteriori image deblurring and image improvement, and may be applied in electron microscopy[92].

VI. SUMMARY

This review has presented some possible applications of different methods of holography in biomedicine. In general, biomedical holographic investigations can be divided into two groups: investigations in vitro and in vivo. To study the behaviour of different object in vitro, cw lasers can be used. In this way some problems in experimental biomechanics, ortopaedics or dentistry can be solved. However, applications of holographic methods in clinical investigations in vivo require the use of Q-switched ruby lasers or pulsed dye lasers. The principal contributions in this field to date deal with examinations in vitro; however, the general goal is for in vivo diagnosis and clinical study.

Results obtained from the holographic examination of biological specimens can support other biomedical studies or clinical investigations. The ability of holographic methods to perform non-destructive evaluation can be exploited more intensively, especially when combined with endoscopic imaging. This can lead in the future to more extended applications of holographic metrology in biomedical research.

REFERENCES

1. R. K. Erf, (ed), "Holographic non-destructive testing", Academic Press (1974)

2. C. M. Vest, "Holographic interferometry", John Wiley and Sons (1979)
3. N. Abramson, "The making and evaluation of holograms", Academic Press (1981)
4. E. Marom, A. A. Friesem and E. Wiener (eds), "Applications of Holography and optical Data Processing", Pergamon Press (1977)
5. J. Ostrovsky, M. Butusov and G. Ostrovskaya, "Interferometry by Holography", Springer Series in Optical Sciences, Springer-Verlag (1980)
6. P. Greguss (ed), "Holography in Medicine", IPC Science and Technology Press (1975)
7. M. Hoke and G. V. Bally, Proc. Symp., 1976 Spec. Res. Area and Int. Conf. on Electrocochl. and Holography in Medicine, Münster (1976)
8. G. V. Bally (ed), "Holography in Medicine and Biology", Springer Series in Optical Sciences, Springer Verlag (1979)
9. G. V. Bally and P. Greguss (eds), "Optics in Biomedical Sciences", Springer Series in Optical Sciences, Springer Verlag (1982)
10. G. V. Bally, "Holography in Biomedicine", SPIE, vol.673, 327 (1987)
11. K. Piwernetz and G. V. Bally, "Holography in Orthopaedics", 7, see in (8)
12. U. Hanser, "Holographische Bestimmung von Verformungen in der experimentellen Biomechanik", Biomedizinische Technik, Ergänzungs Band 23, 186 (1978)
13. U. Hanser, "Quantitative Evaluation of Holographic Deformation Investigation in Experimental Orthopaedics", 27, see in (8)
14. D. Vukicevic, V. Nikolic, S. Vukicevic, J. Hancevic and Z. Sucur, "Holographic investigation of Mechanical Characteristics of the Complex Leg-Foot in Conditions of Lesion and Reconstruction", 34, see in (8)
15. A. Kojima, R. Ogawa, N. Izuchi, T. Matsumoto, K. Iwata and R. Nagata, "Holographic investigation of mechanical properties of tibia fixed with an internal fixation plate", Selected Proceedings of the Fifth Meeting of the European Society of Biomechanics, 243, Martinus Nijhoff Publisher (1987)
16. H. Podbielska and H. Kasprzak, "Biomechanical investigation of external fixing devices by holographic interferometry", 363, Intern. Series on Biomechanics, Biomechanics XI-A, Free University Press, Amsterdam (1988)
17. H. Podbielska, H. Kasprzak and G. V. Bally, "Holographic investigation of different types of surgical fixing devices", Proc. SPIE vol. 952, 843 (1988)
18. P. Jacquot, P. Rastogi and L. Pflug, "Mechanical testing of the external fixator by holographic interferometry", Orthopaedics 7/3, 513, (1984)
19. M. Manley, B. Ovryn and L. Stern, "Evaluation of double-exposure holographic interferometry for biomechanical measurements in vitro", J. of Ortop. Research 5, 144 (1987)
20. B. Ovryn, M. Manley and L. Stern, "Holographic interferometry: a critique of technique and its potential for biomedical measurements", Annals of Biomedical Engineering 15, 67 (1987)
21. H. Bjelkhagen, "Holography in dentistry", 157, see in (8)
22. I. Dirtoft, "Dental Holography", Proc. SPIE vol. 370, 108 (1983)
23. R. Pryputniewicz, C. Burstone and D. Goldin, "Computer aided holographic analysis of displacement of human teeth", J. Dent. Research 60(A), 515 (1981)
24. I. Dirtoft, "Holographic measurement of deformation in complete upper dentures - Clinical Application", 100, see in (9)
25. P. Pavlin, D. Vukicevic and Z. Rajic, "Strain distribution in the facial skeleton arising from orthodontic appliance activity", 177, see in (8)
26. J. Wagner, J. Ebbeni and M. Clemens, "Application de l'interférometrie holographique à l'étude du complexe tibio-péronière charge", Acta Orthop. Belgica 41, 24 (1975)

27. D. Vukicevic et al., "Holographic investigation of the human pelvis", 138, see in (9)
28. K. Piwernetz and R. Röhler, "Elastomechanical properties of trabecular bone from the human vertebral body", 15, see in (8)
29. T. Matsumoto, A. Kojima, R. Ogawa, K. Iwata and R. Nagata, "Deformation Measurement of lumbar vertebra by holographic interferometry", Proc. SPIE vol. 673, 340 (1987)
30. H. Kasprzak, H. Podbielska and N. Sultanova, "Mechanical features of the human thigh bone investigated by means of holographic interferometry", Acta Politechnica Scandinavia 150, 198 (1985)
31. H. Kasprzak, H. Podbielska and G. V. Bally, "Human tibia rigidity examined in bending and torsion loading by using double-exposure holographic interferometry", SPIE vol. 1026, 196 (1988)
32. J. Ebbeni, A. Huybrecht, S. Orloff, "Holographic determination of demineralization of bones", Proc. SPIE vol. 211, 84 (1979)
33. A. Dancer et al., "Holographic interferometry applied to the investigation tympanic membrane displacements in guinea pig ears subjected to acoustic impulses", J. Acous. Soc. of America 58, 223 (1975)
34. G. V. Bally, "Otological investigations in living man using holographic interferometry", 198, see in (8)
35. S. Zivi and G. Humberstone, "Chest motion visualized by holographic interferometry", Med. Res. Eng. 9, 5 (1970)
36. B. Hök, K. Nilson and H. Bjelkhagen, "Imaging of chest motion due to heart action by means of holographic interferometry", Med. and Biol. Eng. and Comp. 16, 363 (1978)
37. R. Pawluczyk et al., "Holographic vibration analysis of the frontal part of the human neck during singing", 131, see in (9)
38. P. Rastogi, L. Pflug and R. Delez, "Noninvasive observation of embryonic behavior in chicks using holographic interference", Appl. Opt., vol.28 no.7, 1378 (1989)
39. K. Harding, "Preliminary study of fracture fixation using holographic interferometry", 307, see in (7)
40. U. Hanser, "Anwendung der holographischen interferometrie in der experimentellen interferometrie", 343, see in (7)
41. G. Häusler et al., "Holographische Deformationsmessungen zur Optimierung von Hüftgelenke Implantaten", 349, see in (7)
42. H. Podbielska, G. V. Bally and H. Kasprzak, "Mechanical reaction of human skull bones to external load examined by holographic interferometry", SPIE vol. 673, 321 (1987)
43. J. Kinder et al.,, "Holographische Untersuchungen des thermischen Verhaltens von Schmelz, Dentin und ausgewälten Dentalstoffen", 301, see in (7)
44. C. Sieger et al., "Measurement of vibration wave forms using temporally modulated holography", 247, see in (9)
45. S. Khanna et al., "Tympanic membrane vibration analysis in cats studied by time averaged holography", J. Acoust. Soc. Am. 51, 1904 (1972)
46. J. Tonndorf et al., "Tympanic membrane vibration in human cadavers ears studied by time-averaged holography", J. Acoust. Soc. Am. 52, 1221 (1972)
47. T. Gundersen et al., "Holographic vibration analysis of the ossicular chain", Acta Otholaryng. 82, 16 (1976)
48. S. Khanna et al., "The vibratory patterns of the round window in cats", J. Acoust. Soc. Am. 50, 1457 (1971)
49. K. Vaughan et al., "Holography of the eye: a critical review", 77, in M. Wolbarsht (ed), "Laser applications in medicine and biology", Plenum Press (1974)
50. J. Calkins, "Fundus camera holography", 85, in (6)
51. J. Atkinson et al., "Measurement of the area of real contact between, and wear of, articulating surfaces using holographic interferometry", 289, see in (4)

52. M. Lalor et al., "Holographic studies of wear in implant materials and devices", 20, see in (8)

53. M. Koukash et al., "The measurement of wear in dental restorations using digital image processing techniques", Proc. IEEE No 265, 63, (1986)

54. J. Atkinson, D. Groves, M. Lalor, D. Cunningham and J. Wiliams, "The measurement of wear in dental restorations using laser dual source contouring", Wear no. 76.91 (1982)

55. O. Lokberg, "The present and future importance of ESPI", SPIE vol. 746, 86 (1987)

56. O. Lokberg et al., "Use of ESPI to measure the vibration of the human eardrum in vivo and other biological movements", 212, in (8)

57. O. Lokberg et al., "Bio-medical applications of ESPI", 154 in (9)

58. O. Lokberg, "Electronic speckle pattern interferometry", in O. Soares, "Optical metrology", Martinus Nijhof Publisher, 542 (1987)

59. D. Hadbawnik, "Holographic endoscopy", Optik 45, in German, 21 (1976)

60. G. Raviv et al., "In vivo holography of vocal cords", Journ. of Surgical Oncology, 20, 213 (1982)

61. G. Raviv, M. Marphic and M. Epstein, "Fiber optics delivery for endoscopic holography", Optics Communication, 55, 261 (1985)

62. G. V. Bally et al., "Gradient index optical system in holographic endoscopy", Appl. Opt., 23, 1725 (1984)

63. G. V. Bally et al., "Holographic endoscopy with gradient index optical system and optical fibers", Appl. Opt., 25, 3425 (1984)

64. T. Dudderar and J. Gilbert, "Fiber optic pulsed laser holography", Appl. Phys. Lett., 43,730 (1983)

65. J. Gilbert, J. Herrick, "Holographic displacement analysis with multimode fiber optics", Exp. Mech., 20, 315 (1981)

66. H. Bjelkhagen, et al., "Holographic interferometry through imaging fibers using cw and pulsed lasers", Proc. SPIE, vol. 746, 201, (1987)

67. U. Grünewald et al., "Interferometric investigations of the rabbit urinary bladder", 147, in G. V. Bally, (ed), "Holography in Biology and Medicine", Springer Verlag (1979)

68. G. V. Bally, "Otoscopic investigations by holographic interferometry", 110, in G. V. Bally and P. Greguss (eds), "Optics in Biomedical Sciences" (1982)

69. P. Greguss, "Thoughts on the future of holography in biology and medicine", Optics and Laser Techn., 253 (1975)

70. H. Ohzu and T. Kawara, "Application of Holography in Ophtalmology", 133, see in (8)

71. R. Bexon et al., "In line holography and the assessment of aerosols", Optics and Laser Techn., 8, 161 (1976)

72. E. Bals, "The principles and new developments in ultra low volume spraying", Proc. of 5th Brit. Insect. Fungic. Conf., 189 (1969)

73. D. Gabor, "New Microscopic Principle", Nature 161, 777 (1948)

74. Z. Antaloczy, I. Bukosza, Z. Fuzessy, F. Guimesi, "Three dimensional contour map holodisplay of the heart's electric field", Appl. Opt. 24, 11:1564 (1985)

75. K. Sugimura et al., "Clinical applications of multiplex holography", SPIE vol. 370, 20 (1983)

76. K. Johnson et al., "Multiple holographic display of CT data", Proc. SPIE vol. 367, 149 (1982)

77. J. Tsuijuchi, "Multiplex holograms and their application in medicine", Proc. SPIE vol. 673, 312 (1987)

78. D. Lacey, "Geometric modelling with image plane integral holography", Proc. SPIE vol. 507, 121 (1984)

79. D. Lacey, "Radiologic Applications of Holography", Proc. SPIE 761:114 (1987)

80. R. Ligten and H. Osterberg, "Holographic microscopy", Nature 211:282 (1966)

81. M. Pluta, "Holographic microscopy" in M. Pluta (ed), "Optical Holography", PWN, Warsaw, in polish (1980)

82. P. Greguss, "Laser as a probe in biomedical research", in W. Waidelich (ed), Proc. Intern. Optoelectronics Cong., Laser 75:155, Munich (1975)

83. R. Ligten, "Holographic microscopy in exobiology", 44, see in (6)

84. G. Ellis, "Holomicrography: Transformation of image during reconstruction a posteriori", Science 154, 1195 (1966)

85. E. Feleppa, "Biomedical Applications of Holography", Physics Today 22, 25 (1969)

86. G. Knox et al., "Holographic motion picture microscopy", Proc. Roy. Soc. London b 174, 115 (1966)

87. G. van der Haagen, "Ein Mikroskop mit holographischer 16-mm-Filmaufzeichnung Laser 2" (1972)

88. S. Almeida et al., "A real-time optical processor for pattern recognition of biological specimens", 573, see in (4)

89. H. Caulfield, "The Application of coherent optical image processing for medicine and biology", 39, see in (6)

90. B. Smolinska, "Holographic testing of human vision", SPIE vol. 673, 412 (1987)

91. W. Komarnicki and B. Smolinska, "Optical correlator matched with human eye", SPIE vol. 673, 414 (1987)

92. G. Stroke et al., "Image improvement in high-resolution electron microscopy using holographic image deconvolution", Optik 41, 319 (1974)

LASER DOPPLER FLOW MEASUREMENT FOR THE EVALUATION

OF CEREBRAL MICROCIRCULATION

V. A. Fasano, R. Urciuoli, P. Bolognese and M. Fontanella

Neurosurgery Institute
University of Turin
Turin, Italy

I. INTRODUCTION

The intraoperative assessment of cerebral blood flow (CBF) variations would be of the utmost value to the neurosurgeon. Intraoperative monitoring of CBF is directed to identify either the impairment of hemodynamic regulating mechanisms (autoregulation, vasoreactivity) produced by a lesion [SAH (sub aracnoidal hemorrhagy), tumor, trauma] or the modifications of local perfusion produced by surgical maneuvers. The ideal intraoperative device would not be invasive and cumbersome; moreover, it would not alter regional CBF during the detection procedures, allowing real time continuous monitoring of CBF; and the operation would not be prolonged by the flow measurement. The laser Doppler (LD) flowmeter has been extensively used to study microcirculation in various tissues.

II. MATERIAL AND METHODS

The LD flowmeter measures CBF by analysing the frequency shift of laser light scattered from Red Blood Cells (RBC's); the blood vessels under study range between 5 and 100 µm in diameter. A continuous 2 mW laser beam (HeNe) is conveyed to the target tissue through an optical fiber. The electromagnetic energy enters the tissue, spreading into a hemisphere; the diameter of this hemisphere [1 mm in the Central Nervous System (CNS)] depends inversely on the absorption.

The laser radiation interacts with the tissue, generating scattering inside the sample volume. If the scattering centres are fixed or moving, the scattered light is respectively unshifted or shifted in frequency. A portion of the light returns to the probe, where it enters two optical fibers and is carried to a couple of photodetectors; here the electromagnetic energy is transformed into electric energy. The row signal is then processed. A linearizer avoids underestimating the blood flow in high perfusion situations, eliminating the effects of multiple frequency shifting. Finally, the flow signal reaches the outputs.

Two different LD flowmeters have been used: the Periflux PF2 and the Periflux PF2b (Perimed, Sweden). Five different probes have been used.

Surface and needle probes are fixed by an authostatic hold fast, with a 90° angle between laser beam and cortical surface. Simultaneous recordings from cortical surface and white matter have been obtained using two LD instruments, driving a double pen recorder. Arterial pressure (AP) and blood gas have been recorded during the measurements. Cerebral microcirculation has been studied intraoperatively in basal condition and after stimulation of various types: Hypotensive drugs - sodium nitroprusside (0.4 μm/kg/m' for 3') and nitroglycerin; Hypocapnia - obtained by means of hyperventilation, controlled through haemogasanalysis; Hypercapnia - CO inhalation at 5%, 10% of concentration; Vasoactive drugs - nipodipine and eupaverine topically administered; Mannitol 20% - 200 ml/m' for 3-5 min. Local perfusion variations have been monitored during different surgical procedures: use of retractors, removal of tumors, temporary clipping of the major cerebral arteries, arterial venous malformation (AVM) removal, hematoma evacuation, cyst emptying and ventricular puncture.

III. RESULTS

II.1. Basal conditions

Laser Doppler flow measurement has been used on 145 patients. The mean flow of the brain is high in comparison with many other tissues (i.e. the skin) in basal conditions. The mean flow decreases to very low levels in pathologic conditions, i.e., in the presence of vasospasm, in the tissues compressed by tumors, and in the areas surrounding AVM's.

Three rhythmical phenomena have been identified during intraoperative monitoring of CBF in Neurosurgery. High frequency rhythmical variations (HFRV's) are visible only using fast time constants (0.2, 0.1, 0.002 s). Their frequency is synchronous with the heart rate. Their amplitude depends on physiologic and especially on pathologic conditions. A digital storage oscilloscope interfaced to a personal computer through a bus cable has been used to display the fastest components of this phenomenon. A steeper anacrotic tract, a catacrotic tract and a dicrotic notch are visible.

Medium frequency rhythmical variations (MFRV's) are visible using fast and medium time constants; in the first case the HFRV's are visible superimposed on the MFRV's. They are always present during automatic ventilation and in all body tissues their frequency is equal to that of the ventilator. The MFRV's are formed by an anacrotic and a catacrotic portion; the top is round; and the relative steepness between the two branches is influenced by the Inspiration/Exhalation ratio (I/E ratio), which is controlled by the ventilator.

Low frequency rhythmical variations (LFRV's) have been described by many authors in several tissues using different terms: vasomotion, fluxmotion etc. Vasomotion is a physiological phenomenon; a slight ischemia increases and decreases the amplitude of CBF in many tissues, while a severe ischemia deletes vasomotion completely. All the available time constants of the LD system can show the LFRV's.

Autoregulation or vasoreactivity and vasomotion are the most important physiological parameters. Autoregulation and CO_2 reactivity were generally intact in patients with a preserved vasomotion, while vasomotion was always absent in patients with a lack of autoregulation and CO_2 reactivity.

III.2. Flow changes after surgical maneuvers

In arterial aneurysm surgery the safety of temporary clipping can be checked by monitoring the CBF.

In AVM surgery after the extirpation, an increase of flow can be detected at the border of the malformation. CBF increasing up to 10 times the preexcision value has been observed, immediately downstream of the feeding artery occlusion. Brain compression with retractors produces a flow reduction. After brain decompression by ventricular puncture the CBF increases immediately, then falls to a lower value, which is always higher than the starting value. The empting of a tumoral cyst causes a CBF increase in the adjacent normal area, and when the tumor is removed the flow further increases.

III.3. Drug induced variations

Mannitol increased the mean flow value in a group of patients, without flow changes in a second group. Eupaverine and Nimodipine, topically administered, produced an increase of blood flow.

IV. DISCUSSION

The proper aim of a LD intraoperative flowmeter is not the analysis of the absolute flow distribution in the different regions of an exposed area, but the evaluation of the basal flow and the monitoring of microcirculatory flow changes, to evaluate modifications of physiological response to stimuli and the immediate local effects of the surgeon's maneuvers. In addition, the method provides a better treatment of the patients.

Intracranial hypertension can be associated either with hyperemia or with ischemia. Monitoring of CBF distinguishes these two situations, allowing a correct therapeutic decision. After tumor removal, changes in local perfusion have been registered in the adjacent areas, which can be connected to the onset of the ischemic edema.

Experiments are now in progress with a new probe which permits CBF postoperative monitoring with LD. The probe is an angle probe, disposable, with a soft plastic stabilizer disk of about 1.5 cm diameter. The probe is removed after 3 days. By means of postoperative monitoring, a continuous recording can be obtained. Any drop in blood flow can be immediately detected and an appropriate treatment can be put into effect in an attempt to prevent ischemia.

LASER SAFETY AND SYSTEM DESIGN

SAFETY ISSUES IN LASER SURGERY

D. H. Sliney

Laser Microwave Division
US Army Environmental Hygiene Agency
Aberdeen Proving Ground, Maryland, USA

I. INTRODUCTION

The introduction of lasers into the medical environment poses several unique problems, as with electrical or electronic equipment. Potential hazards of electrical shock exist requiring appropriate grounding, and other electrical safety procedures are essential. However, there are no particular unique electrical safety problems associated with laser use and biomedical engineers and bioelectronic technicians familiar with safe installation of electrical and electron equipment in hospital and health care environments should have no difficulty in providing guidance for the safe electrical use of laser equipment. As already said in the first chapter, the use of proper radiometric terminology and interaction mechanisms is recommended. Unfortunately, the misuse of radiometric terms such as fluence (for radiant exposure) and fluence rate (for irradiance) continue in the literature. These two terms should only be used for flux densities in tissue. The principal radiometric concepts used to describe laser-tissue interactions and the laser interaction mechanisms used in laser surgery are represented in Figures 1 and 2, respectively the dosimetric terms and the laser tissue interactions; approximate thresholds for laser interactions as a function of pulse duration are shown in Figure 3. Since thresholds actually depend also somewhat upon focal image size, the above values are only approximate and represent values developed from extensive research on laser retinal injury.

Unlike conventional surgical techniques a laser can produce potentially hazardous fumes and vaporized tissues. Vaporized tissue in sufficient quantities must receive special attention, and local exhaust ventilation will be required. The one hazard that is truly unique to the laser and which requires special attention is that resulting from the laser beam itself - the optical radiation hazard. Unlike other light sources the laser beam may be collimated and directed over some distance; hence, the area of potential hazard may not be limited to the immediate surgical site. Unwarranted fears often accompany the introduction of lasers for the first time into the surgical theater or the clinical environment. Therefore a proper appreciation of the real laser beam hazard is necessary for each member of the medical staff so that realistic safety precautions are followed.

Laser hazards depend upon the laser in use, the environment, and the personnel involved with the laser operation (the operator, ancillary per-

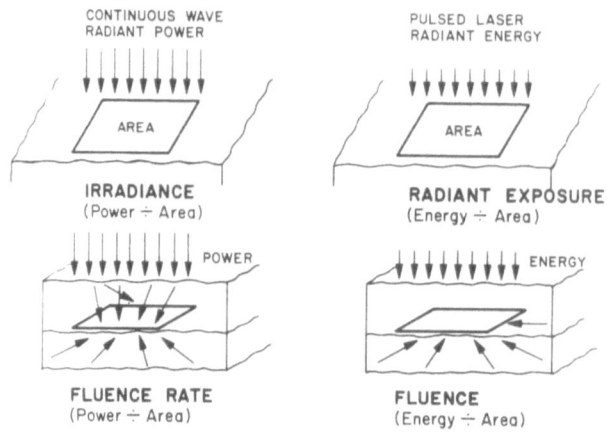

Fig. 1. Dosimetric terms.

sonnel and patient). The laser hazard is roughly defined by the hazard
classification (1 through 4), whereas the other factors must be analyzed in
each situation. A basic understanding of laser biological effects and
hazards is necessary to intelligently assess laser hazards in the operating
room. Once the hazards are understood, the safety measures are obvious.

II. BIOLOGICAL HAZARDS OF LASER BEAMS

II.1. Hazards to the eye

    Because of the special optical properties of the human eye, that organ
is considered the most vulnerable to laser light. Aside from the oral mucosa, the only living tissue exposed to the environment is the cornea and
conjunctiva. Without the comparative protective features of the stratum
cornea of the skin, the eye is exposed to the arsh environment of sun,
wind, dust, ultraviolet radiation and intense light. The eye has a natural
protective mechanism in its lid reflex which limits the exposure to the
retina of very intense visible light or to intense exposure from infrared
rays, which raise the temperature of the cornea. However, some laser beam

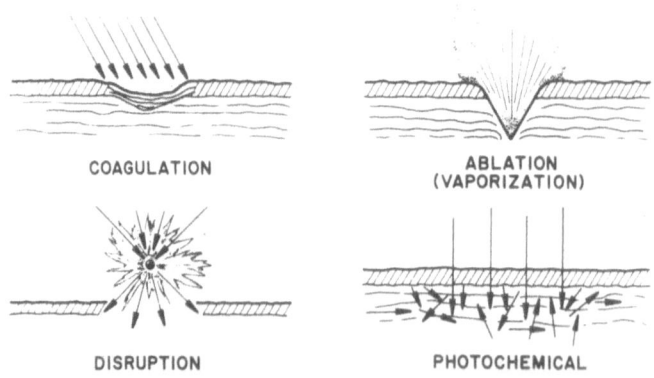

Fig. 2. Laser-tissue interactions.

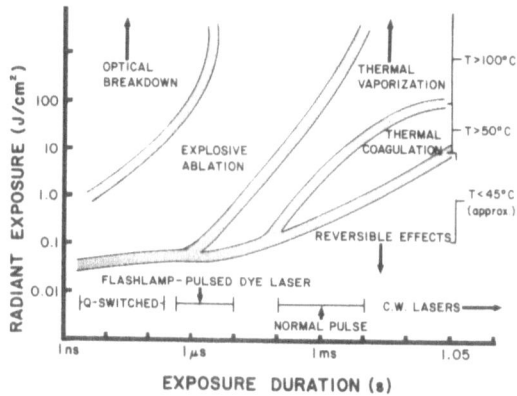

Fig. 3. Thresholds for laser interactions.

intensities are so great that injury can occur faster than the protective action of the lid reflex which occurs between 0.2 and 0.25 sec.

The biological effects of laser light upon the eye depend most predominantly upon wavelength as is shown in Figure 4. Laser light cannot damage tissue unless the light energy is able to penetrate to and be absorbed in that structure. For this reason, rays in the visible and near infrared (visible and IR-A band) which can be transmitted through clear ocular media and be absorbed in the retina can, in sufficient intensity, damage the retina. The high collimation of a laser beam permits the rays to be focused to an extremely small spot on the retina. The image size of such a point at the retina is of the order 10-20 µm (smaller than the diameter of a human hair). For this reason lasers operating etween 400 and 1400 nm are particularly dangerous to the retina. This spectral region is often referred to as the retinal hazard region. The increased concentration of light as it enters the eye is of the order of 100,000. Hence, a collimated beam of 1 W/cm^2 at the cornea will focus to a small spot with an irradiance of 100 kW/cm^2. Although damage to such a small region of the retina may seem insignificant at first thought, it is important to realize that certain parts of the retina, as for instance the central retina, the macula and its fovea (center of the macula) are extremely small areas responsible for critically important high acuity vision. If these areas are damaged by laser radiation, substantial loss of vision can result.

The image area alone may not be the only site of damage, but as a result of heat flow and mechanical (acoustic) transients, the tissue surrounding the image site may also be damaged, leading to more severe consequences upon visual function. For example, it has not been uncommon for an individual to lose almost total function in an eye exposed to a very small amount of energy (several hundred microjoules) when a laser is accidentally imaged on the fovea. Instead of a normal visual acuity of 20/20 (6/6), the visual acuity in such accidental situations has often been recorded as 20/200 (6/60) following the accident. Such low vision would be considered legally blind in most states. Fortunately, in most accidents, only one eye is exposed to a collimated beam. There is normally little recovery of vision since the neural tissue of the retina has very little ability for repair. The visual loss for the most part is permanent.

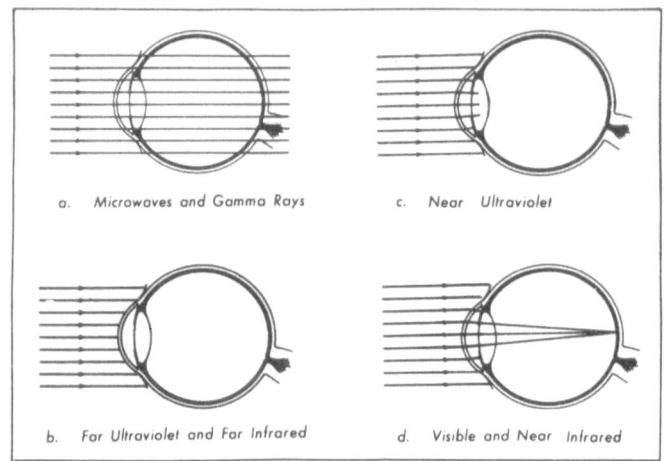

a. *Microwaves and Gamma Rays*
c. *Near Ultraviolet*
b. *Far Ultraviolet and Far Infrared*
d. *Visible and Near Infrared*

Fig. 4.   The biological effects of optical radiation upon
the eye depend upon the absorption properties
which depend in turn upon the spectral region.

At wavelengths outside of the retinal hazard region, in both the ultraviolet and far infrared regions of the spectrum, injury to the anterior
segment of the eye is possible. Certain spectral bands may injure the lens
(notably at wavelengths between 295 and 320 nm and wavelengths between 1
and 2 μm). Injury to the cornea is possible from a wide range of wavelengths in the ultraviolet and most of the infrared at wavelengths beyond
1400 nm. Injury to the cornea is normally very superficial, involving only
the corneal epithelium, and with the cornea's high metabolic rate, corneal
repair occurs within a day or two, and total recovery of vision will occur.
If, however, significant injury occurs in deeper corneal layers in the
stroma or endothelium and in the germinative layers of the cornea, corneal
scars can result, leading to permanent loss of vision unless a corneal
transplant can be effected.

Excimer lasers operating in the ultraviolet are particularly hazardous
to the cornea, and the 308 nm excimer laser can be considered extremely
dangerous as it can cause an immediate cataract of the lens. The argon,
krypton, copper vapor, gold vapor, helium-neon, and neodymium-YAG lasers
are all potentially hazardous to the retina. The erbium:YAG, erbium:YLF,
holmium:YAG, hydrogen-fluoride, carbon-dioxide, and carbon-monoxide lasers
are all potentially hazardous to the cornea because wavelengths which cause
corneal damage are not reconcentrated by the eye as are wavelengths in the
retinal hazard region. The thresholds for injury of the cornea are generally much higher than those which may injure the retina. Table 1 lists
permissible occupational exposure limits for most of the commonly used biomedical lasers[1÷8]. Note that: a) not all standards/guidelines have MPE's
below 200 nm; and, b) to convert MPE's in mW/cm^2 to mJ/cm^2, multiply by
exposure time t in seconds; e.g., the He-Ne or Argon MPE at 0.1 s is
0.32 mJ/cm^2.

II.2. Skin hazards

The skin is generally considered far less vulnerable to injury than
the eye. While this is true in terms of absolute irradiance or exposure
dose because of the protective character of the stratum corneum, one should
remember that the probability of exposure to some part of the skin from a
reflected laser beam is far greater than to the small area occupied by the
eye. Injury to the skin can occur from either photochemical damage mechanisms which are predominant in the ultraviolet end of the spectrum. For

example, erythema results from injury to the epidermis, and to some extent, the dermis as well, and originates from a photochemically initiated event. This type of injury is commonly referred to as "sunburn". First, second and third degree skin burns can also be induced by visible and infrared laser beam exposure producing thermal injury.

The severity of the injury depends upon the length of exposure and the penetration depth of the laser radiation. Generally, if the exposure lasts for a second or more, a pain response elicits a jerk movement to move the exposed tissue away from the laser beam, thereby limiting the exposure duration to a second or less. High power laser beam exposure will not result in a deep tissue burn at CO_2 wavelengths if the exposure time is extremely short, since the penetration depth of the carbon-dioxide laser beam is very shallow (of the order 20 μm) and, in fact, does not penetrate the normal thickness of the stratum corneum. Injury to the epidermis from the CO_2 laser is by heat conduction from the stratum corneum to deeper layers. However, short pulsed exposure to 1064 nm Nd:YAG laser radiation, which penetrates several millimeters into tissue, can cause a deep, severe burn at a radiant exposure just above burn threshold, albeit at a much higher threshold than for a CO_2 laser burn.

Clearly the focal spot of a surgical laser is designed to ablate or vaporize tissue and will be hazardous to skin if located near the focal spot. Skin injury can also occur as a result of ignition of clothing by a reflected laser beam with tragic consequences. Significant skin injuries from accidental exposure to industrial or medical lasers rarely occur; at least, they are rarely reported. Actual thresholds of injury to the skin are normally of the order of joules-per-square centimeter, and this level of exposure does not occur outside of the focal zone of a surgical laser.

Table 1. Selected occupational exposure limits (MPE's) for some common lasers (From References 2, 4, 5).

Laser	Wavelength	Exposure Limit
Argon-fluoride laser	193 nm	3.0 mJ/cm^2 over 8 h
Xenon-chloride laser	308 nm	40 mJ/cm^2 over 8 h
Argon ion laser	488, 514.5 nm	3.2 mW/cm^2 for 0.1 s 2.5 mW/cm^2 for 0.25 s
Helium-neon laser	632.8 nm	1.8 mW/cm^2 for 1.0 s 1.0 mW/cm^2 for 10 s
Krypton ion laser	568, 647 nm	
Neodymium-YAG laser	1064 nm 1334 nm	5.0 μJ/cm^2 for 1 ns to 100 μs No MPE for t < 1 ns 5 mW/cm^2 for 10 s
Carbon-dioxide laser	10.6 μm	100 mW/cm^2 for 10 s to 8 h, limited area
Carbon-monoxide laser	~ 5 μm	10 mW/cm^2 for > 10 s for most of body

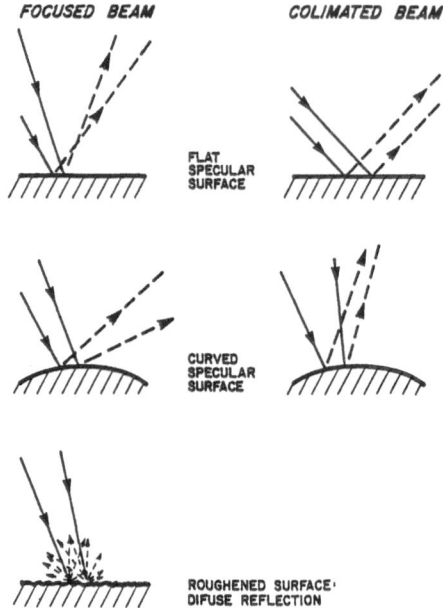

FOCUSED BEAM   COLIMATED BEAM

FLAT
SPECULAR
SURFACE

CURVED
SPECULAR
SURFACE

ROUGHENED SURFACE·
DIFUSE REFLECTION

Fig. 5. Examples of reflections of laser radiation from
specular (mirror like) surfaces (e.g., metallic
instrument surfaces).

III. REFLECTIONS AND PROBABILITY OF EXPOSURE

When one examines the accident history with lasers, it becomes apparent that the source of ocular exposure is a reflection. Figure 5 illustrates the types of laser beam reflections that can occur from ophthalmic
contact lenses and metallic instruments used in other surgical procedures.
Normally the collimated beam is considered the most hazardous type of reflection, but at very close range, a diverging beam may pose a greater
likelihood of striking the eye.

A number of steps can be taken to minimize the potential hazards to
both the patient and surgical staff. Preventive measures will depend upon
the type of laser. The most common type of laser employed today in most
surgical applications is the CO_2 laser. Since the CO_2 laser wavelength of
10.6 μm is in the far-infrared spectral region and invisible, the presence
of hazardous secondary beams could go unnoticed. This added hazard resulting from an infrared laser beam's lack of visibility is common to other
infrared lasers such as the 1064 nm Nd:YAG laser. Because there have been
a number of serious retinal injuries caused by improper attention to safety
with Nd:YAG lasers [2,5], the use of the Nd:YAG laser must be approached with
even greater caution than the CO_2 laser. By contrast, the argon laser and
the second-harmonic Nd:YAG (sometimes referred to as the "KDP") laser emit
highly visible, blue-green, (488, 514.5, and 532 nm) beams, and in some
ways pose a lesser potential hazard.

Most current surgical lasers, such as the CO_2, Nd:YAG, or argon lasers
are continuous-wave (CW), or nearly so; even "super-pulse" is quasi-CW compared to single-pulse laser photodisruptors or some experimental excimer
ablative lasers. The biological effects and potential hazards from high-
peak power pulsed lasers are quite different from those of CW lasers. This
is particularly true of lasers operating in the retinal hazard region of
the visible (400-760 nm) and near-infrared spectrum (IR-A: 760-1400 nm).
The severity of retinal lesions from a visible or near-infrared (IR-A) CW
laser are normally considered to be far less than from a Q-switched or mo-

delocked laser. Another major factor that influences the potential hazard is the degree of beam collimation. Almost all surgical lasers are focused, thereby limiting the hazardous area (referred to as the "nominal hazardous zone" in Reference 2). An exception is the highly collimated beam from many argon laser photocoagulators which may be still concentrated and hazardous at quite some distance from the instrument[10].

Reflections are most serious from flat mirror-like (specular) surfaces characteristic of many metallic surgical instruments. Many surgical instruments now have black anodized or sand-blasted, roughened surfaces to reduce (but not eliminate) potentially hazardous reflections. The surface roughening is generally more effective than the black (ebonized) surface, since the beam is diffused. However, combining a black surface with roughening provides the greatest protection. Furthermore, any metal surface is quite absorbing at the 10.6 μm CO_2 laser wavelength anyway.

One should know that both the surface finish and reflectance seen in the visible spectrum do not indicate those qualities in the invisible far-infrared spectrum. In fact, a roughened surface which appears to be quite dull and diffuse at shorter, visible or IR-A wavelengths, will always be much more specular at far-infrared wavelengths (e.g., the CO_2 laser wavelength). This behavior results from the fact that the relative size of the microscopic structure of the surface relative to the incident wavelength determines whether the beam is reflected as a specular or diffuse reflection. In any case, a specularly reflected beam with only 1% of the initial beam's power can still be quite hazardous. Hence the rougher the surface of an instrument likely to intercept the beam, the safer the reflection. For example, even a 1% reflection of a 40 W laser beam is 400 mW!

It is indeed somewhat surprising that there have been few cases reported of eye injuries to residents and other persons observing Nd:YAG laser surgery without eye protectors. Hazardous specular reflections from a focused laser beam from a surgical instrument or other polished surface are limited in extent because of the focused beam. Figure 6 shows the beam irradiance of a focussed laser beam as a function of distance from a laser. Note the rapid decrease of irradiance beyond the focal point.

Most surgical lasers have a visible alignment beam. Infrared lasers most often make use of a low-power coaxial He-Ne (632.8 nm) red laser. It is desirable for this alignment beam to be 1 mW or less, since the maximum CW, visible laser beam power that can safely enter the eye within the aversion response (i.e., within the blink reflex, etc. of 0.25 sec.) is 1 mW.

III.1. Patient and Personnel Susceptibility

Laser safety regulations do not apply to the exposure of the patient

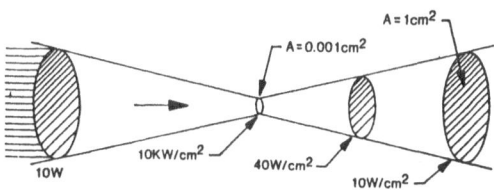

EXAMPLE:
POWER = 10 WATTS
FOCAL DIAMETER = 0.35mm
FOCAL LENGTH = 250mm

Fig. 6. Irradiance in focussed beams.

at the target site for surgery. However accidental exposure to the patient from misdirection of the laser beam should be of concern and can result in injury of eye and skin. This is of particular concern when lasers are used in or near the eye and where exposure of the eye itself is not intended.

III.2. Surgeon's Susceptibility

The surgeon or laser operator is normally not highly susceptible to injury due to proper design of the laser instrument. Normally the surgeon views the target tissue through the optics of an endoscope, an operating microscope, colposcope, slit-lamp biomicroscope, etc., and the reflections are safely attenuated within the optics. However, whit hand-held laser delivery systems, one should remember that the surgeon's hand is the closest to the laser target and therefore it is closest to potentially hazardous reflections from adjacent surgical instruments (e.g., metal retractors).

III.3. Surgical Assistant's Susceptibility

Nurses, other surgical assistants and operating room staff are potentially exposed to misdirected laser beams. Lasers have been accidentally initiated when the beam delivery system was directed other than at the patient, a foot switch was accidentally pressed, or similar errors have occurred, and the beam directed at person. Assistants are also potentially exposed to secondary reflections from surgical devices while the surgeon's eyes are protected by viewing through an optical delivery system. For example, an ophthalmic surgeon normally directs the laser beam into the patient's eye through a slit-lamp microscope, and the surgeon's eyes are protected by filters built into the delivery system. Reflections from the cornea or the contact lens used in ophthalmic surgery may be hazardous to assistants or bystanders in line of view of the contact lens to a distance of 1 to 2 m. The operating microscope used in laser microsurgery by a number of specialties would protect the eyes of the surgeon if properly designed; whereas, assistants and bystanders may be exposed to potentially hazardous reflections from surgical instruments inserted into the beam path.

III.4. Bystander's Susceptibility

Bystanders in the surgical facility or clinical laser facility who are present to observe or to calm the patient (e.g. a patient's relative) may be susceptible to exposure from reflected laser beams in the same manner that a surgical assistant or nurse may be susceptible. In addition, because of lack of training or knowledge about the laser surgical procedure the bystander may be at greater risk by inadvertently placing themselves in a dangerous position.

III.5. Service Personnel Susceptibility

Service personnel are particularly susceptible to laser injury since they often gain access to collimated laser beams from the laser cavity itself or by opening up the beam delivery optics and gaining access to collimated laser prior to the beam focusing optics in the beam delivery system. As previously explained, most surgical lasers employ a focused beam, thereby reducing the hazard distance of the direct beam (Figure 6). Hence, once the laser beam leaves the delivery system and comes rapidly to a focus, it then diverges again. The zone where the beam is concentrated to a level sufficient to pose a severe hazard to the eyes or skin (the "Nominal Hazard Zone", or NHZ), is normally a limited zone of 1-2 m near the beam focal point. However, collimated laser beam, as the raw beam for most laser cavities, or a specular reflection from a turning mirror or Brewster window in the laser console may be emitted from the laser cabinet when the service person gains access. At least two serious eye injuries have occurred to service personnel exposed to secondary, collimated, invisible 1064 nm Nd:YAG laser beams when the service personnel gained access to the laser cavity.

IV. OCCCUPATIONAL EXPOSURE LIMITS

Relevant EL's for lasers of interest are given in Table 1 and are calculated or measured at the cornea. If the laser beam is less than 7 mm in diameter, it is assumed that the entire beam could enter the dark-adapted pupil and one can express the maximal safe power or energy in the beam; it is the EL multiplied by the area of a 7 mm pupil, i.e., 0.45 cm^2. For example, for the visible CW lasers, an exposure limited by the natural aversion response of 0.25 sec is 2.5 mW/cm^2, and this EL multiplied by 0.4 cm^2 results in the limiting power of 1 mW. This 1 mW value has a special significance in laser safety standards, since it is the "Accessible Emission Limit (AEL)" of Class 2, i.e., the dividing line between two laser safety hazard classifications: Class 2 and Class 3.

IV.1. Laser Hazard Classification

As just mentioned, any CW visible laser (400-700 nm) that has an output power less than 1 mW is termed a Class 2 (low-risk) laser, and could be considered more or less equivalent in risk with staring at the sun, at a tungsten-halogen spotlight, or at other bright lights which can cause a photic maculopathy (central retinal injury). Only if one forces themselves to overcome their natural aversion response to bright light, can Class 2 laser pose a real ocular hazard. An aiming beam or alignment laser operation at a total power above 1.0 mW would fall into hazard Class 3, and could be hazardous even if viewed momentarily within the aversion response time. A sub-category of Class 3, termed Class 3a, consists of lasers from 1 to 5 mW in power and these lasers pose a moderate ocular hazard under viewing conditions where most of the beam enters the eye. Class 3b is then the subcategory that comprises, amongst certain pulsed lasers, CW visible lasers that emit between 5 and 500 mW output power. Even momentary viewing of Class 3b lasers is potentially hazardous to the eye.

Only lasers that are totally enclosed or that emit extremely low output powers fall into Class 1 and are safe to view. Any CW laser with an output power above 0.5 W (500mW) falls into Class 4. Class 4 lasers are considered to pose skin or fire hazards as well as hazards if not properly used. The purpose of assigning adequate safety measures, i.e., Class 3a measures are more stringent than Class 3b measures. Virtually all surgical lasers are Class 4.

V. CONCLUSIONS

The potential exposure levels to the eye and skin from scattered laser radiation from most surgical laser applications are substantially below threshold, and only the direct beam or specular reflections are of concern. Only with UV lasers should one be seriously concerned with chronic exposure and delayed effects. The surgical laser user can be assured that today a consensus exists almost worldwide[2÷4, 8, 11] regarding the appropriate laser safety measures to preclude injury from acute or chronic effects.

REFERENCES

1. ACGIH, "TLV's, Threshold Limit Values and Biological Exposure Indices for 1990-1991", American Conference of Governmental Industrial Hygienists, Cincinnati, Ohio (1990)
2. ANSI, "Safe Use of Lasers", Standard Z-136.1 -1986, American National Standards Institute, Laser Institute of America, Orlando, Florida (1986)
3. ANSI, "Safe Use of Lasers in Health Care Facilities", Standard Z-136.3-1988, American National Standards Institute, Laser Institute of America, Orlando, Florida (1988)
4. British Standards Organisation, "Radiation Safety of Laser Products and Systems", Standard BS4803, London, BSI (1984)
5. Deutsche Institut für Normung, "Radiation Safety of Laser Products", Standard VDE 0837, Berlin, DIN/VDE (1984)
6. Health Council of the Netherlands, "Acceptable Levels for Micrometer Radiation", Rijswijk, Gezondheidsraad (1979)
7. IRPA, International Non-Ionizing Radiation Committee, "Guidelines for Limits of Human Exposure to Laser Radiation", Health Physics, 49(5):341-359 (1985)
8. International Electrotechnical Commission, "Radiation Safety of Laser Products, Equipment Classification, and User's Guide", Document WS 825, IEC, Geneva (1984)
9. D. H. Sliney and M. L. Wolbarsht, "Safety with Lasers and Other Optical Sources", Plenum Publishing Corp, New York (1980)
10. D. H. Sliney and M. A. Mainster, "Potentially hazardous reflections to the clinician during photocoagulation" Amer J. Ophthalmol. 103(6):758-760 (1987)
11. World Health Organization (WHO), Environmental Health Criteria No. 23, "Lasers and Optical Radiation", joint publication of the United Nations Environmental Program, the International Radiation Protection Association and the World Health Organization, Geneva (1982)

LASER SURGERY OF THE EYE: PHOTOTOXICITY CONCERNS

D. H. Sliney

Laser Microwave Division
US Army Environmental Hygiene Agency
Aberdeen Proving Ground, Maryland, USA

I. INTRODUCTION

Lasers have been employed extensively in ophthalmology for a variety of procedures and have the longest history of large-scale clinical use. In most applications there has been a concern for undesired side effects resulting from phototoxicity. This concern has been accentuated by recent interest in applying ultraviolet (UV) lasers to corneal refractive surgery and phakoablation of the lens. However, retinal phototoxicity resulting from scattered light from argon laser retinal photocoagulation is also a realistic potential hazard.

Photochemical injury mechanisms are normally of greatest concern at short wavelengths (UV and short-wavelength blue light) where photon energies are greatest, and also will be most readily observed for lengthy exposure durations. Because of the reciprocal relationship of irradiance (dose-rate in W/cm^2) and exposure duration (in seconds, s) to achieve a threshold photochemical radiant exposure (dose in J/cm^2), the cumulative exposures of adjacent tissues to scattered light from repeated pulses of laser energy directed at target tissue will be additive. For example, when an argon laser is used for pan-retinal photocoagulation, the dominant tissue interaction mechanism at the target retinal tissue is thermal. However, the small amount of energy scattered toward the macular area with each pulsed laser exposure will add, such that after 500 focal burns, the accumulated macular dose may produce photoretinitis.

Although for each laser wavelength there is a specific target tissue which is the primary site of concern, there are some wavelengths capable of causing injury to more than one structure in the human eye. Corneal surgery with the excimer laser operates at 193 nm and, as Figure 1 illustrates, all wavelengths less than 280 nm are entirely absorbed in the cornea. This is not true at longer UV wavelengths and mid and near-UV wavelengths, the primary injury site is not only the cornea, but injury to the lens and the retina may also be possible. As an example, the 308 nm excimer laser wavelength deserves particular note. At this wavelength, a collimated beam striking the eye will penetrate to the retina and because of its absorption properties can cause damage to the cornea lens, and retina all in one exposure. Ocular structures are susceptible to damage at this wavelength for only small amounts of 308 nm laser radiation. The injury thresholds for the retina and lens are comparable. Since the corneal injury threshold is

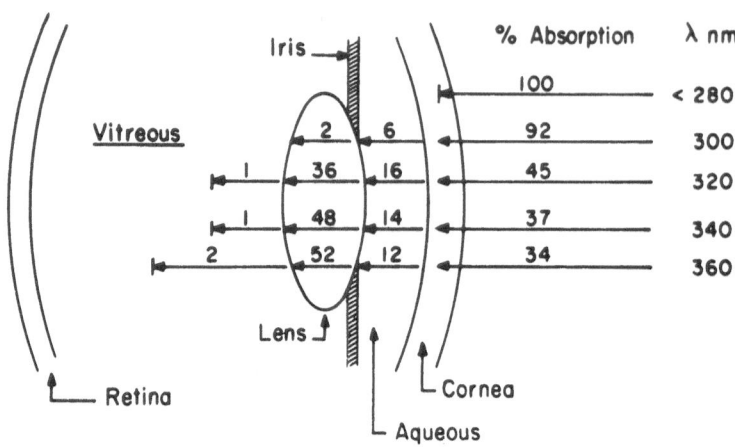

Fig. 1. Spectral properties of the human eye in
the ultraviolet spectrum.

lower than for either lens or retina, laser safety standards limit exposure
based upon the corneal injury threshold.

One should also recognize that the aphakic or pseudophakic observer
would be particularly at risk from retinal injury when exposed to laser
radiation longer that 300 nm. The crystalline lens normally absorbs in excess of 99% of UV radiation from 315 to 400 nm.

Unlike most laser wavelengths, chronic exposure is of particular concern when working with ultraviolet radiation sources. UV radiation (UVR)
with a wavelength from 280 to 315 nm (referred to as UV-B) has carcinogenic
potential. Ocular damage from this range of mid-UVR is thought to be limited largely to cataractogenesis, although the potential for macular damage
has also been posed as a possibility. Although biological proof of cataractogenesis and retinal degeneration from chronic exposure to the near UV
has yet to be solidly proven, it is clear that individuals working with
excimer lasers must exercise extreme caution and always wear eye protection
at wavelengths between 280 to 340 nm. They should also greatly limit skin
exposure where feasible.

II. ULTRAVIOLET RADIATION HAZARDS

Ultraviolet radiation (UVR) biological effects have been studied for
nearly a century[1-5]. Biological studies have been performed at the molecular and cellular level as well as at the organ level.

When considering UVR bioeffects, it is useful to employ the convention
of the International Commission on Illumination (CIE) for spectral bands.
The CIE has designated 315-320 to 400 nm as UV-A, 280 to 315-320 nm as
UV-B, and 100-280 nm as UV-C (CIE, 1970). Wavelengths below 180 nm (vacuum
UV) are of little practical significance since they are readily absorbed in
air. The 308 nm UV wavelength is therefore in the UV-B spectral region.
UV-C wavelengths are more photochemically active, because these wavelengths
correspond to the most energetic photons, are strongly absorbed in certain
amino acids and therefore by most proteins[6-9]; whereas, UV-B wavelengths
are less photochemically active, but are more penetrating in most tissues.
UV-A wavelengths are far less photobiologically active, but are still more
penetrating than UV-B wavelengths. UV-A wavelengths play an interactive
(sometimes synergistic) role when exposure occurs following UV-B exposure[10]. UV-B radiation has been shown to alter enzyme activity in the
lens[11].

UV-C wavelengths are strongly absorbed in proteins, are very photo-chemically interactive, and have the least penetration into biological tissue. In this regard, the ArF excimer laser wavelength of 193 nm is the only wavelength shown very clearly to produce extremely sharp ablative edges[12], and small corneal tissue fragments (suggestive of photochemical photodecomposition) during corneal ablation studies[13]. Figure 2 shows the absorption spectrum of representative proteins/amino acids. Note that all are very weakly absorbing, if at all, at the shortest UV wavelengths and at 308 nm (after Reference 8).

With regard to photocarcinogenesis, it is generally agreed that only a very narrow UV-B wavelength band is generally considered very effective in producing skin carcinogenesis[14,15]; and, for that matter, severe sunburn[16], and cataractogenesis in humans[4,17]. Even though we are concerned with delayed effects upon ocular tissues of UVR, one can obtain a deeper insight by reviewing the known biological effects upon the skin as well as the eye in this wavelength region.

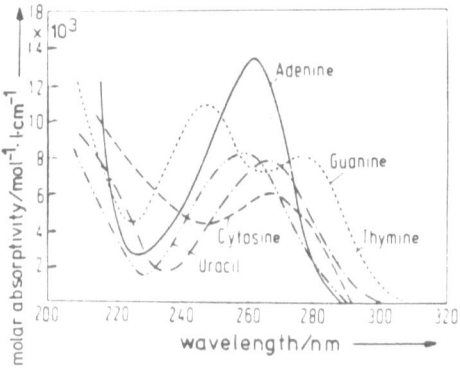

Fig. 2. Absorption spectra of representative proteins/ amino acids.

II.1. Effects upon the skin

Sunburn and erythema have been studied more extensively than UV photo-keratoconjunctivits and other UV ocular effects, and therefore probably offer some insight into UVR effects upon any human tissue. It should be noted that erythema produced by the longer UV-B wavelengths (295-315 nm) is more severe and persists longer[16]. The increased severity and time-course of the erythema results from deeper penetration of these wavelengths into the epidermis. Maximum sensitivity of the skin occurs at approximately 295 nm [18,19], with much less (approximately 0.07) sensitivity[20] at 308 nm. Although erythema can be elicited at UV-A wavelengths, radiant exposures of more than a thousand-fold greater than at 295 nm are generally required[21]. The minimal erythemal dose (MED) related to 295 nm reported in more recent studies for untanned, lightly pigmented skin range[22÷24] from 6-30 mJ/cm^2. The MED at 193 nm is very difficult to extrapolate, but should be of the order of 400 mJ/cm^2, or as high a 1.0 J/cm^2. Skin pigmentation and tanning, and most importantly, thickening of the stratum corneum can increase this MED by at least one order of magnitude. The current occupational exposure limit (EL) at the 308 nm wavelength is: 120 mJ/cm^2, which is near the threshold for photokeratitis[17].

Chronic exposure to sunlight - especially the UV-B component - accelerates the aging of the skin and increases the risk of developing skin cancer[1, 3, 11, 25÷27]. Several epidemiologic studies have shown that the incidence of skin cancer is strongly correlated with latitude, altitude and sky cover which correlate with UVR exposure[28, 29]. The solar spectrum is greatly attenuated by the earth's ozone layer limiting terrestrial UV to wavelengths greater than 290-295 nm. The UV-B irradiance at ground level is a function of elevation and the atmospheric path length, i.e., time of day[30, 31].

Exact quantitative dose-response relationships for human skin carcinogenesis have not yet been established, although fair-skinned individuals, particularly those of Celtic origin, are much more prone to develop skin cancer. Nevertheless, it must be noted that the UVR exposures necessary to elicit skin tumors in animal models may be delivered sufficiently slowly that erythema is not produced, and the relative effectiveness of 308 nm (relative to the peak at 302 nm) reported in those studies[14, 15] is approximately 0.022.

II.2. Effects upon the eye

Actinic ultraviolet radiation (UV-B and C) is strongly absorbed by the cornea and conjunctiva. Overexposure of these tissues causes photokeratoconjunctivitis commonly referred to as welder's flash, arc-eye, or snowblindness. The action spectrum and time-course of photokeratitis in the human, rabbit and monkey cornea, is reported in Reference 32. The latent period varies inversely with the severity of exposure ranging from 1.5 to 24 hours, but usually occurs within 6-12 hours, and discomfort usually disappears within 48 hours. Conjunctivitis follows and may be accompanied by erythema of the facial skin surrounding the eyelids. Of course, UVR exposure rarely results in permanent ocular injury. Threshold data for photokeratitis in humans for 10 nm wavebands from 220 to 310 nm is reported in Reference 33. The maximum sensitivity of the cornea was found to occur at 270 nm, differing markedly from the maximum for the skin. Presumably, 270 nm radiation is more effective because of the lack of a stratum corneum to attenuate the dose to the corneal epithelium tissue at shorter UVR wavelengths. The wavelength response, or action spectrum, did not vary as greatly as did erythemal action spectra, with thresholds varying from 4-14 mJ/cm^2 at 270 nm. The threshold reported at 308 nm was approximately 100 mJ/cm^2.

It is well to remember that repeated exposure of the eye to potentially hazardous levels of UVR does not increase the protective capability of the cornea as does skin exposure which leads to tanning and to tickening of the stratum corneum. The UVR absorption properties of the cornea and aqueous as well as the effects of UV-B radiation upon the corneal epithelium, corneal stroma, and the corneal endothelium are reported in References 34 ÷ 39. Electron microscope studies showed remarkable repair and recovery properties of corneal tissue. Although one could readily detect significant damage to all of these layers, and apparently initially appearing in cell membranes, morphological recovery was complete after a week. The destruction of keratocytes in the stromal layer was apparent, and endothelial recovery was pronounced despite the normal lack of rapid cell turnover in the endothelium. Endothelial damage has been found[40] that was persistent if the UVR exposure was persistent. The corneal endothelium following

UV-B exposure is reported in Reference 41 and it is concluded that severe, single insults were not likely to have delayed effects; however, it is also concluded that chronic exposure could accelerate changes in the endothelium related to ageing of the cornea.

Wavelengths above 295 nm can be transmitted through the cornea and are almost totally absorbed by the lens. It has been shown[17] that catarats can be produced in rabbits by wavelengths in the 295-320 nm band. Thresholds for transient opacities ranged from 0.15 to 12.6 J/cm^2 depending on wavelength, with a minimum threshold at 300 nm. Permanent opacities required greater radiant exposures. No lenticular effects were noted[17,42] in the wavelength range 325 to 395 nm even with much higher radiant exposures of 28 to 162 J/cm^2. These studies clearly illustrate the particular hazard of the 300-315 nm spectral band as would be expected because of the penetration of these wavelengths together with the photon energy to efficiently produce photochemical damage.

Using a Xe-Cl 308 nm laser, rabbit eyes have been irradiated[43] at a range of exposure levels from 0.02 to 5.1 J/cm^2 to study effects upon the corneal endothelium. A corneal thickening was noted, which was maintained for two weeks following the exposure, even at the lowest exposure levels. The ocular effects of 308 nm laser exposure showing similar results are reported in Reference 44.

Epidemiologic evidence has been provided[45] that UV-B in sunlight was an etiologic factor in senile cataract, but showed no correlation of cataract with UV-A exposure. Although once a popular belief because of the strong absorption of UV-A by the lens, the hypothesis that UV-A can cause cataract has not been supported by either experimental laboratory studied or epidemiologic studies. From the laboratory experimental data which showed that thresholds for photokeratitis were lower than for cataractogenesis, one must conclude that levels less than that required to produce photokeratitis on a daily basis should be considered hazardous to lens tissue. Even if one were to assume that the cornea is exposed to a level nearly equivalent to the threshold for photokeratitis, one would estimate that the daily UVR dose to the lens at 308 nm would be less than 120 mJ/cm^2 for 12 hours out-of-doors[46]. Indeed, a more realistic average daily exposure would be less than half that value.

In determining the action spectrum for photoretinitis produced by UV radiation from 320 to 400 nm, it has been showed[47] that thresholds in the visible spectral band, which were 20-30 J/cm^2 at 440 nm, were reduced to approximately 5 J/cm^2 for a 10 nm band centered at 325 nm. The action spectrum was monotonically increasing with decreasing wavelength. We should therefore conclude that levels well below 5 J/cm^2 at 308 nm, if reaching the retina, should produce retinal lesions, although these lesions would not become apparent for 24-48 hours after the exposure. There are no published data for retinal injury thresholds below 325 nm, and one can only expect that the pattern for the action spectrum for photochemical injury to the cornea and lens tissues would apply to the retina leading to an injury threshold of the order of 0.1 J/cm^2.

Although UV-B radiation has been clearly shown to be mutagenic and carcinogenic to the skin, the extreme rarity of carcinogenesis in the cornea and conjunctiva is quite remarkable. There appears to be no scientific evidence to link UV exposure with any cancers of the cornea or conjunctiva in humans, although the same is not true of cattle[48,49]. This would suggest a very effective immune system operating in the human eye, since there are certainly outdoor workers who receive a UV exposure comparable to cattle. This conclusion is further supported by the fact that individuals suffering from a defective immune response, as in xeroderma pigmentosum, frequently develop neoplasias of the cornea and conjunctiva[50].

II.3. Safety standards

Occupational exposure limits for UVR have been developed and include
an action spectrum curve which envelopes the threshold data for acute effects obtained from studies of minimal erythema and keratoconjunctivi-
tis[51, 52]. This curve does not differ significantly from the collective
threshold data considering measurement errors and variations in individual
response. The curve is well below the UV-B cataractogenic thresholds. The
Figure 3 (adapted from Reference 51) shows the thresholds for ultraviolet
acute injury of the cornea and skin. Note the lack of available data in
the spectral region below 254 nm.

The EL for incoherent UVR at 308 nm is J/cm^2, whereas, the EL for
longpulse or CW laser exposure of the cornea at 308 nm depend upon pulse
duration. This EL is 0.04 J/cm^2 integrated over a 24 hour period for pulse
durations greater than 0.026 ms (26 μs) and varies as the fourth root of
the duration for shorter pulse durations[53÷55]. For a pulse duration of
10 ns, the EL is 5.6 mJ/cm^2; and for pulse durations of 100 ns, 1 μs and
10 μs, the EL is 10, 17 and 31 mJ/cm^2, respectively.

Regardless of pulse duration, the total daily exposure to 308 nm laser
radiation is still limited to the long-exposure EL of 40 mJ/cm^2. In other
words, if the unprotected cornea were exposed to 1000 pulses of scattered
laser radiation from a 10 ns, 308 ns Xe-Cl laser in a laboratory work environment, the limit on an averaged per-pulse basis would be only 40 μJ/cm^2.

III. LASER PHAKOABLATION

The appeal of using a laser to vaporize the nucleus and cortex of the
lens in cataract surgery is the potential to remove all of the contents of
the cloudy lens from within the capsule through a very small opening created for a fiber optic delivery system. With a trasparent capsule remaining, it could be filled with a viscous trasparent optical medium to permit
ocular accommodation as experienced in youth.

It is somewhat difficult to accurately estimate the exposure of the
corneal endothelium, ciliary body, iris and retina during an endocapsular
procedure whit a 308 nm laser. The energy levels reported[56] to have been
required to ablate the cortex and nucleus of enucleated eyes was of the

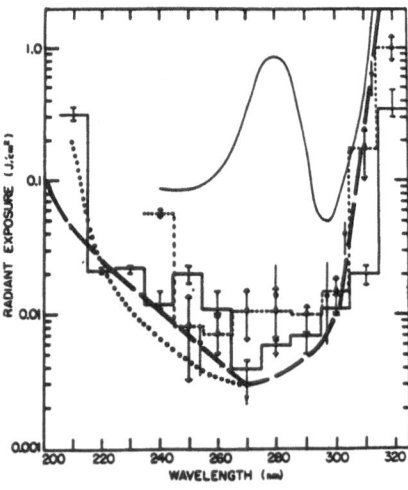

Fig. 3. Thresholds for ultraviolet acute injury and skin.

order of 200 J. If one were to assume only 10% of the total energy to be scattered, this would be 20 J of scattered radiation. As a first estimate, one could assume the averaged scatter for the entire procedure to be reasonably uniform (isotropic). Some of the energy would be attenuated by absorption in the remaining media, but this would be minimal except for absorption in the remaining lens tissue. The greatest attenuation would result from reduction of irradiance by the inverse square law. Radiant exposure doses would range from 0.5 to 10 J/cm^2 depending upon the distance from the lens.

The orientation of the fiber tip and the surgical approach could clearly limit the dose to other structures. For example, ablating the nucleus first, taking advantage of the absorption of scattered radiation by the cortex and aiming away from the retina, would limit retinal exposure. In any case, exposure of the ciliary body would probably be greatest, followed by exposures to the cornea and retina. The exposure to the cornea would probably exceed the threshold for photokeratitis. Because the lens capsule is nearly 90% transmitting[57] the retina would experience its greatest exposure during the end of the procedure and could exceed the probable retinal injury threshold at 308 nm, even though the retina probably would not be exposed to levels exceeding the threshold for photomaculopathy[47] of about 5 J/cm^2 applicable at 325 nm. Therefore, retinal exposure would probably pose the greatest concerns, since delayed effects of retinal exposure to UV-B have not really been studied.

The potential risk of injury to the ciliary is difficult to predict. In all likelihood the threshold exposure dose in J/cm^2 for injury of that structure would probably be similar as that for the retina since the tissues have similar embryological origins. Indeed, one should recognize that the thresholds of most tissues studied (cornea, lens and epidermis) are comparable and are of the order of 0.1 J/cm^2. Of particular concern is that the ciliary body would normally not be exposed to UV-B radiation and therefore would not be expected to have any adaptive/protective mechanism.

Although the ciliary body is normally well shielded from external UV-B exposure from sunlight, the retina does receive some exposure[58]. In the adult human eye approximately 1 to 3% of the 308 nm laser radiation reaches the retina and this percentage decreases with age[4]. Incidents of apparent photic maculopathy from UV-B radiation in young persons indicate that there is a possibility that UV-B may cause retinal injury at lower levels than previously thought possible[50]. The thresholds for acute injury from UV-B exposure of the ciliary body, the lens capsule and the retina may not actually be exceeded. If animal experiments whit the 308 nm laser show no inflammatory response, one is left with the question of whether molecular damage could lead to delayed effects. Our knowledge of delayed effects even to the retina is extremely limited.

IV. LASER CORNEAL REFRACTIVE SURGERY

The argon-fluoride (ArF) 193 nm excimer laser has been shown[13,60-64] experimentally to produce corneal incisions and surface ablations with ablated surfaces which are superior to those attainable by use of other laser wavelengths. However, since 193 nm UV-C photons are sufficiently energetic (6.4 eV) to break most biologically significant chemical bonds, concern has centered on questions of mutagenicity of the remaining unablated corneal tissue. Furthermore, fluorescence produced by laser photoablation contains UV-C and UV-B photons which could penetrate to other ocular structures and pose potential hazards, e.g., to the lens and retina.

To determine the potential for delayed effects from UV fluorescence and direct collateral tissue damage from the 193 nm wavelength, photokeratitis thresholds have been studied by the author. The threshold for photokeratitis at 193 nm was obtained for the rabbit cornea using an ArF excimer laser. The value was 1-2 J/cm^2 for superficial epithelial cell damage and approximately 15 J/cm^2 for full epithelial keratitis involving Bowman's layer (judged by fluorescein staining). Because ablation occurs at a level below that for photokeratitis, it was necessary to expose the cornea to a lengthy series of low-energy exposures. It was concluded that classical photokeratitis occurs from the fluorescence emitted at the corneal epithelial absorption site. Indeed, an intact tear film helps to protect the cornea from low-level, scattered 193 nm laser radiation. The most superficial epithelial haze created at the lower threshold was probably due to photodecomposition of the cell membranes facing the environment.

V. RETINAL PHOTOCOAGULATION

Photomaculopathy from blue laser light has been shown[47] to be a photochemical injury mechanism with an action spectrum which peaks at a wavelength of approximately 440 nm for the phakic eye. In theory, scattered light from photocoagulation could produce photic maculopathy from repeated paramacular exposures of the argon laser (488 nm in particular). In recent years there has been an effort to eliminate the blue argon laser line from commercial laser photocoagulators, partially from this concern. Although only a small fraction of the laser beam energy is scattered in each exposure, a panretinal grid of 500 to 1000 exposures could produce a cumulative dose to the macula that would exceed the 20-30 J/cm^2 threshold for photochemical injury. The macular injury might go unnoticed by the clinician, since it will not appear until one or two days after the exposure session. Figure 4 shows how the small fraction of scattered light illuminates retinal areas of the bean spot. Note that only a small fraction of scattered light illuminates retinal areas outside of the spot, but cumulatively from many exposure, it may pose a significant hazard to the macula.

To calculate the cumulative exposure to the macula from scattered light requires a knowledge of the angular scatter function. This has been measured both by psychophysical and by optical methods for the normal eye[4]. However, patients undergoing panretinal laser photoablation (photocoagulation) frequently have compromised ocular media and scattering is significantly greater than in a normal human or primate eye for which we have the scatter function. Even if we use the typical scatter function for a normal eye, it has been calculated that 500 exposures at 488 nm should produce a threshold dose for photomaculopathy.

VI. CONCLUSIONS

Molecular damage from ultraviolet exposure occurs constantly and repair mechanisms exist to deal with the exposure of skin and ocular tissues

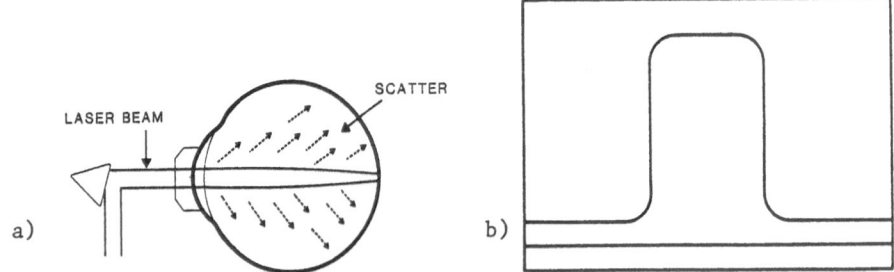

Fig. 4. a) Intraocular scatter and, b) Retinal light profile.

to ultraviolet radiation. Only when these repair mechanisms are overwhelmed, does acute biological injury become apparent[9]. For these reasons, any clinical application of ultraviolet laser in ophthalmology must carefully examine the potential for injury to adjacent and remote sites resulting from either the scattered primary beam energy or tissue fluorescence. Most materials have a characteristic fluorescence when exposed to UV and emit longer wavelength photons, whose damage potential must be evaluated. UVR laser photoablation at 193 nm appears to be safe when taken in this context; however, the risks of using 308 nm laser radiation are brought into serious question in any ocular tissue.

We are left with the clear need to explore the use of lasers with wavelengths other than 308 nm for ocular surgical procedures. For phakoablation, the IR-B wavelength band is far more promising than UV-C or UV-B. Since the optimum infrared laser wavelength cannot be conclusively determined based purely on the peak absorption of water, it is recommended that a tunable infrared laser (e.g., magnesium-calcium-fluoride) be employed to scan over the 2-3 μm spectral region to determine the clinically most suitable laser ablation wavelength. One does not need to face the reasonable questions of benefit vs. risk with UV-B exposures unless the 308 nm laser wavelength were shown to be far more effective than an IR-B laser wavelength.

REFERENCES

1. F. Urbach (Ed), "The Biologic Effects of Ultraviolet Radiation", Pergamon Press, New York (1969)
2. J. A. Parrish, R. R. Anderson, F. Urbach and D. Pitts, "UV-A, Biological Effects of Ultraviolet Radiation with Emphasis on Human Responses to Longwave Radiation", Plenum Press, New York (1978)
3. W. F. Passchier and B. F. M. Bosnjakovic (Eds.), "Human Exposure to Ultraviolet Radiation: Risks and Regulations", New York, Excerpts Medica Division, Elsevier Science Publishers (1987)
4. D. H. Sliney and M. L. Wolbarsht, "Safety with Lasers and Other Optical Sources", Plenum Publishing Corp., New York (1980)
5. F. Urbach and R. W. Gange (Eds.), "The Biological Effects of UV-A Radiation", Praeger Publishers, Westport, Connecticut, USA (1986)
6. J. J. Harding and K. J. Dilley, "Structural proteins of the mammalian lens: a review with emphasis on changes in development, aging and cataract", Exp. Eye Res., 22(1):1-73 (1976)
7. L. I. Grossweiner, "Photochemistry of proteins: a review", Curr. Eye Res., 3(1):137-144 (1984)
8. F. Hillenkamp, "Laser interactions with biological tissues", in F. Hillenkamp, C. A. Sacchi and F. T. Arecchi, Eds., Lasers in Biology and Medicine, Plenum Press, New York (1980)
9. K. C. Smith, "The Science of Photobiology", Plenum Press, New York (1988)
10. I. Willis, A. Kligman and J. Epstein, "Effects of long ultraviolet rays on human skin: photoprotective or photoaugmentative", J. Invest. Dermatol., 59:416-420 (1972)
11. W. H. Tung, K. T. Chylasck and U. P. Andley, "Lens hexokinase deactivation by near-uv radiation", Curr. Eye Res., 7(3):257-263 (1988)
12. J. Marshall, S. Trokel, S. Rothery and H. Schubert, "An ultrastructural study of corneal incisions induced by an excimer laser at 193 nm", Ophthalmology, 92:749-758 (1985)
13. C. A. Puliafito, K. Wong and R. F. Steinert, "Quantitative and ultrastructural studies of excimer laser ablation of the cornea at 193 and 248 nm", Laser Surg. Med., 1:155-159 (1987)
14. C. A. Cole, D. F. Forbes and P. D. Davies, "An action spectrum for UV photocarcinogenesis", Photochem. Photobiol., 43(3):275-284 (1986)

15. H. J. C. M. Sterenborg and J. C. van der Leun, "Action spectra for tumorigenesis by ultraviolet radiation" in: W. F. Passchier and B. F. M. Bosnjakovic (Eds), "Human Exposure to Ultraviolet Radiation: Risks and Regulations", pp. 173-191, New York, Excerpta Medica Division, Elsevier Science Publishers (1987)

16. K. W. Hausser, "Influence of wavelength in radiation biology", Strahlentherapie 28:25-44 (1928)

17. D. G. Pitts, A. P. Cullen and P. D. Hacker, "Ultraviolet Effects from 295 to 400 nm in the Rabbit Eye", Contract CDC-99-74-12, Nat. Inst. for Occup. Safety and Health, DHEW Pub. No. (NIOSH) 77-175, Cincinnati, Ohio (October 1977); see also: "Ocular effects of ultraviolet radiation from 295 to 365 nm", Invest. Ophthal. Vis. Sci., 16(10): 932-939 (1977)

18. M. L. Luckiesh, L. Holladay and A. H. Taylor, "Reaction of untanned human skin to ultraviolet radiation", J. Opt. Soc. Am., 20:423-432 (1930)

19. W. R. Coblentz, R. Stair and J. M. Hogue, "The spectral erythemic relation of the skin to ultraviolet radiation", Proc. Nat. Acad. Sci. US, 17:401-403 (1931)

20. A. F. McKinlay and B. L. Diffey, "A reference action spectrum for ultraviolet induced erythema in human skin", in: W. F. Passchier and B. F. M. Bosnjakovic (Eds), "Human Exposure to Ultraviolet Radiation: Risks and Regulations", pp. 83-87, New York, Excerpta Medica Division, Elsevier Science Publishers (1987)

21. J. A. Parrish, K. F. Jaenike and R. R. Anderson, "Erythema and melanogenesis action spectra of normal human skin", Photochem. Photobiol. 36(2):187-191 (1982)

22. M. A. Everett, R. L. Olsen and R. M. Sayer, "Ultraviolet erythema", Arch. Dermatol. 92:713-719 (1965)

23. R. S. Freeman, D. W. Owens, J. M. Knox and H. T. Hudson, "Relative energy requirements for an erythemal response of skin to monochromatic wavelengths of ultraviolet present in the solar spectrum", J. Invest. Dermatol., 47:586-592 (1966)

24. D. Berger, F. Urbach and R. E. Davies, "The action spectrum of erythema induced by ultraviolet radiation", Preliminary Report XIII Congressus Internationalis Dermatologiae (München 1967), pp.1112-1117, W. Jadassohn and C. G. Schirren, Eds., Springer-Verlag, New York (1968)

25. T. B. Fitzpatrick, M. A. Pathak, L. C. Harber, M. Seiji and A. Kukita (Eds), "Sunlight and Man, Normal and Abnormal Photobiologic Responses", University of Tokyo Press, Tokyo, Japan (1974)

26. P. D. Forbes and P. D. Davies, "Factors that Influence Photocarcinogenesis", in: J. A. Parrish, M. L. Kripke and W. L. Morison, Eds., Chapter 7, "Photoimmunology", Plenum Publishing Corp., New York (1982)

27. World Health Organization (WHO), Environmental Health Criteria No.14, Ultraviolet Radiation, jont publication of the United Nations Environmental Program, the International Radiation Protection Association and the World Health Organization, Geneva (1979)

28. J. Scotto, T. R. Fears and G. B. Gori, "Measurements of ultraviolet radiations in the United States and comparisons with skin cancer data", US Department of Health, Education and Welfare Publication No. (NIH)80-2154, Government Printing Office, Washington (1980)

29. World Health Organization (WHO), Environmental Health Criteria No.23, Lasers and Optical Radiation, joint publication of the United Nations Environmental Program, the International Radiation Protection Association and he World Health organization, Geneva (1982)

30. D. H. Sliney, "Physical factors in cataractogenesis: ambient ultraviolet radiation and temperature", Invest. Ophthalmol. Vis. Sci., 27(5): 781-790 (1986)

31. D. H. Sliney, "Estimating the solar ultraviolet radiation exposure to an intraocular lens implant", J. Cataract. Refract. Surg., 13(5): 296-301 (1987)

32. D. G. Pitts, "The human ultraviolet action spectrum", Am. J. Optom. Physiol. Optics, 51(12):946-960 (1974)

33. D. G. Pitts and T. J. Tredici, "The effects of ultraviolet on the eye", Am. Ind. Hyg. Assoc. J., 32(4):235-246 (1971)

34. A. Ringvold, "Cornea and ultraviolet radiation", Acta Ophthalmol., 58:63-68 (1980)

35. A. Ringvold, "Aqueous humour and ultraviolet radiation", Acta Ophthalmol. 58:69-82 (1980)

36. A. Ringvold, M. Davanger and E. G. Olsen, "Changes of the cornea endothelium after ultraviolet radiation", Acta Ophthalmologica, 60:41-53 (1982)

37. A. Ringvold and M. Davanger, "Changes in the rabbit corneal stroma caused by UV-radiation", Acta Ophthalmologica, 63:601-606 (1985)

38. A. Ringvold, "Damage of the cornea epithelium caused by ultraviolet radiation", Acta Ophthalmologica, 61:898-907 (1983)

39. E. G. Olsen and A. Ringvold, "Human cornea endothelium and ultraviolet radiation", Acta Ophthalmologica, 60:54-56 (1982)

40. A. P. Cullen, B. R. Chou, M. G. Hall and S. E. Jany, "Ultraviolet-B damages corneal endothelium", Am. J. Optom. Physiol. Opt., 61(7):473478 (1984)

41. M. V. Riley, S. Susan, M. I. Peters and C. A. Schwartz, "The effects of UV-B irradiation on the corneal endothelium", Curr. Eye Res., 6(8):1021-1033 (1987)

42. J. A. Zuclich and J. S. Connolly, "Ocular damage induced by near-ultraviolet laser radiation", Invest. Ophthalm. 15(9):760-764 (1976)

43. S. Takise, S. Horiguchi, I. Karai, S. Matsumura, M. Harima, T. Miki, S. Yoshikawa and H. Yamashita, "Effects of ultraviolet laser beam irradiation on rabbit cornea and lens", Sangyo Igaku, 30(2):112-120 (1988)

44. G. A. Peyman, J. R. Kuszak and K. Weckstrom, "Effects of Xe-Cl excimer laser on the eyelid and anterior segment structures", Arch. Ophthalmol., 104:118-122 (1986)

45. H. R. Taylor, S. K. West, F. S. Rosenthal, B. Munoz, H. S. Newland, H. Abbey and E. A. Emmett, "Effect of ultraviolet radiation on cataract formation", New Engl. J. Med. 319:1429-1433 (1988)

46. D. H. Sliney, "Unintentional exposure to ultraviolet radiation: Risk reduction and exposure limits", in: "Human Exposure to Ultraviolet Radiation: Risks and Regulations" (Eds: W. F. Passchier and B. F. M. Bosnjakovic), New York, Excerpta Medica Divsion, Elsevier Science Publishers, pp. 425-437 (1987)

47. W. T. Ham, H. A. Mueller, J. J. Ruffolo, D. Guerry III and R. K. Guerry, "Action spectrum for retinal injury from near ultraviolet radiation in the aphakic monkey", Am. J. Ophthalmol., 93(3):299-306 (1982)

48. K. E. Kopecky, G. W. Pugh Jr, D. E. Hughes, G. D. Booth and N. F. Cheville, "Biological effect of ultraviolet radiation on cattle", Am. J. Vet. Res., 40(12):1783-1788 (1979)

49. M. F. Lavin, P. A. Jennings and D. J. Hughes, "Bovine ocular squamous cell carcinoma: UV sensitivity in lymphocytes", Photochem. Photobiol. 35(5):685-689 (1982)

50. S. Stenson, "Ocular findings in xeriderma pigmentosum: report of two cases", Ann. Ophthalmol., 14(6):580-585 (1982)

51. D. H. Sliney, "The merits of an envelope action spectrum for ultraviolet radiation exposure criteria", Am. Ind. Hyg. Assoc. J., 33:644-653 (1972)

52. International Radiation Protection Association (IRPA): "Proposed change to the IRPA 1985 guidelines limits of exposure to ultraviolet radiation", Health Physics, 56(6):971-972 (1989)

53. American Conference of Governmental Industrial Hygienists (ACGIH), Threshold Limit Values (TLV's) and Biological Exposure Indices for 1989-1990, ACGIH, Cincinnati, Ohio, USA (1989)

54. American National Standard Institute (ANSI), Safe Use of Lasers, Standard Z-136.1-1986, American National Standard Institute, New York, published by Laser Institute of America, Toledo (1986)

55. International Radiation Protection Association (IRPA), "Guidelines for limits of human exposure to laser radiation", Health Physics, 49(2):341359 (1985); and change: "Recommendations for minor updates to the IRPA 1985 guidelines on limits of exposure to laser radiation", Health Phys. 54(5):573-573 (1988)

56. N. Mueller-Stoltzenberg, G. Mueller and N. Stange, "Retinal UV exposition, during endocapsular phacoablation at 308 nm", Lasers and Light in Ophthalmology, 2(3): 197 (1989)

57. R. H. Keates, D. E. Genstler and S. Tarabichi, "Ultraviolet light transmission of the lens capsule", Ophthalmic Surg. 13(5):374-376 (1982)

58. M. A. Mainster, "Spectral transmission of intraocular lenses and retinal damage from intense light sources", Am. J. Ophthalmol. 85:167170 (1978)

59. L. A. Yanuzzi, Y. L. Fisher, A. Krueger and J. Slakter, "Solar retinopathy, a photobiological and geophysical analysis", Tr. Am. Ophthalmol. Soc., 85:120-158 (1987)

60. S. L. Trokel, R. Srinivasan and B. Braren, Excimer laser surgery of the corneas", Am. J. Ophthalmol., 96:710-715 (1983)

61. C. A. Puliafito, R. F. Steinert, T. F. Deutsch, F. Hillenkamp, E. J. Dehm and C. M. Adler, "Excimer laser ablation of the cornea and lens", Ophthalmology, 92(6): 741-748 (1985)

62. J. Marshall, S. L. Trokel, S. Rothery and R. R. Krueger, "Photoablative reprofiling of the cornea using an excimer laser: photorefractive keratectomy", Lasers in Ophthalmol., 1:21-48 (1986)

63. J. Marshall, S. L. Trokel, S. Rothery and R. R. Krueger, "A comparative study of corneal incisions induced by diamond and steel knives and two ultraviolet radiations from an excimer laser", Br. J. Ophthalmol. 70:482-501 (1986)

64. J. Marshall, S. L. Trokel, S. Rothery and R. R. Krueger, "Long term healing of the central cornea after photorefractive keratectomy using an excimer laser", Ophthalmology, 95:1411-1421 (1988)

SURGICAL ENDOSCOPY UNITS: ROOM INSTALLATION

F. Longhi

Unità Sanitaria Locale N. 29
Servizio Attività Tecniche
Ospedale di Bellaria, Bologna, Italy

I. CLINICAL APPLICATION OF LASER ENDOSCOPY

The increasing demand for endoscopic treatment by laser techniques accompanied with other techniques as endo/electrosurgery in the same room or alternatively during the same operation, is a challenge in organizing a single surgery room. Furthermore, in our case, at Bellaria Hospital of Bologna, we have to face the necessity of sharing laser service between the two Divisions of Bronchology and Gastroenterology.

The clinical application of lasers in endoscopy requires an appropriate environment. An example of an operating unit arrangement is shown in Figure 1, with reference to some international safety standards. The existing standards and general rules for safety in our country do not require particular application in the present case. The implementation of a few general rules as in the case shown, could be a step towards issuing future standards.

II. SAFETY STANDARDS

During an initial period of operations performed in a traditional operating room which was "borrowed" while waiting for the new one, we observed the following issues to be faced:

Microclimate: The room was not designed for the use of anaesthetic gases, but the fumes produced by burning tissues were sufficiently objectionable to require localized aspiration. As many as ten persons might be present in the operating room in some cases.

Electrical safety: Invasive techniques require both a network fed through a continuously monitored isolation transformer and an equipotential wiring system for grounding. Some critical steps during intervention require uninterrupted power and light. The electronic monitoring systems require a well stabilized power source.

Protection against laser radiation: Although the laser source, classified as Class 4 according to ANSI standards, was equipped with the necessary protective devices for the machine and the personnel (such as microswitches, interlocks, and glasses for operators), the environment had to be adapted to that purpose.

Regulations in force in our country could easily be applied to cover the first two issues described above. For the third one, the environment, we had to refer to other standards as ANSI (American National Standards Institute), ACGIH (American Conference of Governmental Industrial Hygienists) and LIA (Laser Institute of America), and we were able to secure cross-reference and confirmation from studies published in Italy by CNR (Consiglio Nazionale delle Ricerche) and I.S.S. (Istituto Superiore di Sanità). Other authorities concerned with this matter are FDA (Food and Drug Administration), BRH (Bureau of Radiological Health), NCDRH (National Center for Devices and Radiological Health) and OSHA (Occupational Safety and Health Administration).

It is not our intention to reiterate the possible hazards for skin and eyes or to quote the TLV (threeshold limit values), which can be easily obtained from the literature. We want to show only the results of the application of standards and recommendations as implemented in our installation. We merely note that Class IV is the most high and dangerous level actually classified for safety purposes, and that our laser is a Nd:YAG type with 150 Watt output power.

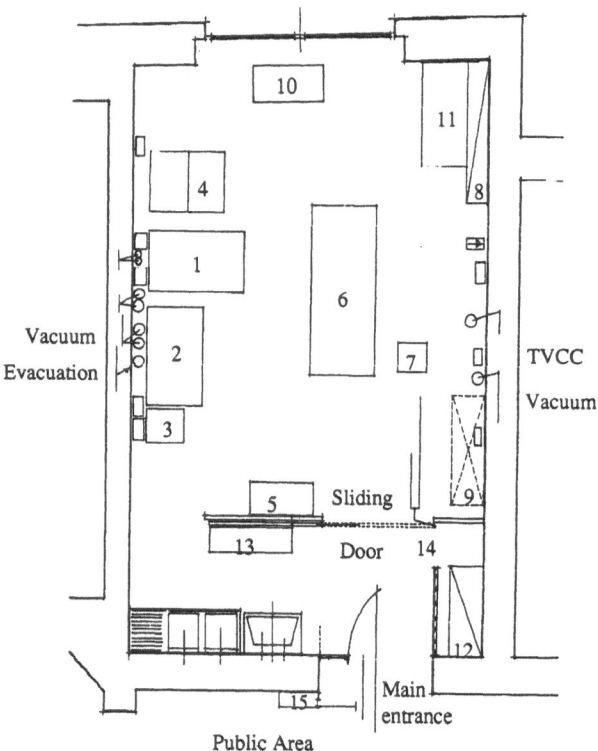

Fig. 1. Plan of the operating room at Bellaria Hospital of Bologna, Pad. "D" (Floor level), Laser Department: 1) Laser Nd YAG; 2) EVE light source and video processor; 3) Anesthesia instruments; 4) Laser; 5) Electrosurgery Unit; 6) Operating table; 7) Monitor; 8) Film negative scope; 9) Defibrillator Unit; 10) Cabinet; 11) Desk; 12) Switch board & isolation transformer; 13) Sterilization.

III. OPERATING ROOM

With reference to Figure 1 showing the plan of our operating room, we can note the following:

- The small area available had to be organised as a single, independent unit, without any accessory room.
- Since access is directly from a corridor open to the public, we have created an internal barrier with a wall and a sliding door which is kept closed during the intervention: a small passage without a door on the left is for service purposes only and creates a labyrinth for the laser beam.
- In the area between the entrance and the wall we have located the washing and sterilizing facilities. In this way, exit is always possible from the restricted area while we keep the sliding door closed and controlled with two devices. The first is a microswitch interlocked with the laser firing circuit: the laser can be in stand-by but cannot be operated without the microswitch signal which is given only when the sliding door is closed. The second device is an electromagnetic lock which is activated to keep the sliding door closed when the microswitch is already closed and the laser fires. In this way firing is possible only if the sliding door is closed, and no interruption can be made when surgeon fires, because an accidental opening of the sliding door is inhibited by the second device. In addition, we also have flashing warning lights in the corridor, with ANSI warning signals.

Hazards due to scattering and reflection are reduced by employing rough/opaque surfaces on walls, furniture and wherever possible even on the plumbing and gas connections, to avoid reflections in case of rupture of the fiber or accident during servicing. The window has been screened by a vertically foldable curtain of aluminium strips, covered with an opaque paint which is non-reflecting and will not produce fire or smoke. We use a well lighted room to avoid excessive dilatation of the pupil, thereby making the eyes less vulnerable. Finally, two large red emergency switches on a yellow background can be pulsed to switch off the power source in case of emergency.

The room organisation is determined by the minimal space available: since the bed is on wheels and interchangeable, we left room enough for entrance and for other wheel-mounted instruments. The main supplies of electricity, water for laser cooling, oxygen, vacuum, fumes/exhausting are on the left side. Two separated and interlocked power supplies permit allocating two different lasers for separate kinds of therapies. All other outlets are fed by a separate network through an isolation transformer whose function is double: i) to separate the local network from the main one, keeping the dispersion currents under the safety TLV in case a failure occurs in some electrical device, and ii) although failure is possible, to continue the work without danger because no switch-off is necessary. An audible and visual alarm warns about the occurring failure, by a system which monitors the isolation status of the network.

The laser power supplies are protected by differential switches because their high power is not compatible with the isolation transformer. Furthermore in case of lack of power by the distribution network, we can provide power from an autonomous source within 10 secs.

To protect the safety system against microshock, hazards, all the metal structures and the grounding outlets have been connected to a so-called junction "equipotential" and then, from here, to the grounding system according to our regulations in force CEI 64-4. For lighting we have provided the room with supplementary battery-operated neon lamps, which switch on in a few msec in case of failure of the main line, in order to guarantee the illumination necessary during the intervention.

Finally, in order to satisfy the increasing demand for audio-visual communication, including live broadcasts, we provided a large number of electrical outlets all around the walls, all transformer-protected, to avoid interference from non-medical apparatus fed by other power lines. Particular precautions were taken in the layout of the cable and wiring system using wall brackets and supports to eliminate the floor distribution of any kind of wire or tube.

CONSIDERATIONS IN THE DESIGN OF MEDICAL

LASER SYSTEMS AND INSTRUMENTATION

R. A. Kirschner

The Institute for Applied Laser Surgery, Inc.
Suburban General Hospital
Bala Cynwyd, Pennsylvania, U.S.A.

I. MEDICAL LASER SYSTEMS

In reviewing considerations of medical laser systems, it becomes important to examine the factors that exist in high-tech industries. High technology has grown in an exponential rate for the past three and a half decades. In Photonics we have seen areas of technology spawn three generations within the past decade. The Medical Laser Industry has been particularly volatile and difficult to predict.

The cyclic nature of high technology must be understood in order to learn how best to deal with it. Over the past thirty five years Americans have excelled in the early phases of technologic development and have faltered in reaping many of the economic benefits that should have been forthcoming. The same scenario is true for many NATO countries.

Technology may be seen to develop in stages:

i. Theory. In this stage a theory is developed that is not proven. It may be partially or completely correct or completely erroneous. The length of this stage is very variable and is dependent upon many factors.

ii. Discovery. This phase is marked by the realization that parts or all of the theory are correct. Support for research will arise from a variety of sectors of the economy: a) Military; b) Industry; c) Foundations; and, d) Government Agencies.

iii. Germination. This is a period in which portions of the theory are being tested and possible applications are proposed. The more useful and profitable the possible applications are, the greater the funding will be and the shorter thus period will last.

iv. Application and Proliferation. During this period applications are developed and commercialization follows. This period is punctuated by successive generations of improved technology and growing applications. This represents industrial growth with the development of manufacturing, marketing and sales.

v. Saturation. During the period of application, the outlets for this technology have become progressively saturated. At the saturation point, sales are past their peak and have turned down to the point where they represent replacement purchases only. As newer and possibly better technology becomes available, even these sales will decrease.

High technology is expensive, almost by definition. Medical Hi-Tech is even more so. Many factors come into play that are not germaine to other industries. Technology has to be refined and rendered fool-proof as possible before it can be used in the field. Equipment must undergo extensive laboratory testing and clinical evaluation before it can pass regulatory muster.

The laser has truly altered many aspects of patient care today. Less than a decade ago, shepticism for this technology was rampant. This predicament is shared by almost all new technology. In spite of this, the laser has filled an important niche in the surgical armamentarium. The companies that have provided this technology, however, have not done well from a business stand-point. There are certain factors that are necessary to improve the financial posture of this industry.

If a medical laser company is to succeed with a cycle such as the above, it must have a technology with an extremely long phase of "Application and Proliferation" or at least several products at various stages of this cycle. Due to the small size of most laser companies, it has been difficult for most of them to devote adequate funds for R&D. With the rapid change of technology, a company can be rendered impotent in rather short order.

It is difficult with any new modality to predict the overall size of the market that will develop. Additionally, with the acceptance of most new medical laser modalities we have seen the emergence of competition. Any new technology with high market potential will stimulate interest by a number of potential competitors. This makes "share of market" predictions quite difficult. It becomes readily apparent that new products must have a certain "uniqueness" to make success possible and that a very ordered strategy must be followed to ensure that success.

The ingredients to succeed in the medical laser industry for the nineties are summarized in the following five points:

i. continued flow of new technology, a) that offers new surgical capabilities, b) that decreases overall treatment cost, c) that is less expensive, d) that is less invasive and safer, e) that reduces treatment time, f) that offers a reasonable lead time over competitive technology, and, g) that has adequate market potential.

ii. Adequate financial resources, a) to minimize expensive financing, b) to permit ongoing R&D, c) to adequately support facilities providing work on clinical protocols, d) to minimize product development time, e) to provide the opportunity to proceed without compromise in crucial areas, and, f) to pursue development of concurrent technologies.

iii. Development of scientifically designed strategic marketing to match the sophistication of the technology involved.

iv. To work with regulatory agencies rather than challenging them.

v. Vision. Einstein has stated that "Imagination is more important than knowledge." His legacy included both the theory this technology is based upon and this philosophy without which there would be no "Hi-Tech".

II. LASER INSTRUMENTATION

The medical laser has brought many new capabilities to the surgeon. At the present time, almost all surgical specialties have adopted the laser for their use. As each new application developed, so has the need for specific peripheral instrumentation. This instrumentation has developed in response to the need:

i) To provide safe operation in the O.R.: a) Safety goggles (- specific wavelength, - multi-wavelength); b) Flame retardent drapes; c) Safety lenses for endoscopes; d) Laser endotracheal tubes.

ii) To extend the capabilities of lasers: a) Contact tips; b) Laser endoscopes; c) Micromanipulators.

iii) In order to get both i) and ii): a) Non-reflective instrumentation.

The development of non-reflective instrumentation was a necessity from the standpoints of safety and extension of capability. The dangers of a reflected beam are sometimes even more dangerous than a direct beam. This is because the surgeon and the operating room (OR) personnel usually pay very close attention to the aiming beam.

Actually the term non-reflective is a relative term. A more exact terminology would be relatively non-reflective. Many attempts at non reflectivity have been made, including these:

i. Sand blasting. Sand blasting breaks a smooth surface into many smaller ones. These irregularities tend to disperse coherent light into non-coherant light. The sand blasting process that was utilized on our original carbon dioxide laser instrumentation utilized two sizes of glass beads.

ii. Ebonizing. Ebonizing creates a surface that increases light absorption and decreases reflection. The biggest problem with this surface treatment is that many ebonizing treatments deteriorate with routine use in OR. If the surface treatment deteriorates, areas that will reflect light strongly become exposed. The ebonizing process that was originally utilized on our instruments was applied in an electrolytic process after a copper coating was applied to the instrument.

iii. The combination of sand blasting and ebonizing. This maximized the attributes of i) and ii).

iv. Gold. Aesculap combines a rough surface with a gold color. This process maximized reflection of the coherent light as non-coherent light.

There are problems attendant with each of these surface treatments. The sand-blasting process creates a rough instrument that is sometimes rough on tissue. A secondary solution to this problem is to have the roughened surface limited to the areas that require decreased reflectivity and have smooth surfaces where tissue contact occurs. Ebonizing also creates problems. When the metal is ebonizing it reflects less and absorbs more. This absorption is converted to heat. This can be injurious to tissue by itself. If the instruments are small they can actually glow as the laser energy is absorbed. A rough gold finish is excellent except that the high level of reflected non-coherent light can obscure the surgeon's visualization.

III. COMPUTER ASSISTED MICROTELESCOPIC SURGERY (CAMS)

CAMS is the result of a variety of Hi-Tech developments. About four years ago J. Richards, F. Brown and myself introduced a prototype telescope for microtelescopic surgery. This instrument had the following features: a) Low cost, b) Semi-disposable, and, c) Brighter image. Further work on this technology has added the ability to bring the illuminating light through the same channel as the viewing optics.

This technology will be used with the following laser modalities: A. The Luxar CO_2 waveguide; and, B. The KDG-Rotem CW Holmium laser.

The Luxar waveguide was introduced to the market during the first week of April, 1990. The CW holmium laser will be undergoing laboratory testing and modifications for medicine at the Philadelphia College of Osteopathic Medicine in June, 1990. The waveguide will transmit carbon dioxide laser radiation through a semi-flexible conduit. This obviates the need for articulated arms for many applications and allows the design of many new peripheral instruments.

The CW holmium laser cuts in a manner that is quite similar to that of CO_2. It also will coagulate almost as well as Nd:YAG. This laser light will also be able to be transmitted through quartz fibreoptics. This laser has the potential to deliver the theoretically ideal wavelength.

One difficulty with some types of surgery that are performed with the telescope is the ability to determine accurate positioning. This is a problem even with the superb optics of the instrumentation. ISG Technology, Toronto, Canada has developed a superb system for reconstructing information obtained from both MRIs and CAT scans. These reconstructions will be available on an operating room workstation. The tip of the microtelescope will now act as a digitizer. This will permit the surgeon to accurately determine the position of the scope. This is an example of cutting edge technologies being married in design for tomorrow's state of the art medical application. There is a large potential market for this technology however, all of the factors discussed previously will apply to each of the components.

RADIATION SAFETY OF LASER PRODUCTS:

EUROPEAN AND ITALIAN REQUIREMENTS

R. Barbini

ENEA - Dipartimento Sviluppo Tecnologie di Punta
Frascati, Rome, Italy

I. INTRODUCTION

This chapter describes the requirements on laser safety which have been developed by the CEI-CT76 committee, in accordance with the corresponding specifications and coordination documents issued by International Institutions such as IEC and CENELEC. In the framework of the CT-76 committee, five User Guides have been prepared, according to specific laser applications:

Guide-A : material working
Guide-B : industrial, civilian and environmental uses
Guide-C : telecommunications and fiberoptics
Guide-D : medical applications
Guide-E : scientific research laboratories.

In particular, I will stress the contents of the last Guide, which is addressed to personnel who have a high degree of expertise in the laser field, but who often face situations which require a special knowledge of the hazards which can be encountered in laboratory practice. The Guide recommends that a Laser Safety Officer be appointed for properly advising personnel in non standard situations, and specific training courses on laser safety. Finally, procedures for hazard control according to the AEL (Accessible Emission Limit) classification scheme for laser products will be reviewed.

Some definitions used throughout this chapter are listed below:
<u>Accessible Emission Limit</u> (AEL). The maximum accessible emission level permitted within a particular class.
<u>Maximum Permissible Exposure</u> (MPE). The maximum level of laser radiation to which, under normal circumstances, persons may be exposed without suffering adverse effects at eyes or at skin, immediately or after a long time. The MPE levels are related to the wavelength of radiation, the pulse duration or exposure time, the tissue at risk annd, for visible and near infrared radiation in the range 400 to 1400 nm, the size of the retinal image.
<u>Irradiance</u>. At a point on a surface, the radiant flux incident on an element of a surface containing the point divided by the area of that element. Units: Watts/m^2.
<u>Radiant exposure</u>. At a point on a surface, the radiant energy incident on

an element of a surface divided by the area of that element. Units: Joules/m^2.

II. USER GUIDE-E

On the basis of existing regulations[1,3,4], the CEI (Comitato Elettrotecnico Italiano) CT-76 committee issued the Italian Standard on laser safety in December 1989, accompanied by five specialized user's guides; among them, the User Guide-E, in particular, is specifically devoted to: i) safety precautions to be taken, ii) hazards incidental to laser operation which can occur, and, iii) procedures for hazard control to be adopted in scientific laboratories where basic and applied research on laser sources and systems is carried out, with the aim of reducing the possibility of exposure to hazardous levels of laser radiation, and other associated hazards. Whenever the application of any one or more control measures reduces the possible exposure to a level at or below the applicable Maximum Permissible Exposure (MPE) (see Figure 1), then the additional control measures should not be necessary.

According both to international specifications and to a public enquiry carried out among Italian laser specialists, the CT-76 committee of CEI has issued the following recommendations[2].

In those laboratories where lasers of Class greater than Class 3A are operated a Laser Safety Officer (LSO) should be appointed with the responsibility to review the laser safety precautions and to designate the appropriate controls to be implemented.

The complementary responsibilities of personnel involved in basic and applied research upon lasers and laser systems can be summarized as follows: a) the LSO advises upon safety measures and appropriate controls to be taken; b) the laboratory Director is responsible for the acquisition and implementation of the required protective measures; and, c) the Qualified Operators actually involved in the research are personally responsible of using the laser-system in accordance with the relevant safety precautions.

III. THE CLASSIFICATION SCHEME

Because of the wide range possible for the wavelength, energy content and pulse characteristics of a laser beam, the hazards arising in laser use vary widely. It is impossible to regard lasers as a single group to which common safety limits can apply. Laser products are therefore grouped into four general classes for each of which Accessible Emission Limits (AEL's) are specified, as shown in Figure 2.

Various situations can arise, according to the specific research activity carried out upon or with lasers and laser systems. When using commercial laser systems, for instance, users can take advantage of the manufacturer's classification to determine the laser system class, thus avoiding all measurements. When building lasers in the laboratory, the classification must be determined by the person or organization responsible for the apparatus.

Finally, when modifications are performed upon commercial systems which affect any aspects of the product's performance or intended functions within the scope of this standard (Guide E), the person or organization performing any such modification is responsible for ensuring the reclassification or relabelling of the laser product.

Exposure time t (s) → Wavelength λ (nm) ↓	$<10^{-9}$	10^{-9} to 10^{-7}	10^{-7} to 1.8×10^{-5}	1.8×10^{-5} to 5×10^{-5}	5×10^{-5} to 10	10 to 10^3	10^3 to 10^4	10^4 to 3×10^4
200 to 302.5	3×10^{10} W.m^{-2}	30 J.m^{-2}						
302.5 to 315	3×10^{10} W.m^{-2}	C_1 J.m^{-2} ($t<T_1$)			C_2 J.m^{-2} ($t>T_1$)		C_2 J.m^{-2}	
315 to 400	5×10^{6} W.m^{-2}	C_1 J.m^{-2}				10^{4} J.m^{-2}	C_2 J.m^{-2}	
400 to 550	5×10^{6} W.m^{-2}	5×10^{-3} J.m^{-2}				100 J.m^{-2}	10 W.m^{-2}	10^{-2} W.m^{-2}
550 to 700	5×10^{6} W.m^{-2}	5×10^{-3} J.m^{-2}		$18\,t^{0.75}$ J.m^{-2}		100 J.m^{-2} ($t<T_2$)	$t>T_2$ $C_3\times10^{2}$ J.m^{-2}	$C_3\times10^{-2}$ W.m^{-2}
700 to 1050	$5\times C_4\times10^{6}$ W.m^{-2}	$5\times10^{-3}\times C_4$ J.m^{-2}		$18\times C_4\,t^{0.75}$ J.m^{-2}			$3.2\times C_4$ W.m^{-2}	
1050 to 1400	5×10^{7} W.m^{-2}	5×10^{-2} J.m^{-2}		$90\times t^{0.75}$ J.m^{-2}			16 W.m^{-2}	
1400 to 10^6	10^{11} W.m^{-2}	100 J.m^{-2}	$5600\times t^{0.25}$ J.m^{-2}				1000 W.m^{-2}	

Fig. 1. Maximum permissible exposure (MPE) at the cornea for direct ocular exposure to laser radiation (intrabeam viewing). The C's and T's parameters are defined in the various spectral regions as:

$C_1 = 5.6 \times 10^3\, t^{0.25}$ (from 302.5 to 400 nm); $T_1 = 10^{0.8(\lambda-295)}\times10^{-15}$ S (from 302.5 to 315 nm);

$C_2 = 10^{0.2}\,(\lambda-295)$ (from 302.5 to 315 nm); $T_2 = 10\,(\lambda-550)$ S (from 550 to 700 nm);

$C_3 = 10^{0.015}\,(\lambda-550)$ (from 550 to 700 nm); $C_4 = 10^{(\lambda-700)/500}$ (from 700 to 1050 nm).

Diameter of limiting apertures shall be for: 1 mm, 200<λ<400 nm; 7 mm, 400<λ<1400; 1 mm, 1400<λ<10^5 nm; 11 mm, 10^5<λ<10^6 nm.

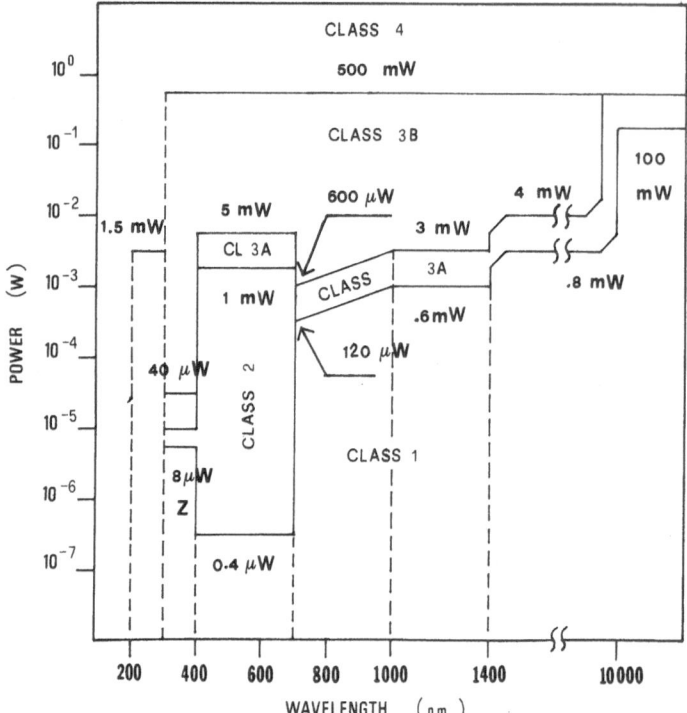

Fig. 2. Accessible emission limits for Classes 1, 2, 3A, 3B, 4 versus wavelength for CW lasers [1÷3].

It is a duty of the laboratory Administration, either to adequately train internal personnel to be employed as a LSO, or to appoint external consultants as LSO.

IV.   SAFETY PROTECTIONS APPLIED TO THE LASER SYSTEMS

For Class 3B and Class 4 lasers, the following safety protections should be applied to the laser source itself and at the laser location:

a) Warning signs should be attached at entrances to areas containing such laser products.

b) Protective enclosures should be used wherever practicable. In laser research laboratories however, it may be necessary to operate laser systems without protetive enclosures. In such case, the LSO will determine the hazards and designate the appropriate controls to be implemented (upon access, by area, physical barriers, attenuators, eye protection, training) to ensure the safe operation of the laser system.

c) Warning labels should be placed upon removable parts of protective enclosures or at service connections where a hazard is introduced by their removal or by disconnections.

d) Remote interlock connector should be connected to an emergency master disconnect interlock located near (< 5 m) the experimental area, or to room, door or fixture interlocks. The person in charge may be permitted momentary override of the remote interlock connector to allow access to other authorized persons if it is clearly evident that there is no optical radiation hazard at the time and point of entry.

e) Key control. When it is not in use, each laser product should be protected against unauthorized use by removal of the control key.

f) Beam stop to be attached permanently to the laser product, in order to prevent output emission in excess of the appropriate MPE level when the laser product is on stand-by.

g) Beam path terminations and enclosures. The emitted beam should be terminated at the end of its useful path by a diffusely-reflecting material of appropriate reflectivity and thermal properties or by absorbers. Open laser beam paths should be located above or below eye level where practicable. Laser beams should be enclosed (e.g. within a tube) where practicable.

h) Specular reflections hindrance. The maximum attention should be exercised to prevent the unintentional specular reflection of radiation from the laser.

V. PROTECTION OF THE PERSONNEL

When personnel may be exposed to potentially hazardous laser radiation (Class 3B and Class 4 laser products), adequate personal protection should be provided.

The need to use personal protection against the hazardous effects of laser operation should be kept to a minimum by administrative controls, engineering design and by beam enclosure.

V.1. General procedures for hazard control

Three aspects of the use of lasers need to be considered in the evaluation of hazards and application of control measures: i) the capability of the laser or laser system to injure personnel (any possibility of human access to the main exit port or any subsidiary port has to be considered); ii) the environment in which the laser is used; and, iii) the level of training of the personnel who operate the laser or who may be exposed to its radiation.

In this regard, the classification scheme of lasers is a very useful tool for deciding the safety precautions to be applied to a particular laser installation, since it provides a practical method for evaluation and control of laser radiation hazards. For those situations not specifically covered by the Guide E, the Laser Safety Officer is responsible for providing informed judgements.

V.2. Eye protection

A protective eyewear designed to provide adequate protection against specific laser radiation should be used in all hazard areas where Class 3B or Class 4 lasers are in use. Exceptions to this are: i) when engineering and administrative controls are such as to eliminate potential exposures in excess of the applicable MPE; and, ii) when, due to the unusual operating requirements, the use of eye protection is not practicable. Such operating procedures should only be undertaken with the approval of the LSO.

In the choice of protective eyewear the following considerations are to be taken into account: a) wavelength(s) of operation; b) radiant exposure or irradiance; c) Maximum Permissible Exposure (MPE); d) eyewear optical density at laser output wavelength(s); e) visible light transmission requirements; f) radiant exposure or irradiance at which eyewear damage occurs; g) need for prescription glasses; h) comfort and adequate ventilation (to avoid fogging); i) degradation or modification of absorbing media even if temporary or transient; j) strength of materials (resistance to shock); and, k) peripheral vision requirements (wide field of view).

As far as the optical density D of laser protective eyewear is concerned, we recall here that it is normally highly wavelength dependent. Where the eyewear is required to cover a band of radiation, the minimum value of D measured within the band shall be quoted. The value of D required to give eye protection can be calculated from the formula:

$$D = \log_{10} (H_0 / MPE) \qquad (1)$$

where H_0 is the expected unprotected eye exposure level.

Finally, all laser protective eyewear shall be clearly labelled with information adequate to ensure the proper choice of eyewear with particular lasers.

Besides the eye protection, where personnel may be exposed to levels of radiation that exceed the MPE for the skin, suitable protective clothing should be provided. Class 4 lasers especially are a potential fire hazard and protective clothing worn should be made from a suitable flame and heat resisting material.

V.3. Personnel training

Operation of class 3B and class 4 laser systems can represent a hazard not only to the use but also to other people over a considerable distance. Because of this, only persons who have received training at an appropriate level should be placed in control of such systems. The technical personnel operating in laboratories for laser research and development must be provided with adequate information in order to minimize the professional hazard connected with this type of research. To this aim, appropriate laser safety training should be given to researchers, by organizing for them specific training courses to be run on a national basis by an approved organization.

Work is in progress in this connection, within the CEI CT-76 Committee, devoted to the preparation of a schedule of training courses, to be carried out with the collaboration of Italian Institutions such as Universities, Research Institutions and Ministries. The training should include, but is not limited to: a) familiarization with system operating procedures; b) the proper use of hazard control procedures, warning signs etc.; c) the need for personal protection; d) accident reporting procedures; and, e) bioeffects of the laser upon eye and skin.

V.4. Medical supervision

The following recommendations should be considered: a) pre-, interim and post employment, ophthalmic examinations should be carried out by a qualified specialist and should be confined to workers using Class 3B and Class 4 lasers; and, b) a medical examination by a qualified specialist should be carried out immediately after an apparent or suspected injurious ocular or dermatological exposure. Such an examination should be supplemented with a full biophysical investigation of the circumstances under which the accident occurred. Pre-, interim and post employment ophthalmic and dermatological examinations upon workers using Class 3B and Class 4 lasers have value for medical-legal reasons only and are not a necessary part of a safety programme.

VI. HAZARDS INCIDENTAL TO LASER OPERATION

Depending on the type of laser used, associated hazards involved in laser operations may include the following:

a) Gases from laser systems such as bromine, chlorine, fluorine, hydrogen cyanide, etc. flowing in gas laser systems or arising from the by-products of laser reactions.

b) Gases or vapours from cryogenic coolants

c) UV collateral radiation. There may be a considerable hazard from UV radiation associated with flash-lamps or CW laser discharge tubes, especially when UV transmitting tubing or mirrors (such as quartz) are used.

d) VIS-IR radiation. The visible and near infrared radiation emitted from flashtubes and pump sources and target re-radiation may be of sufficient radiance to produce a potential hazard.

e) Electrical hazards. Most lasers make use of high voltages (> 1 kV). Pulsed lasers are especially dangerous because of the stored energy in the capacitor banks. Unless properly shielded, circuit components such as electronic tubes working at anode voltages greater than 5 kV may emit X-rays.

f) Fire and combustion due to beam interaction with flammable materials. Burning particles may be emitted during laser machining operations.

g) Other hazards. Explosions of the capacitor banks or optical pumping systems are possible during the operation of high power lasers. Explosive reactions of chemical laser reagents or other gases used within the laboratory are also possible.

VII. LABORATORY LASER INSTALLATIONS

The following specifications are devoted to those laser research activities and installations located indoors:

i) Class 1 laser products. Use without protective measures.
ii) Class 2 laser products: a) Prevent continuous viewing of the direct beam. A momentary (0.25 s) exposure as would occur in accidental viewing situations is not considered hazardous, because of the eye blinking reflex. b) The laser beam should not be intentionally aimed at people.
iii) Class 3A laser products. The use of optical viewing aids (e.g. binoculars) with these laser products may be hazardous and should not be authorized without the agreement of the LSO: a) A laser warning sign should be posted in areas where such lasers are used. b) Wherever practicable, mechanical or electronic means should be used to assist in the alignment of the laser. c) The laser beam should be terminated at the end of its useful path (outside the working volume): in all cases if the NOHD (Nominal Ocular Hazard Distance) extends beyond the controlled area. d) The laser beam path should be located well above or below eye level, wherever practicable. e) Avoid unintentionally directing the laser beam at mirror-like (specular) surfaces (most importantly, at flat mirror-like surfaces). f) When not in use, the laser should be stored in a location where unauthorized cannot gain access.
iv) Class 3B laser products are potentially hazardous if a direct beam or specular reflection is viewed by the unprotected eye (intrabeam viewing). Accordingly, the following precautions should be taken in addition to those for Class 3A lasers: a) The laser should only be operated in a controlled

area; b) Carefully avoid unintentional specular reflections; c) Terminate, where possible, the laser beam at the end of its useful path (outside the working volume) by a material that is diffuse and of such a colour and reflectivity as to make beam positioning possible while still minimizing the reflection hazards (note that conditions for safe viewing of a diffuse reflection of Class 3B visible laser are: minimum viewing distance of 13 cm between screen and cornea and a maximum viewing time of 10 sec., whilst other viewing conditions require a comparison of the diffuse reflection radiance with the MPE); and, d) Eye protection is required if there is any possibility of viewing either the direct or specularly reflected beam, or of viewing a diffuse reflection not complying with the conditions of the previous item.

v)   Class 4 laser products can cause injury from both the direct beam or its specular reflections and from diffuse reflections. They also present a potential fire hazard. The following controls should be employed in addition to those for Class 3B to minimize these risks: a) Beam paths should be enclosed whenever practicable. Access to the laser environment during laser operation should be limited to technical personnel wearing proper protective eyewear and clothing. b) These lasers should be operated by remote control whenever practical, thus eliminating the need for personnel to be physically present in the laser environment. c) Good room illumination is important in areas where laser eye protection is worn. Light-coloured diffuse wall surfaces help to achieve this condition. d) Fire is a principal hazard associated with high power lasers (CO_2, HF, DF). Fire bricks or other refractory material should be provided as a backstop for the beam. The maximum caution must be exercised with these materials, which can suffer surface glazing with prolonged exposure, giving rise to specular reflections. Cooled non-flat metal targets (cones, absorbers) are preferred. e) Special precautions may be required to prevent unwanted reflections in the invisible spectrum from FIR laser radiation, and the beam and target area should be surrounded by a material opaque to the laser wavelength (even dull metal surfaces may become highly specular at the CO_2 wavelength of 10.6 µm).

The development and use outdoor of lasers systems, such as LIDARS, has been increasing in recent years. Accordingly, the laser safety specification given below should be followed.

i)   Class 1 laser products. Use without protective measures
ii) Class 2 and Class 3A laser products. Use without aiming at people, and beam stop wherever possible. Prescriptions valid for Class 3B and 4 lasers may be followed whenever possible.
iii) Class 3B and Class 4. The hazard potential of these lasers may extend over a considerable distance. The range from the laser at which the irradiance or radiant exposure falls below the appropriate MPE is termed the Nominal Ocular Hazard Distance (NOHD). The area within which the beam irradiance or radiant exposure exceeds the appropriate MPE is called the Nominal Ocular Hazard Area (NOHA). This area is bounded by the limits of traverse, elevation and pointing accuracy of the laser system and extends either to the limit of the NOHD or to the position or any target or backstop. The exact NOHA will also depend on the nature of any material within the beam path, e.g. specular reflectors. The NOHD is dependent on the output characteristics of the laser, the appropriate MPE, the type of optical system used and atmospheric effects on beam propagation. Class 3B and Class 4 lasers in outdoor and similar environments should only be operated by personnel adequately trained in their use and approved by the Laser Safety Officer. The following precautions should be considered when using Class 3B and Class 4 laser products outdoors: a) Wherever possible, a restricted zone should be established as shown in Figure 3, in order to prevent access to the hazardous area. Personnel must be excluded from the beam path at all where the beam irradiance or radiant exposures exceed the

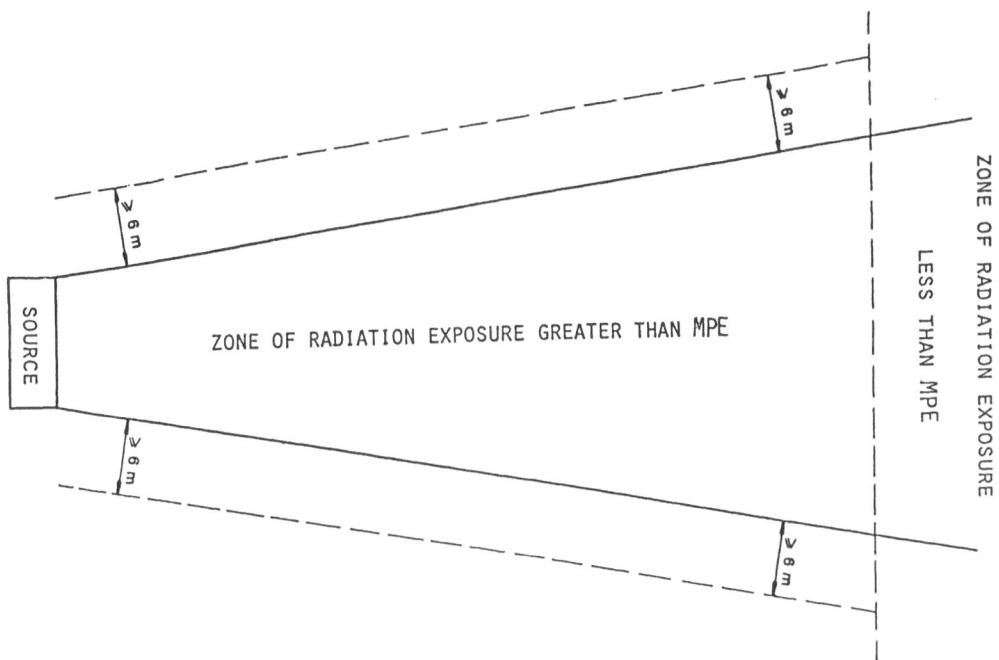

Fig. 3. Installation of laser products of classes 3B and 4 for outdoor
measurements: layout of the controlled area.

MPE unless they are wearing appropriate protective eye shields and clothing. Engineering controls such as physical barriers, interlocks limiting
the beam traverse and elevation, etc. should be used wherever practicable
to augment administrative controls. b) The intentional tracking of non-
target vehicular traffic or aircraft should be prohibited within the NOHD.
c) The beam paths should, whenever practicable, be cleared of all surfaces
capable of producing unintended reflections that are potentially hazardous,
or the hazard area should be extended appropriately.

REFERENCES

1. "Radiation safety of laser products, equipment classification, requirements and users guide", in International Electrotechnical Commission
   Publication IEC-825, First Edition, CEI Geneve (1984)
2. "Apparecchi laser: sicurezza dalle radiazioni, classificazione dei ma-
   teriali, prescrizioni e Guide per l'utilizzazione", in Pubblicazione
   italiana CEI 76-2 - Fascicoli 1283-1284 G (1989)
3. "American national standard for the safe use of lasers", in ANSI
   Publication Z 136.1, 1986, Rev. ANSI Z 136.1 (1980)

4. "Safety in universities: notes of guidance, Part 2:1 - Lasers", in The
   Committee of vice-chancellors and principals of the universities of
   the United Kingdom - Second Edition (1987)